W9-DGE-261

THE ABSOLUTE DIFFERENTIAL CALCULUS

QA
433
L4
1977

THE ABSOLUTE DIFFERENTIAL CALCULUS

(Calculus of Tensors)

BY

TULLIO LEVI-CIVITA
Late Professor of Rational Mechanics in the University of Rome
Fellow of R. Accademia Nazionale dei Lincei

EDITED BY

DR. ENRICO PERSICO

Authorized Translation by

MARJORIE LONG
Late Scholar of Girton College, Cambridge

DOVER PUBLICATIONS, INC.
NEW YORK

All rights reserved under Pan American and International Copyright Conventions.

Published in Canada by General Publishing Company, Ltd., 30 Lesmill Road, Don Mills, Toronto, Ontario.

Published in the United Kingdom by Constable and Company, Ltd., 10 Orange Street, London WC2H 7EG.

This Dover edition, first published in 1977, is an unabridged and unaltered republication of the English translation by Marjorie Long, first published by Blackie & Son Limited, London and Glasgow, in 1926. The Dover edition is published by special arrangement with Blackie & Son Limited, Bishopbriggs, Glasgow G64 2NZ, Scotland.

International Standard Book Number: 0-486-63401-9
Library of Congress Catalog Card Number: 76-27497

Manufactured in the United States of America
Dover Publications, Inc.
180 Varick Street
New York, N.Y. 10014

PREFACE

The present volume contains a complete translation, made in consequence of a suggestion by my eminent friend, Professor E. T. Whittaker, F.R.S., of the Italian text of my *Lezioni di calcolo differenziale assoluto*.[1] Two new chapters have been added, which are intended to exhibit the fundamental principles of Einstein's General Theory of Relativity (including, of course, as a limiting case, the so-called Special or Restricted Theory) as an application of the Absolute Calculus.

I have already had occasion to remark in the Preface to the Italian edition that we possess various systematic and well-written expositions of Relativity by celebrated authors. The short treatment which is offered in the two new chapters of the present work presents some distinctive features which it may be well to point out explicitly.

In the first place, in order not to increase the size of the book unduly, I have thought it expedient to confine myself to tracing the relativistic evolution of Mechanics (properly so called) and of Geometrical Optics, and to developing its most important consequences. In this treatment the whole of Electromagnetism is sacrificed. The sacrifice is certainly regrettable, since Electromagnetism was historically related in the most intimate way to Einstein's conception, having served indeed as the support and model for Restricted Relativity. Furthermore, Electromagnetism, in common with every other physical phenomenon, now comes within the ambit of General Relativity. Much as the omission of Electromagnetism is to be regretted, it has the advantage of reducing the programme to subjects belonging to the pure Newtonian tradition (or to its developments); and it allows us to take a clearer and more exact view of the transition from the classical scheme of Mechanics to the relativistic one.

[1] Compiled by Dr. Enrico Persico (Rome, Stock, 1925).

For this reason I have followed the method—which I have adopted sometimes in lectures or articles on special subjects—of taking the classical laws as the starting point and then of trying to find inductively what modifications—negligible in ordinary circumstances—should be introduced in order to take account of Einstein's ideas; and in the first place, naturally, to take account of his Principle of Relativity, that is to say, the invariant behaviour of these laws under all transformations of space and time, an auxiliary four-dimensional ds^2 being duly employed. This method has seemed to me to be preferable to the procedure of enunciating the postulates of relativistic Mechanics in abstract tensorial form, which is so comprehensive in physical content as to be almost inaccessible to ordinary intuition, except with ample comment and illustration.

A further characteristic of our exposition is that we make extensive use not only of geometrical representation but also of the differential properties pertaining to the space-time continuum; attention is drawn also to the special importance of the Einsteinian statics, the treatment being rigorous in some cases, while in others which involve fields variable with the time it is approximate.

In closing this introduction to Chapters XI and XII I would add that they were prepared, still in collaboration with Professor Persico, at the suggestion of Mr. F. F. P. Bisacre, M.A.

In connexion with the whole of the English edition, I must warmly thank the translator, Miss Marjorie Long, formerly Scholar of Girton College, who with double competence, scientific and linguistic, has known how to combine scrupulous respect for the text with its effective adaptation to the spirit of the English language.

I owe hearty thanks also to Dr. John Dougall, who, while revising the proofs, has checked the analysis throughout, detected some oversights, and made many useful suggestions for improvement. I wish finally to thank my English publishers, who have not only acceded to, but almost always anticipated, my wishes in regard to symbols and the typography of the book.

T. LEVI-CIVITA.

ROME, *October, 1926.*

PREFACE TO THE FIRST (ITALIAN) EDITION

Riemann's general metric and a formula of Christoffel constitute the premises of the absolute differential calculus. Its development as a systematic branch of mathematics was a later process, the credit for which is due to Ricci, who during the ten years 1887–1896 elaborated the theory and worked out the elegant and comprehensive notation which enables it to be easily adapted to a wide variety of questions of analysis, geometry, and physics.

Ricci himself, in an article published in Volume XVI of the *Bulletin des Sciences Mathématiques* (1892), gave a first account of his methods, and applied them to some problems in differential geometry and mathematical physics. Later on other interesting applications, made by himself or his students (to which group I had the privilege of belonging), suggested the desirability of preparing a general account of the whole subject, including methods, results, and a bibliography. This was the origin of the memoir "Méthodes de calcul différentiel absolu et leurs applications", which was compiled by Professor Ricci and myself in collaboration, on the courteous invitation of Klein, and appeared in Volume 54 of *Math. Ann.* (1901).

There is a chapter on the foundations of the absolute calculus, with special reference to the transformation of the equations of dynamics, in Wright's Tract, *Invariants of Quadratic Differential Forms* (Cambridge University Press, 1908); apart from this, while special researches based on the use of this method were continued after 1901 by a limited number of mathematicians, yet general attention was not again directed to it until the great renaissance of natural philosophy, due to Einstein, which found in the absolute differential calculus the necessary instrument for formulating the new ideas mathematically and for the subsequent numerical work.

Einstein's discovery of the gravitational equations was announced by him in the famous note "Zur allgemeinen Relativitätstheorie" [1] in the following words: "Sie bedeutet einen wahren Triumph der durch Gauss, Riemann, Christoffel, Ricci . . . begründeten Methoden des allgemeinen Differentialkalculus."

In an earlier memoir Einstein had given a new exposition of those elements and formulæ of the absolute calculus which more specifically served his purposes. A similar standpoint was subsequently adopted by the most distinguished workers in the field of general relativity, in particular by Weyl,[2] Laue,[3] Eddington,[4] and Birkhoff,[5] all of whom made conspicuous original contributions, both of idea and of method, to the physical theories, in addition to useful and elegant developments of the tensor calculus. Similar statements can be made for Carmichael,[6] Marcolongo,[7] Kopff,[8] Becquerel [9]—to mention, from the vast literature on the subject, only the books I have myself had occasion to consult —while de Donder [10] has avoided the notation of the absolute calculus and used instead the theory of integral invariants.

In recent years there have been some general treatises devoted to the absolute calculus; for instance, those of Juvet,[11] Marais,[12] and Galbrun.[13] Lastly, there is another calculus, in a new order of ideas, not less comprehensive and perhaps even more general, invented by Schouten, and developed with the collaboration of Struik.[14]

In face of this plentiful and valuable literature a new discussion of Ricci's methods might seem to be superfluous; and conceptually this is perhaps true.

In fact, of the improvements and additions to the scheme of 1901 (the memoir in *Math. Ann.*), derived mainly from the notion of parallelism [15] and on this basis introduced by me into two courses of lectures given at the University of Rome during the sessions 1920–1921 and 1922–1923, all, or almost all, will be found as independent discoveries of the authors already cited, in one or other of their books.

For instance, the definition of a tensor, and some algebraic anticipations of the results intended to simplify the proofs, are to be found in Weyl, Laue, and Marais, all of whom, like Eddington, establish a more or less intimate connexion between covariant differentiation and parallelism. A thorough discussion of the latter is also given by Juvet and Galbrun. But the association with the algebraico-tensorial notation and with the elements of differential geometry is always less detailed and systematic than what I tried to establish in my lectures. The line of argument followed in them has a particular unity, which may perhaps justify their appearance in print at this juncture.

The manuscript was edited with great care and intelligence by Dr. Enrico Persico, from notes of the lectures. I wish to express my thanks to him for his valuable help, and to my publisher, Signor Stock (who also attended the lectures), to whose continued encouragement the existence of the book is due.

TULLIO LEVI-CIVITA.

ROME, *December. 1923.*

REFERENCES

[1] *Sitzungsberichte der Preuss. Ak. der Wissenschaften*, 11th Nov., 1915, pp. 778–786.

[2] *Raum, Zeit, Materie.* Fifth edition. Berlin, Springer, 1923.

[3] *Die Relativitätstheorie*, Vol. II. Brunswick, Vieweg, 1921.

[4] *The Mathematical Theory of Relativity.* Cambridge, University Press, 1923.

[5] *Relativity and Modern Physics.* Cambridge (Mass.), Harvard University Press, 1923.

[6] *The Theory of Relativity.* Second edition. New York, Wiley, 1920.

[7] *Relatività.* Second edition. Messina, Principato, 1923.

[8] *I Fondamenti della Relatività Einsteiniana.* Italian edition, with preface by G. ARMELLINI. Milan, Hoepli, 1923.

[9] *Le principe de relativité et la théorie de la gravitation.* Paris, Gauthier-Villars, 1923.

[10] *La gravifique einsteinienne.* Paris, Gauthier-Villars, 1921.

[11] *Introduction au calcul tensoriel et au calcul différentiel absolu.* Paris, Blanchard, 1922.

[12] *Introduction géométrique à l'étude de la relativité.* Paris, Gauthier-Villars, 1923.

[13] *Introduction à la théorie de la relativité. Calcul différentiel absolu et géométrie.* Paris, Gauthier-Villars, 1923.

[14] Cf. in particular D. J. STRUIK: *Grundzüge der mehrdimensionalen Differentialgeometrie* (Berlin, Springer, 1922), which also contains a full and accurate bibliography; and also SCHOUTEN: *Der Ricci-Kalkul* (same publisher, 1924). Alongside the extension of methods there is a further point which I was unable to deal with in the lectures, but should refer to here, namely, the extension of geometry beyond the already very wide boundaries laid down by Riemann. I allude to the physical speculations of WEYL and EDDINGTON, which have recently

culminated in a further generalization—the work of Einstein—of the relativistic scheme. The spatial structure in relation to Weyl's new concept (affine geometry) has been the subject of study by WEYL himself, whose results have been published in the volume *Mathematische Analyse des Raumproblems* (Berlin, Springer, 1923), of further important systematic research by CARTAN, and of numerous notes on special problems by BERWALD, BLASCHKE, DIENES, EISENHART, KASNER, VEBLEN, and others.

[15] *Rendiconti del Circolo Matematico di Palermo,* fascicolo XLII, 1917, pp. 173–215.

NOTE TO SECOND ENGLISH IMPRESSION

Advantage has been taken of a reprint to correct a few typographical errors and to add references to some recent work (see p. 441).

T. L.-C.

ROME, *November, 1928.*

PUBLISHER'S NOTE

Professor Tullio Levi-Civita died in Rome on the 29th of December, 1941.

An appreciation of his work was published in the *Atti della Accademia Nazionale dei Lincei*, Serie Ottava, Vol. I, Fascicolo 11, November, 1946, with a list of his 204 scientific publications.

This volume includes the Author's last revisions of the English Version.

April, 1947.

CONTENTS

PART I—INTRODUCTORY THEORIES

CHAPTER I

FUNCTIONAL DETERMINANTS AND MATRICES

CHAPTER II

SYSTEMS OF TOTAL DIFFERENTIAL EQUATIONS

CHAPTER III

LINEAR PARTIAL DIFFERENTIAL EQUATIONS COMPLETE SYSTEMS

CHAPTER IV

ALGEBRAIC FOUNDATIONS OF THE ABSOLUTE DIFFERENTIAL CALCULUS

CHAPTER V

GEOMETRICAL INTRODUCTION TO THE THEORY OF DIFFERENTIAL QUADRATIC FORMS

(a) *The Line Element on a Surface*

PART II

THE FUNDAMENTAL QUADRATIC FORM AND THE ABSOLUTE DIFFERENTIAL CALCULUS

CHAPTER VI

COVARIANT DIFFERENTIATION; INVARIANTS AND DIFFERENTIAL PARAMETERS; LOCALLY GEODESIC CO-ORDINATES

CHAPTER VII

RIEMANN'S SYMBOLS AND PROPERTIES RELATING TO CURVATURE; RICCI'S AND EINSTEIN'S SYMBOLS; GEODESIC DEVIATION

CHAPTER VIII

RELATIONS BETWEEN TWO DIFFERENT METRICS REFERRED TO THE SAME PARAMETERS; MANIFOLDS OF CONSTANT CURVATURE.

CHAPTER IX

DIFFERENTIAL QUADRATIC FORMS OF CLASS ZERO AND CLASS ONE

CHAPTER X

SOME APPLICATIONS OF INTRINSIC GEOMETRY

PART III.—PHYSICAL APPLICATIONS

CHAPTER XI

EVOLUTION OF MECHANICS AND GEOMETRICAL OPTICS; THEIR RELATION TO A FOUR-DIMENSIONAL WORLD ACCORDING TO EINSTEIN

CHAPTER XII

THE GRAVITATIONAL EQUATIONS AND GENERAL RELATIVITY

THE ABSOLUTE DIFFERENTIAL CALCULUS

PART I
Introductory Theories

CHAPTER I

Functional Determinants and Matrices

1. Geometrical terminology.

In analytical geometry it frequently happens that complicated algebraic relationships represent simple geometrical properties. In some of these cases, while the algebraic relationships are not easily expressed in words, the use of geometrical language, on the contrary, makes it possible to express the equivalent geometrical relationships clearly, concisely, and intuitively. Further, geometrical relationships are often easier to discover than are the corresponding analytical properties, so that geometrical terminology offers not only an illuminating means of exposition, but also a powerful instrument of research. We can therefore anticipate that in various questions of analysis it will be advantageous to adopt terms taken over from geometry.

For this purpose it is essential to adopt the fundamental convention of using the term *point of an abstract* n-*dimensional manifold* (n being any positive integer whatever) to denote a set of n values assigned to any n variables $x_1, x_2, \ldots x_n$. This is an obvious extension of the use of the term in the one-to-one correspondence which can be established between pairs or triplets of co-ordinates and the points of a plane or space, for the cases $n = 2$ and $n = 3$ respectively. For the case of n variables we can thus also speak of a *field* of points (rather than of values assigned to the x's), and of the *region round* a specified point x_i $(i = 1, 2, \ldots n)$.

If the x's are n functions $x_i(t)$ of a real variable t, then when t varies continuously between t_0 and t_1 we get a simply infinite succession of points, the aggregate of which (as for $n = 2$ and $n = 3$) is called a *line*, and more precisely an *arc* or *segment of a line*.

2. **Functional determinants and change of variables.**

Let there be n functions of n variables:

$$u_i(x_1, x_2, \ldots x_n),$$

the functions and their derivatives to any required degree being supposed finite and continuous in the field considered.

To simplify the notation, let x (without a suffix) represent not only (as is usual) any one of the n variables $x_1, x_2, \ldots x_n$, but also (as is sometimes done) the whole set of them; and similarly for other letters which will be used farther on. With this convention the given functions can be written in the abridged form:

$$u_i(x).$$

With the usual notation, the *functional determinant* or *Jacobian* of the u's is the determinant of the nth order whose terms are the first derivatives of the u's; i.e.

$$D = \begin{vmatrix} \frac{\partial u_1}{\partial x_1} & \frac{\partial u_1}{\partial x_2} & \ldots & \frac{\partial u_1}{\partial x_n} \\ \frac{\partial u_2}{\partial x_1} & \frac{\partial u_2}{\partial x_2} & \ldots & \frac{\partial u_2}{\partial x_n} \\ \ldots & \ldots & \ldots & \ldots \\ \frac{\partial u_n}{\partial x_1} & \frac{\partial u_n}{\partial x_2} & \ldots & \frac{\partial u_n}{\partial x_n} \end{vmatrix}.$$

Such a determinant is sometimes represented by the abridged notation

$$\begin{pmatrix} u_1 \, u_2 \ldots u_n \\ x_1 \, x_2 \ldots x_n \end{pmatrix}$$

analogous to that used for fractions and substitutions, the set of functions u representing the numerator and the set of variables x the denominator of a fraction. The analogy of form is justified by the analogy of properties, as can be seen by considering the effect on a functional determinant of a change of variables. For let the x's be functions of n variables y,

$$\left.\begin{aligned} x_1 &= x_1(y_1, \ldots y_n), \\ &\cdot \quad \cdot \quad \cdot \quad \cdot \quad \cdot \quad \cdot \\ x_n &= x_n(y_1, \ldots y_n), \end{aligned}\right\} \quad \cdot \quad \cdot \quad \cdot \quad \cdot \quad \cdot \quad (1)$$

and suppose further that these equations represent a reversible transformation, i.e. that they also define the y's as functions of the x's, or, in other words, that they are soluble with respect to the y's. If then the u's are considered as functions of the y's (being given in terms of the x's, which are functions of the y's), and the corresponding functional determinant

$$D_1 = \begin{pmatrix} u_1 & \ldots & u_n \\ y_1 & \ldots & y_n \end{pmatrix}$$

is formed, it will be found, as will be shown below in § 4, that $D_1 = D$ multiplied by the determinant of the functions defined by equations (1), i.e. by

$$\Delta = \begin{pmatrix} x_1 & \ldots & x_n \\ y_1 & \ldots & y_n \end{pmatrix}.$$

3. The fundamental theorem on implicit functions.

Before proving the theorem just referred to, we must recall a fundamental theorem relating to implicit functions. It is known that a relation between two variables of the type

$$f(y, x) = 0$$

defines y as a function of x, provided certain suitable qualitative conditions are satisfied.[1] A classical form of the conditions sufficient for solubility is as follows. Let x^0, y^0, be a point at which f vanishes, f being finite in a (plane) region I round the point. Let $\dfrac{\partial f}{\partial y}$ exist in I and be not zero for $x = x^0$, $y = y^0$. Then

[1] When an equation is said to be "soluble", this will not necessarily mean that the process of finding an algebraic solution can be carried out.

in a certain (linear) region round the value x^0 the given equation defines a continuous function $y(x)$ such that $f(y(x), x)$ vanishes identically.

For implicit functions of several variables the following theorem, which is a generalization of the one just stated, holds.

Let there be given n equations between n variables y and any number of variables x of the form

$$f_i(y \mid x) = 0 \qquad (i = 1, 2, \ldots n).$$

Let there be a set of values x^0, y^0, which satisfy these equations; in a region round the point x^0, y^0, let the f's and their derivatives with respect to the y's be continuous, and let the determinant

$$\begin{pmatrix} f_1 \ldots f_n \\ y_1 \ldots y_n \end{pmatrix}$$

be not zero. Then the given equations define the y's as functions of the x's in a region round the set of values x^0.

It will be seen that from a certain point of view the functional determinant of several functions of the same number of variables constitutes a natural generalization of the derivative of a function of one variable. This will follow explicitly from the applications of the following section.

4. Effect on a functional determinant of a change of variables.

Consider first the (sufficient) condition of solubility of the set of equations (1). Write the equations in the form

$$x_i(y_1, \ldots y_n) - x_i = 0 \qquad (i = 1, 2, \ldots n),$$

and suppose that there exists at least one set of values of the y's and the x's which satisfy them and for which the functions $x_i(y)$ and their derivatives are continuous. Then, to apply the preceding theorem, we must calculate the partial derivatives of the left-hand side of each equation with respect to the y's, and form their determinant. But these derivatives are the terms $\frac{\partial x_i}{\partial y_j}$ $(j = 1, 2, \ldots n)$, and hence the condition of solubility with respect to the y's is

$$\Delta = \begin{pmatrix} x_1 \ldots x_n \\ y_1 \ldots y_n \end{pmatrix} \neq 0.$$

Now take the theorem stated in § 2, and suppose $\Delta \neq 0$. Multiply together the two determinants D and Δ, i.e., interchanging rows and columns in Δ, form the product

$$\begin{vmatrix} \frac{\partial u_1}{\partial x_1} & \frac{\partial u_1}{\partial x_2} & \cdots & \frac{\partial u_1}{\partial x_n} \\ \frac{\partial u_2}{\partial x_1} & \frac{\partial u_2}{\partial x_2} & \cdots & \frac{\partial u_2}{\partial x_n} \\ \cdots & \cdots & \cdots & \cdots \\ \frac{\partial u_n}{\partial x_1} & \frac{\partial u_n}{\partial x_2} & \cdots & \frac{\partial u_n}{\partial x_n} \end{vmatrix} \times \begin{vmatrix} \frac{\partial x_1}{\partial y_1} & \frac{\partial x_2}{\partial y_1} & \cdots & \frac{\partial x_n}{\partial y_1} \\ \frac{\partial x_1}{\partial y_2} & \frac{\partial x_2}{\partial y_2} & \cdots & \frac{\partial x_n}{\partial y_2} \\ \cdots & \cdots & \cdots & \cdots \\ \frac{\partial x_1}{\partial y_n} & \frac{\partial x_2}{\partial y_n} & \cdots & \frac{\partial x_n}{\partial y_n} \end{vmatrix}.$$

Applying the ordinary rule, the product by rows gives as the typical element a_{rs} of the resulting determinant the expression

$$\sum_{1}^{n}{}_{i} \frac{\partial u_r}{\partial x_i} \frac{\partial x_i}{\partial y_s} = \frac{\partial u_r}{\partial y_s}$$

(remembering the rule for differentiating a function of one or more functions). Hence, as already stated, the product is the determinant D_1. This result is expressed by the formula

$$\begin{pmatrix} u_1 & \ldots & u_n \\ x_1 & \ldots & x_n \end{pmatrix} \times \begin{pmatrix} x_1 & \ldots & x_n \\ y_1 & \ldots & y_n \end{pmatrix} = \begin{pmatrix} u_1 & \ldots & u_n \\ y_1 & \ldots & y_n \end{pmatrix}, \qquad (2)$$

which justifies the use of this notation for the functional determinants.

5. The necessary and sufficient conditions for the independence of n functions of n variables.

If therefore the functional determinant of n functions of n variables does not vanish identically, it follows that this property still holds when the original variables are replaced by others related to the first set by the transformation (1) (with the condition $\Delta \neq 0$); in other words, this is an *invariant* property. The following definition may therefore be given:

DEFINITION.—n *functions of* n *variables are said to be independent when their functional determinant does not vanish identically.*

The reason for applying the word "independence" to this property is shown by the following theorem.

THEOREM.—*Given* n *functions* u *of* n *variables* x, *the necessary and sufficient condition for the non-existence of any (differentiable) relation between them of the type*

$$f(u_1, u_2, \ldots u_n) = 0 \quad . \quad . \quad . \quad . \quad (3)$$

involving only the u's *and not the* x's, *is that their functional determinant does not vanish identically.*

We shall first show that the condition is sufficient; then that it is also necessary, but for the moment confining the proof to a particular case; the theorem in its general form will be shown farther on to be itself only a particular case of another still more general theorem (cf. § 7).

Suppose the condition satisfied

$$D = \begin{pmatrix} u_1 \ldots u_n \\ x_1 \ldots x_n \end{pmatrix} \neq 0. \quad . \quad . \quad . \quad . \quad (4)$$

We shall then show that no relation of the type (3) can exist. (Identities are of course not considered; i.e. we exclude the case where equation (3) is satisfied when arbitrary values are assigned to the u's, as it would not then represent any relation between the u's.) Suppose that such a relation does exist. Differentiating with respect to $x_1, \ldots x_n$, we should get n equations

$$\sum_{1}^{n}{}_{\alpha} \frac{\partial f}{\partial u_\alpha} \frac{\partial u_\alpha}{\partial x_i} = 0 \qquad (i = 1, 2, \ldots n),$$

linear and homogeneous in the derivatives $\dfrac{\partial f}{\partial u_\alpha}$. Now since by hypothesis f is a true function, not zero identically, these derivatives are not all zero. Hence the determinant of the coefficients of this group of equations vanishes; i.e. $D = 0$, which is contrary to our hypothesis. The condition (4) is therefore sufficient to secure the non-existence of any relation of the type (3).

To prove that condition (4) is necessary, we shall show that if it is not satisfied, i.e. if

$$D = 0, \quad . \quad . \quad . \quad . \quad . \quad . \quad (5)$$

then the u's are connected by a relation (at least one) of the type (3). For the moment the only case considered will be that in which at least one of the minors of order $n - 1$ of the determinant D does not vanish. This minor will in general be of the type

$$D' = \begin{pmatrix} u_{q_1} & u_{q_2} & \dots & u_{q_{n-1}} \\ x_{p_1} & x_{p_2} & \dots & x_{p_{n-1}} \end{pmatrix}$$

where $p_1, \dots p_{n-1}$ and $q_1, \dots q_{n-1}$ represent any two arrangements of $n - 1$ integers chosen without repetitions from the numbers $1, 2, \dots . n$. But since the order in which the x's and u's are made to correspond to the numbers $1, 2, \dots n$ is immaterial, we can, without loss of generality, suppose numbers assigned to the variables in such a way that D' is the minor formed by the first $n - 1$ rows and $n - 1$ columns; we thus get

$$D' = \begin{pmatrix} u_1 & \dots & u_{n-1} \\ x_1 & \dots & x_{n-1} \end{pmatrix} \neq 0. \quad . \quad . \quad . \quad (6)$$

This condition expresses the fact that no relation exists between the first $n - 1$ functions.

Now we know that if a reversible transformation is applied to the x's, it follows from hypothesis (5) that the determinant of the u's with respect to the new set of variables y is also zero. Let the relation between the x's and y's be given by the following equations:

$$\left.\begin{aligned} y_1 &= u_1(x_1, \dots x_n), \\ &\dots\dots\dots\dots \\ y_{n-1} &= u_{n-1}(x_1, \dots x_n), \\ y_n &= x_n. \end{aligned}\right\} \quad . \quad . \quad . \quad . \quad (7)$$

We may note that these formulæ define a reversible transformation, since the functional determinant of the y's with respect to the x's is

$$\begin{vmatrix} \frac{\partial u_1}{\partial x_1} & \dots & \frac{\partial u_1}{\partial x_{n-1}} & \frac{\partial u_1}{\partial x_n} \\ \dots & \dots & \dots & \dots \\ \frac{\partial u_{n-1}}{\partial x_1} & \dots & \frac{\partial u_{n-1}}{\partial x_{n-1}} & \frac{\partial u_{n-1}}{\partial x_n} \\ 0 & \dots & 0 & 1 \end{vmatrix},$$

and expanding this from the last row, it is seen to be equal to D', which by hypothesis is not zero.

Now consider the u's as functions of the y's; using equations (7) we get

$$\left.\begin{aligned} u_1 &= y_1, \\ &\cdot\ \cdot\ \cdot\ \cdot\ \cdot\ \cdot \\ u_{n-1} &= y_{n-1}, \\ u_n &= u_n(y_1, \ldots y_{n-1}, y_n). \end{aligned}\right\} \quad . \quad . \quad . \quad (8)$$

Expressing the fact that the determinant of the u's with respect to the y's is zero, we get

$$\begin{vmatrix} 1 & 0 & \ldots & 0 & 0 \\ 0 & 1 & \ldots & 0 & 0 \\ \cdot & \cdot & \cdot & \cdot & \cdot \\ 0 & 0 & \ldots & 1 & 0 \\ \dfrac{\partial u_n}{\partial y_1} & \dfrac{\partial u_n}{\partial y_2} & \ldots & \dfrac{\partial u_n}{\partial y_{n-1}} & \dfrac{\partial u_n}{\partial y_n} \end{vmatrix} = \frac{\partial u_n}{\partial y_n} = 0.$$

It follows that the last of the equations (8) does not contain y_n; substituting in it from the remaining equations, it becomes

$$u_n = u_n(u_1, \ldots u_{n-1}),$$

i.e. a relation between the u's which does not contain any of the x's.

Hence from the hypotheses (6) and (5) it follows that there exists *one* relation of the type (3), which is such that u_n can be expressed in terms of the other u's. This relation is *unique*, because if there were another, then eliminating u_n between them we should get a relation between $u_1, \ldots u_{n-1}$; but this, as already pointed out, is incompatible with hypothesis (6).

6. Functional matrices. Definition of the independence of m functions of n variables.

We shall now examine the more general case in which the number m of the functions u is not equal to the number n of the variables x. For this purpose we must consider the *functional matrix* of the given functions, i.e the following matrix of m rows and n columns:

$$\begin{Vmatrix} \frac{\partial u_1}{\partial x_1} & \frac{\partial u_1}{\partial x_2} & \cdots & \frac{\partial u_1}{\partial x_n} \\ \cdot & \cdot & \cdot & \cdot \\ \frac{\partial u_m}{\partial x_1} & \frac{\partial u_m}{\partial x_2} & \cdots & \frac{\partial u_m}{\partial x_n} \end{Vmatrix}.$$

In what follows it will be denoted by M; but it must be noted that no numerical value is attached to the symbol, and therefore that M does not represent a quantity, but is an abbreviation for the arrangement of terms under consideration.

The *characteristic* of a matrix is the order of the non-vanishing determinants of highest order which can be constructed from it; it can therefore obviously not be greater than the number of rows or the number of columns, whichever is the less.

We now give a definition, which will be justified in the following section.

DEFINITION.—m *functions of any number of variables are said to be independent when the characteristic of their functional matrix is* m. It follows immediately that if the number of functions is greater than the number of variables, the functions cannot be independent; while if the two numbers are equal, the definition coincides with that already given, since the matrix becomes a determinant of order m, and if its characteristic is m this is equivalent to saying that the determinant does not vanish.

7. Theorem.

Given m *functions* u *of any number of variables* x, *if the characteristic of their functional matrix is* k, *then there are* m — k *relations (and not more) between the* u*'s which do not involve the* x*'s.*

It will follow immediately as a corollary that if the functions are independent (the case $k = m$) there exists no relation between them.

The theorem just stated has been proved above (§ 5) for the particular cases in which the number of functions is equal to the number of variables and in addition $k = m$ or $k = m - 1$. We proceed to prove it in general, taking various cases in turn, as follows:

(1) $k = m$ (and $\therefore m \leqslant n$), the case of independence;

$$(2)\ k < m: \begin{cases} (2a)\ k = n, \\ (2b)\ k < n. \end{cases}$$

Case (1): $k = m$. This hypothesis is equivalent to saying that there exists a minor of order m which is not zero; remembering the remark made on p. 7, we may suppose without loss of generality that

$$\begin{pmatrix} u_1 & \ldots & u_m \\ x_1 & \ldots & x_m \end{pmatrix} \neq 0.$$

Applying the theorem of p. 6 it follows that the u's are not connected by any relation which does not involve any of the x's.

Case (2a): $k < m$, $k = n$. There is therefore a minor of order n which is not zero. We may arrange the suffixes of the u's and the x's so that the minor in question is that formed by the first n rows and n columns, and we shall have

$$D = \begin{pmatrix} u_1 & \ldots & u_n \\ x_1 & \ldots & x_n \end{pmatrix} \neq 0.$$

We shall now show that $u_{n+1}, u_{n+2}, \ldots u_m$ can all be expressed in terms of the remaining u's, without using the x's, so that we shall have $m - n$ (which is the same as $m - k$) relations between the u's. For since $D \neq 0$ we may change the variables. Let the new variables be given by the equations

$$\begin{aligned} u_1 &= u_1(x_1, \ldots x_n), \\ &\ldots\ldots\ldots \\ u_n &= u_n(x_1, \ldots x_n). \end{aligned}$$

Solving these equations with respect to the x's, and substituting the expressions so obtained in $u_{n+1}, \ldots u_m$, these will be expressed as functions of $u_1, \ldots u_n$; hence the theorem is true for this case.

Case (2*b*): $k < m, k < n$. The hypothesis is that there exists a determinant of order k which is not zero, and that every determinant of higher order vanishes. Let us arrange the u's and the x's so that

$$D = \begin{pmatrix} u_1 & \dots & u_k \\ x_1 & \dots & x_k \end{pmatrix} \neq 0. \quad . \quad . \quad . \quad . \quad (9)$$

We shall show that any function u_h ($h = k + 1, \dots m$) can be expressed in terms of the first k functions u, without involving any of the x's. For this purpose, consider the determinant Θ formed by bordering D with the $(k + 1)$th column and hth row of the matrix; since it is of order $k + 1$, it is zero by hypothesis, i.e.

$$\Theta = \begin{pmatrix} u_1 & \dots & u_k & u_h \\ x_1 & \dots & x_k & x_{k+1} \end{pmatrix} = 0. \quad . \quad . \quad . \quad (10)$$

Now applying the theorem stated on p. 6, it follows from this equation and the inequality (9) that u_h can be expressed as a function of $u_1, \dots u_k$, which does not involve $x_1, \dots x_k, x_{k+1}$; i.e. since we are not yet able to say anything about the remaining x's,

$$u_h = \phi(u_1, \dots u_k \mid x_{k+2}, \dots x_n). \quad . \quad . \quad (11)$$

The next step is to show that $x_{k+2}, \dots x_n$ do not in fact occur in this expression. If $n = k + 1$, there is no need to consider $x_{k+2}, \dots x_n$, and therefore the formula (11) represents the expression we are in search of, giving u_h in terms of $u_1, \dots u_k$ alone. If this is not so, let x_j denote any of the variables $x_{k+2}, \dots x_n$, and consider the determinant Θ', obtained from Θ by replacing x_{k+1} by x_j, so that

$$\Theta' = \begin{pmatrix} u_1 & \dots & u_k & u_h \\ x_1 & \dots & x_k & x_j \end{pmatrix} = 0,$$

Θ' vanishing because it is a minor of order $k + 1$ taken from the matrix. Expanding it, substituting from equation (11), and making certain transformations, we can easily show that it involves the vanishing of $\dfrac{\partial \phi}{\partial x_j}$; whence it follows that ϕ does not contain x. In fact, representing compactly by the letter D the square matrix of those elements of Θ which form the determinant D, we have

$$\Theta' = \left| \begin{array}{ccc|c} & & & \dfrac{\partial u_1}{\partial x_j} \\ & D & & \cdots \\ & & & \dfrac{\partial u_k}{\partial x_j} \\ \hline \dfrac{\partial u_h}{\partial x_1} & \cdots & \dfrac{\partial u_h}{\partial x_k} & \dfrac{\partial u_h}{\partial x_j} \end{array} \right| .$$

Using equation (11) the elements of the last row are given by

$$\frac{\partial u_h}{\partial x_i} = \sum_1^k {}_l \frac{\partial \phi}{\partial u_l} \frac{\partial u_l}{\partial x_i} \qquad (i = 1, 2, \ldots k);$$

$$\frac{\partial u_h}{\partial x_j} = \frac{\partial \phi}{\partial x_j} + \sum_1^k {}_l \frac{\partial \phi}{\partial u_l} \frac{\partial u_l}{\partial x_j}.$$

Multiplying the elements of each of the first k rows in turn by $\dfrac{\partial \phi}{\partial u_1}, \ldots \dfrac{\partial \phi}{\partial u_k}$, and subtracting the sum of these products from the elements of the last row (which does not change the value of the determinant) the last row becomes

$$0 \ldots 0 \qquad \frac{\partial \phi}{\partial x_j},$$

and therefore, expanding from this row, we get

$$\Theta' = \frac{\partial \phi}{\partial x_j} D.$$

Since by hypothesis $D \neq 0$, it follows that $\dfrac{\partial \phi}{\partial x_j} = 0$, which proves the assertion.

The theorem enunciated at the beginning of this section is thus completely proved. Applying it to the particular case $m = n$, it coincides with the theorem of p. 6, which is therefore now shown to hold without any restriction.

CHAPTER II

Systems of Total Differential Equations

1. Preliminary remarks.

The reader may first be reminded of some general considerations on differential expressions.

Given a function $f(x_1, x_2, \ldots x_n)$, the expression

$$df = \sum_1^n {}_i \frac{\partial f}{\partial x_i} dx_i$$

is called the *total differential* of the function f; it is equal (except for infinitesimals of higher order) to the increment of f in passing from the point $x_1, x_2, \ldots x_n$ to the infinitely near point $x_1 + dx_1$, $x_2 + dx_2, \ldots x_n + dx_n$.

Given n functions X_i of the x's, which, together with their first derivatives, we shall suppose finite and continuous, the expression

$$\psi = \sum_1^n {}_i X_i (x_1, x_2, \ldots x_n) \, dx_i \quad . \quad . \quad . \quad (1)$$

is called a *differential*, or *Pfaffian, expression.*

An expression of this form is not always an exact differential; i.e. there does not always exist a function $f(x_1, x_2, \ldots x_n)$ such that the given Pfaffian is its total differential. The necessary and sufficient condition for the existence of such an f, i.e. for the integrability of an equation of the type

$$df = \sum_1^n {}_i X_i \, dx_i, \quad . \quad . \quad . \quad . \quad . \quad . \quad (2)$$

is that the following $\frac{1}{2}n(n-1)$ conditions should be satisfied:

$$\frac{\partial X_i}{\partial x_j} = \frac{\partial X_j}{\partial x_i} \qquad (i, j = 1, 2, \ldots n). \quad . \quad . \quad (3)$$

If these conditions are satisfied in a certain field, the integral calculus shows how to construct the most general function f which has the required property; i.e. it shows how to *integrate* the given differential expression. All the possible f's differ from one another by a constant. If we follow the procedure usual in elementary treatises, and consider not the whole field but a suitably restricted region round a point x arbitrarily fixed in advance, then in this region each of the f's is a *uniform* function (i.e. one-valued, like all the functions we are considering) of the arguments $x_1, x_2, \ldots x_n$.

We now proceed to discuss a more general problem than this. Let there be m unknown functions u of n independent variables x, and let there be given a set of relations between their differentials which define the du's in terms of the dx's, in the form

$$du_\alpha = \sum_{1}^{n}{}_i X_{\alpha|i}(x \mid u)\, dx_i \qquad (\alpha = 1, 2, \ldots m), \quad . \quad (4)$$

where the X's are mn arbitrarily assigned functions (finite and continuous, together with their first derivatives).

A group of relations of the type (4) is called a system of *total differential equations*[1]; equation (2) is obviously only a particular case. It may be remarked that equation (2) is itself equivalent to the system of n equations

$$\frac{\partial f}{\partial x_i} = X_i(x) \qquad (i = 1, 2, \ldots n), \quad . \quad . \quad (2')$$

and that the equations (4) are analogously equivalent to the system of mn equations

$$\frac{\partial u_\alpha}{\partial x_i} = X_{\alpha|i}(x \mid u) \qquad \begin{pmatrix} \alpha = 1, 2, \ldots m \\ i = 1, 2, \ldots n \end{pmatrix}. \quad . \quad (4')$$

Both are problems of partial differential equations, and are soluble only under specific conditions; but if these are satisfied,

[1] In a system of this kind the group of variables to be considered independent is fixed in advance. The late Professor G. Ricci in a recent work has considered instead a system of l equations of the type

$$\sum_{1}^{n} a_{rs}(x)\, dx_s = 0 \qquad (r = 1, 2, \ldots l),$$

determining the conditions that the n variables x may be considered functions of any number $p\,(< n)$ of independent variables, and indicating the steps necessary to find the solution (cf. *Atti del Reale Ist. Ven.*, Vol. XXXI, 1922–3, pp. 179–183).

An account of the general theory of Pfaffian systems, with recent developments due mainly to von Weber, Cartan, and Goursat, is given in the last-named author's *Leçons sur le problème de Pfaff* (Paris, Hermann, 1922).

we shall see that the integration reduces to that of ordinary differential equations.

2. Conditions necessary for integrability. Completely integrable, or complete, systems.

When the problem is stated in the form (4′), it is obvious (from the symmetry of the second derivatives of the u's) that a necessary condition for the existence of solutions is that the following conditions shall be satisfied:

$$\frac{dX_{\alpha|i}}{dx_j} = \frac{dX_{\alpha|j}}{dx_i} \qquad \begin{pmatrix} \alpha \;\; = 1, 2, \ldots m \\ i, j = 1, 2, \ldots n \end{pmatrix}. \quad . \quad (5)$$

The symbol denoting total differentiation has been used as a reminder that in differentiating it is necessary to take into account that the arguments u also depend on the x's, i.e. that

$$\frac{dX_{\alpha|i}}{dx_j} = \frac{\partial X_{\alpha|i}}{\partial x_j} + \sum_1^m{}_\beta \frac{\partial X_{\alpha|i}}{\partial u_\beta} \frac{\partial u_\beta}{\partial x_j}$$

$$= \frac{\partial X_{\alpha|i}}{\partial x_j} + \sum_1^m{}_\beta \frac{\partial X_{\alpha|i}}{\partial u_\beta} X_{\beta|j}. \quad . \quad . \quad (6)$$

Using this result, the equations (5) take the form of $\frac{1}{2}mn(n-1)$ relations of the type

$$F(x \mid u) = 0. \quad . \quad . \quad . \quad . \quad . \quad (5')$$

These, it will be seen, in general contain not only the x's but also the u's (unlike the equations (3)); and we must suppose the u's replaced by those unknown functions of x which satisfy the given system of equations. The conditions of integrability cannot therefore be given explicitly without knowing beforehand the solutions of the system. This difficulty did not arise for the equation (2), since the X's, and therefore their derivatives, did not contain the unknown function.

But it may happen—and this is the most interesting case—that the equations (5) are not only satisfied for those particular u's which form a solution of the system, but are true identically, i.e. for any set of values whatever of the u's and of the x's. In this case, as we shall see, these conditions are not only necessary, but also sufficient, for the integrability of the system, which is then said to be *completely integrable*, or *complete*.

3. **The integration of a mutually consistent system can always be reduced to that of a complete system.**

We shall now show that whenever a system of total differential equations is integrable (in the sense that there exists at least one set of m functions $u_\alpha(x_1, x_2, \ldots x_n)$ which satisfy the system), the integration reduces to that of a complete system; we shall thus be able to confine our subsequent discussions to systems of the latter kind.

As we have already said, there are $\frac{1}{2}mn(n-1)$ conditions of integrability (5′), while there are m u's. Now for $n > 2$, $m < \frac{1}{2}mn(n-1)$. In general, therefore, there cannot be m functions u which satisfy these conditions, and therefore the system can certainly not admit of solutions. If exceptionally these conditions are mutually consistent it may happen either that m of them are independent, so that there is then one single set of values for the u's which satisfies these m conditions, and it only remains to test whether these u's also satisfy the given system of equations; or that they are all satisfied identically (and then the system is complete); or that—the most general case—they reduce to a number $\nu < m$ of mutually consistent and independent equations. In the latter case, ν of the unknowns can be found in finite terms, expressed in terms of the x's and the remaining $m - \nu = \mu$ unknowns. Arranging the u's in a suitable order, we may suppose that the equations (5′) give us the last ν of the functions u, viz. the functions

$$u_{\mu+1}, u_{\mu+2}, \ldots u_m$$

in terms of the x's and the remaining u's,

$$u_1, u_2, \ldots u_\mu.$$

For greater clearness, we shall denote these first μ functions u by u'_α $(\alpha = 1, 2, \ldots \mu)$, and the last ν by $u''_\beta = u_{\mu+\beta}$ $(\beta = 1, 2, \ldots \nu)$. Using this notation, the equations (5′) can be put in the form resolved with respect to the u''_β, namely

$$u''_\beta = f_\beta(x \mid u') \qquad (\beta = 1, 2, \ldots \nu). \quad . \quad (5'')$$

Next, suppose the system of equations (4) divided into two groups; one consisting of the first μ:

$$du'_\alpha = \sum_1^n{}_i X_{\alpha|i}(x \mid u)\, dx_i \qquad (\alpha = 1, 2, \ldots \mu); \quad . \quad (4a)$$

and the other of the remaining ν:

$$du_\alpha = \sum_{1}^{n}{}_i X_{\alpha|i}(x \mid u)\, dx_i \qquad (\alpha = \mu + 1, \mu + 2, \ldots m = \mu + \nu).$$

The latter group, putting $\alpha = \mu + \beta$, we shall write in the form:

$$du''_\beta = \sum_{1}^{n}{}_i X_{\mu+\beta|i}(x \mid u)\, dx_i \qquad (\beta = 1, 2, \ldots \nu). \quad . \quad (4b)$$

Substituting from the equations (5″) and (4a), the two sides of this last equation become linear expressions in the differentials dx_i, with coefficients which depend solely on the x's and the u's. Since the coefficients on both sides must be the same (the differentials dx_i being independent), the equations (4b) reduce to equations in finite terms, $n\nu$ in number, between the u''s and the x's.

If all these reduce to identities, we need only consider the system of equations (4a), in which the functions u'' are to be considered as replaced by their values as given by the equations (5″), so that we have a total differential system, of the same form as the original system (4), involving only the u''s, μ in number, where $\mu = m - \nu < m$. The essential result in the case under consideration is that the system (4a) so reduced is necessarily complete. In fact, it consists of a part of the original system (4) with the additional relations (5″) between the u's. The condition of integrability of the whole system (4) (where *a priori* the u's were treated as so many unknowns) consisted of the equations (5), or, we may say, of the equivalent equations (5″). For the system of equations (4a) the analogous conditions will consist of a part of the conditions (5″) (or combinations of these), with the proviso that every u'' is to be replaced by the corresponding expression given by the equations (5″) themselves. This process obviously leads to mere identities; hence, as stated, the system (4a) is complete.

If on the other hand the equations (4b) give rise to non-identical relations in finite terms between the u''s and the x's, we shall have to associate them with the equations (4a) and treat this whole system of equations in μ unknowns (including some total differential equations and some equations in finite terms) as we have already treated the system of equations (4) and the conditions (5).

Proceeding in this way, we shall reach a stage where either the conditions are found to be mutually inconsistent, when we must conclude that the given system has no solution, or else the problem reduces to the integration of a complete system (with a number of unknowns which is certainly less than m). Q.E.D.

In consequence we shall now confine our attention solely to complete systems.

4. Bilinear covariants and the resulting form for the conditions of complete integrability.

We have expressed the condition of complete integrability by means of the equations (5), which are supposed to hold for arbitrary values of the u's and of the x's. We shall now express this condition in a more concise form.

For this purpose take two different systems of infinitesimal increments of the x's, denoted by dx_i and δx_i respectively; the corresponding increments of a generic function u of the x's will then be denoted by du and δu respectively, and will be given by

$$\left.\begin{aligned} du &= \sum_1^n {}_i \frac{\partial u}{\partial x_i} dx_i, \\ \delta u &= \sum_1^n {}_i \frac{\partial u}{\partial x_i} \delta x_i. \end{aligned}\right\} \quad . \quad . \quad . \quad . \quad . \quad (7)$$

Now the dx's are arbitrary infinitesimals, on which we can *a priori* impose any hypotheses we please; we shall consider them as infinitesimal functions of the x's. With this hypothesis the increments of these dx's, corresponding to the increments δx of the variables, will naturally be denoted by δdx; with a similar interpretation for $d\delta x$. The increment du will also be an infinitesimal function of the x's, and we shall thus have to consider δdu; $d\delta u$ will be similarly defined. We shall next obtain the explicit expression of these two *second differentials* of u, in order to show that a slight restriction on the arbitrariness of the second differentials of the independent variables will be sufficient to ensure the result $\delta du = d\delta u$, whatever the function u may be.

Applying the symbol of operation δ to the first of the equations (7), we get (without any restrictive hypothesis)

$$\delta du = \sum_1^n{}_i \delta\left(\frac{\partial u}{\partial x_i}\right) dx_i + \sum_1^n{}_i \frac{\partial u}{\partial x_i}\delta dx_i$$

$$= \sum_1^n{}_i \sum_1^n{}_j \frac{\partial^2 u}{\partial x_i \partial x_j} dx_i \delta x_j + \sum_1^n{}_i \frac{\partial u}{\partial x_i}\delta dx_i. \quad . \quad . \quad (8)$$

The expression for $d\delta u$ will evidently be similar, with d and δ interchanged. Now the first part of the formula is unaltered by this interchange, while in the second δdx_i is replaced by $d\delta x_i$. If therefore we impose on the arbitrary functions dx and δx of the x's the condition

$$d\delta x_i = \delta dx_i \qquad (i = 1, 2, \ldots n), \quad . \quad . \quad (9)$$

which represents a very small loss of generality, the second part of the formula (8) will also be unaltered when d and δ are interchanged; we shall therefore have, for any function whatever $u(x_1, x_2, \ldots x_n)$,

$$d\delta u = \delta du. \quad . \quad . \quad . \quad . \quad . \quad . \quad (10)$$

It may be noted incidentally that in the differential calculus it is usual to impose a hypothesis involving considerably greater restrictions than the conditions (9); the usual convention is that the second differentials of the independent variables are zero, or that the dx's are not functions of the x's, but constants.

We shall now consider, along with the increments of the independent variables, not a function u with its differentials, but a generic Pfaffian

$$\psi_d = \sum_1^n{}_i X_i dx_i,$$

in which the X's are given functions of the x's.

The suffix d has been inserted as a reminder that the Pfaffian refers to the increments dx_i; the same Pfaffian relative to the increments δx_i will be conveniently distinguished by the analogous notation

$$\psi_\delta = \sum_1^n{}_i X_i \delta x_i.$$

Both ψ_d and ψ_δ will naturally be functions of the x's. Calculating $\delta\psi_d$ we thus get

$$\delta\psi_d = \sum_1^n{}_i\, \delta X_i\, dx_i + \sum_1^n{}_i\, X_i\, \delta dx_i = \sum_1^n{}_i \sum_1^n{}_j \frac{\partial X_i}{\partial x_j} dx_i\, \delta x_j + \sum_1^n{}_i\, X_i\, \delta dx_i;$$

or with the abridged notation which can be used when several summations between the same limits are applied to the same general term,

$$\delta\psi_d = \sum_1^n{}_{ij} \frac{\partial X_i}{\partial x_j} dx_i\, \delta x_i + \sum_1^n{}_i\, X_i\, \delta dx_i.$$

Interchanging d and δ we get $d\psi_\delta$. Using the relation (9), the difference $\delta\psi_d - d\psi_\delta$ reduces to

$$\sum_1^n{}_{ij} \frac{\partial X_i}{\partial x_j} dx_i\, \delta x_j - \sum_1^n{}_{ij} \frac{\partial X_i}{\partial x_j} \delta x_i\, dx_j.$$

But the value of a sum is plainly unaffected by the particular letters of the alphabet which we choose to assign to the suffixes with respect to which the summation is to be made. We may therefore interchange i and j in the second part of the preceding formula, so that we can now write the equation in the form

$$\delta\psi_d - d\psi_\delta = \sum_1^n{}_{ij} \left(\frac{\partial X_i}{\partial x_j} - \frac{\partial X_j}{\partial x_i}\right) dx_i\, \delta x_j. \quad . \quad (11)$$

The expression $\delta\psi_d - d\psi_\delta$ is called the *bilinear covariant* relative to the given Pfaffian. The use of the term " bilinear " is sufficiently justified by the expression just found, which is linear in the arguments dx and also in the arguments δx. The name " covariant " is due to the circumstance that the numerical value and formal structure of the two sides of equation (11) always remain the same when the independent variables x vary in any way whatever. But we shall return to this point farther on (cf. Chapter VI, p. 144) in connexion with the general idea of invariants (functions or differential forms).

Meanwhile it may be noted that if the Pfaffian ψ_d is an exact differential, i.e. if the conditions (3) are satisfied, the right-hand side of equation (11) becomes zero, and we reach a result which has already been found (cf. formula (10)).

We may now return to the examination of the system of equations (4), and the conditions of complete integrability. Consider the m Pfaffians which constitute the right-hand sides of the equations (4):

$$\psi_d^{(a)} = \sum_1^n {}_i X_{a|i}\, dx_i,$$

and construct their bilinear covariants. We shall show that the two conditions: (a) that these covariants vanish identically, however dx and δx are chosen; and (b) that the equations (5) are identically true whatever values are assigned to the u's, are completely equivalent, so that the condition of complete integrability may be written in the form

$$\delta\psi_d^{(a)} - d\psi_\delta^{(a)} = 0 \qquad (a = 1, 2, \ldots m), \quad . \quad (12)$$

it being understood that this equation must hold for arbitrary values of the increments dx and δx.[1]

To prove this, take the explicit expression of these bilinear covariants, in the form given by equation (11). In differentiating it must be remembered that the X's must be considered as functions of the x's, both directly, and also indirectly as functions of the u's. Using the convention already adopted, the derivatives can therefore be denoted by the symbol for total differentiation; equation (12) thus becomes

$$\sum_1^n {}_{ij}\left(\frac{dX_{a|i}}{dx_j} - \frac{dX_{a|j}}{dx_i}\right) dx_i\, \delta x_j = 0. \quad . \quad (12')$$

Now if the conditions (5) for complete integrability are satisfied, the coefficients of this bilinear form (i.e. the expressions in parentheses in equation (12′)) are all zero, and therefore the equation is satisfied however the dx's and δx's are chosen. Vice versa, suppose that the equation is satisfied however the dx's and δx's are chosen. Then all the coefficients must necessarily be zero. For if we take all the dx's and δx's as zero, except one pair, e.g. dx_i, δx_j, where i, j, are two arbitrarily chosen but definite

[1] As a matter of fact we have imposed the restrictions (9) on the second differentials δdx_i, $d\delta x_i$, but the infinitesimal increments dx_i, δx_i to be assigned to the x_i's at the generic point under consideration are still entirely arbitrary.

integers of the series 1, 2, . . . n; then the sum in equation (12′) reduces to the single term

$$\left(\frac{dX_{\alpha|i}}{dx_j} - \frac{dX_{\alpha|j}}{dx_i}\right) dx_i \, \delta x_j,$$

which cannot vanish unless

$$\frac{dX_{\alpha|i}}{dx_j} - \frac{dX_{\alpha|j}}{dx_i} = 0.$$

We therefore conclude that the conditions (5) can be written in the more concise form (12).

5. Morera's method of integration.[1]

We shall now show that the conditions of complete integrability are sufficient for integrability, or more precisely that if they are satisfied there exists one and only one set of m functions $u(x)$ which satisfy the given system of equations and have values arbitrarily fixed in advance at a point also fixed in advance. Considering these initial values of u as arbitrary constants (as evidently they may be considered to be), we can say more shortly that the general integral depends on m arbitrary constants, or that there are ∞^m integrals.

For the proof, we first fix a generic point $P_0(x_1^0, x_2^0, \ldots x_n^0)$, in the field of variation of the x's in which the X's are defined. Let $P_1(x_1^1, x_2^1, \ldots x_n^1)$ be another arbitrary point in the field, and suppose it joined to P_0 by a line T which does not leave the field. T will be defined by parametric equations

$$x_i = \phi_i(t) \qquad (i = 1, 2, \ldots n), \quad . \quad . \quad (13)$$

where t is a parameter which has the value t_0 at P_0 and the value t_1 at P_1. We shall provisionally confine our investigation to the points of this line, so that for the present any functions u of the x's are to be considered as functions of the variable t alone (via the x's and the equations (13)). Their derivatives will be

$$\frac{du_\alpha}{dt} = \sum_1^n{}_i \frac{\partial u_\alpha}{\partial x_i} \frac{dx_i}{dt} \qquad (\alpha = 1, 2, \ldots m),$$

[1] "Zur Integration der vollständigen Differentiale", in *Math. Ann.*, Vol. 27, 1886, pp. 403–411. Cf. also SEVERI: "Sul metodo di Mayer per l'integrazione delle equazioni lineari ai differenziali totali", in *Atti del R. Ist. Veneto*, Vol. LXIX, 1910, pp. 419–425.

or, denoting differentiation with respect to t by a dot, and substituting from equation (4′),

$$\frac{du_\alpha}{dt} = \sum_1^n {}_i X_{\alpha|i}\,\dot{x}_i, \quad . \quad . \quad . \quad . \quad . \quad . \quad . \quad (14)$$

$$\text{or} \qquad \frac{du_\alpha}{dt} = \frac{\psi_d^{(\alpha)}}{dt} \qquad (\alpha = 1, 2, \ldots m). \quad . \quad (14')$$

The x_i's are known functions of t given by equations (13); hence the equations (14′) are of the type

$$\frac{du_\alpha}{dt} = U_\alpha(t \mid u_1, u_2, \ldots u_m) \qquad (\alpha = 1, 2, \ldots m), \quad (14'')$$

i.e. they form a system of ordinary differential equations, in the normal form. Now given m arbitrary constants $u_1^0, u_2^0, \ldots u_m^0$, it is known from the calculus that—subject to qualitative conditions of continuity and existence of derivatives, which we suppose satisfied—there exist m functions $u_\alpha(t)$ which satisfy the system (14″), and which are equal to the given constants when $t = t_0$. If, therefore, the u's are given any arbitrary set of values at P_0, they are defined at all points of the line T, and therefore also at P_1. It may however happen—and does in general—that if the points P_0 and P_1 are joined by another line instead of T, different values will be found for the u's at P_1. But we shall now show that if the conditions of complete integrability are satisfied, the values of the u's at P_1, found by the method just described, are independent of the line T, so that these u's will be functions only of the co-ordinates of P_1, that is, functions of position; they will satisfy the given system of equations not only along a line, but along all the infinite number of lines which can be drawn in the given field, or, in other words, in the whole of this field. They will therefore constitute the required solutions of the total differential system (4), as we shall show later on.

We shall simplify our task by considering infinitesimal displacements; i.e. by showing in the first place that the values of the u's at P_1 remain unaltered if the line T undergoes an infinitesimal deformation; it will follow that they will be the same for any line which can be obtained from T by a succession of infinitesimal deformations, i.e. by a continuous deformation of T. If then we suppose the field such that every line joining P_0 and

P_1 can be obtained in this way, we shall have all that is required. Such fields (e.g. a triangle or a circle in a plane, a cube or a sphere in space) are called *simply connected.*

Consider therefore a line T' infinitely close to T; we may think of it as obtained by displacing each point P of T, of co-ordinates x_i, to a point P' of co-ordinates $x_i + \delta x_i$, and the infinitesimal increment δx_i may be taken in the form $\epsilon\chi_i$, for example, where every χ_i is a finite quantity varying from point to point of the curve (and therefore a function of t), and ϵ is an infinitesimal factor taken as constant, and therefore independent of t. With these conditions the parametric equations of the curve T' will be

$$x_i + \delta x_i = \phi_i(t) + \epsilon\chi_i(t). \quad . \quad . \quad . \quad (15)$$

The functions χ_i may be considered as arbitrary, except for the condition of vanishing for $t = t_0$ and for $t = t_1$, so that the lines T and T' may have the same extremities. We shall adopt the natural convention of using the operator δ to denote the increment of a generic quantity (scalar or vector) in passing from the point P of T to the corresponding point P' of T'.

Now suppose the equations (14″) integrated along T'; we shall get functions of t, $u_a + \delta u_a$, satisfying the equations

$$\frac{d}{dt}(u_a + \delta u_a) = \frac{1}{dt}(\psi_d^{(a)} + \delta\psi_d^{(a)}) \qquad (a = 1, 2, \ldots m),$$

or
$$\frac{d\delta u_a}{dt} = \frac{1}{dt}(\delta\psi_d^{(a)});$$

using hypothesis (12), expressing the complete integrability of the system, we can also write the equations in the form

$$\frac{d\delta u_a}{dt} = \frac{d\psi_\delta^{(a)}}{dt}. \quad . \quad . \quad . \quad . \quad . \quad . \quad . \quad . \quad . \quad (16)$$

From the theorem of the existence of integrals of ordinary differential systems (already referred to in connexion with equations (14′)), it follows that the quantities δu_a are uniquely determined by these equations together with the condition of vanishing at P_0. Now the equations (16) are obviously satisfied by taking

$$\delta u_a = \psi_\delta^{(a)}, \quad . \quad . \quad . \quad . \quad . \quad . \quad . \quad (17)$$

(i.e. assuming for the quantities δu_a the expressions appropriate to the case where the u's are in fact functions of the x's). These expressions vanish with the δx_i's, i.e. at P_0 (so satisfying the initial conditions which, together with the equations (16), determine them uniquely), and also at P_1; which proves the required result.

It is thus proved that in order to construct the functions u whose total differentials are the assigned Pfaffians $\psi_d^{(a)}$ (satisfying identically the equations (12) or the original equations (5)) and which have given values u_a^0 at a given point P_0, we need only join P_0 to any point P_1 by any line T, and integrate the system of ordinary differential equations along T.

To complete the proof, we must now show that the differentials of the functions of the co-ordinates of P_1 obtained in this way are in fact the functions $\psi_d^{(a)}$. Consider a point P_2 infinitesimally close to P_1; to construct the values of the u's at P_2 take the broken line made up of T and the small segment P_1P_2. It is then obvious that integrating the equations (14) along this line we get, in passing from P_1 to P_2, the increment $du_a = \psi_d^{(a)}$.

6. **Note on Mayer's method.**

The method followed in the preceding section to show the existence of the integrals of a complete system of total differential equations, is due to Morera.

There was an earlier method, proposed by Mayer, by no means so clear, and seemingly dependent on a purely formal device. Morera's method, which is inspired by geometrical intuition, brings out the true reason for the success of Mayer's device, and provides a criterion for its validity.

Mayer's method is to join the points P_0 and P_1 by a segment of a straight line, instead of by any line T, so giving the equations (13) the form

$$x_i = x_i^0 + (x_i^1 - x_i^0)t \qquad (i = 1, 2, \ldots n);$$

the proof consists of a series of purely algebraic operations, instead of the proof developed above almost without calculations. In addition, while Morera's method can be applied if we merely suppose that the field in which the given equations hold is simply connected, Mayer's method, on the contrary, obviously requires a much more restrictive hypothesis, namely, that any two points in the field can be joined by a straight line which lies wholly in the field. This property is expressed by saying that the field is *convex.*

7. **Application.**

Given a generic Pfaffian

$$\psi = \sum_{1}^{n}{}_i X_i(x)dx_i,$$

we shall investigate whether it is possible to find a relation between x's of the type

$$f(x_1, x_2, \ldots x_n) = C \qquad (C \text{ constant}), \quad . \quad (18)$$

which shall be an integral of the equation

$$\psi = \sum_{1}^{n}{}_i X_i dx_i = 0, \quad . \quad . \quad . \quad . \quad (19)$$

in the sense that the relation produced by differentiating equation (18), namely,

$$df = \sum_{1}^{n}{}_i \frac{\partial f}{\partial x_i} dx_i = 0, \quad . \quad . \quad . \quad . \quad (18')$$

is equivalent to the equation (19).

For this it is plainly both necessary and sufficient that the derivatives of the unknown function f should be proportional to the given functions X_i. We therefore need some test to apply to the X_i's themselves which will show whether they are proportional to the derivatives of a single function not known in advance.

This problem, which also occurs in geometrical questions (as we shall see in particular on pp. 263–265), reduces at once to a particular case of a total differential system. In fact, given that ψ does not vanish identically, and therefore has at least one of its coefficients not equal to zero, we may legitimately suppose that X_n does not vanish identically. We can thus write equation (19) in the form

$$dx_n = -\sum_{1}^{n-1}{}_i \frac{X_i}{X_n} dx_i. \quad . \quad . \quad . \quad . \quad (19')$$

In order that this may be equivalent to the equation (18′), we must have $\frac{\partial f}{\partial x_n} \neq 0$ in the latter. From this condition it follows that the equation in finite terms (18) defines a function

$$x_n = u(x_1, x_2, \ldots x_{n-1}, C), \quad . \quad . \quad (18'')$$

which makes equation (18) an identity, and therefore also equation (18′), as well as the equivalent equations (19) and (19′). This last equation is evidently a particular case of systems of the type (4) consisting of one equation and one unknown function x_n; it must therefore be completely integrable, having as integral the function given by formula (18″), which depends on the arbitrary constant C. Reciprocally, if (19′) is completely integrable, then there will be a solution (18″) depending on an arbitrary constant C; solving with respect to C, this becomes an integral relation of the desired form (18). The problem therefore reduces to expressing the completeness of equation (19′).

Applying formula (5), the required conditions of completeness are

$$\frac{d}{dx_j}\frac{X_i}{X_n} = \frac{d}{dx_i}\frac{X_j}{X_n} \qquad (i, j = 1, 2, \ldots n-1;\ i \neq j).$$

Expanding the derivatives, these relations are easily put in the following form:

$$\left.\begin{array}{c} X_n\left(\frac{\partial X_i}{\partial x_j}-\frac{\partial X_j}{\partial x_i}\right)+X_i\left(\frac{\partial X_j}{\partial x_n}-\frac{\partial X_n}{\partial x_j}\right)+X_j\left(\frac{\partial X_n}{\partial x_i}-\frac{\partial X_i}{\partial x_n}\right)=0 \\ (i, j = 1, 2, \ldots n-1;\ i \neq j). \end{array}\right\} \quad (20).$$

Introduce for the moment the restriction that all the other functions X, as well as X_n, are different from zero. We can then write

$$p_{rs} = \frac{1}{X_r X_s}\left(\frac{\partial X_r}{\partial x_s} - \frac{\partial X_s}{\partial x_r}\right) \qquad (r, s = 1, 2, \ldots n) \quad (21)$$

whatever r and s may be, so that the conditions of integrability take the more concise form

$$p_{ij} + p_{jn} + p_{ni} = 0 \quad (i, j = 1, 2, \ldots n-1;\ i \neq j). \quad (22)$$

The conditions (22) are $\frac{1}{2}(n-1)(n-2)$ in number, this being the number of ways of choosing two distinct integers, i, j, from the series $1, 2, \ldots n-1$. They represent *all* the conditions of integrability. Now the choice of the variable x_n to be expressed as a function of the remaining x's was arbitrary (subject only to the condition $X_n \neq 0$); hence in general the relations

$$p_{ij} + p_{jk} + p_{ki} = 0 \quad . \quad . \quad . \quad . \quad (22')$$

must be satisfied, where i, j, k, are any three integers, no two of which are the same, chosen from the series $1, 2, \ldots n$. Such a triplet can be chosen in $\frac{1}{6}n(n-1)(n-2)$ ways; this is therefore the number of relations of the form (22′). But these are of course not all independent, since the conditions (22) (which form only part of (22′)) are sufficient for the complete integrability of the expression under consideration. In fact, it is easy to show directly that only $\frac{1}{2}(n-1)(n-2)$ of the equations (22′), e.g. those given by formula (22), are essential, the others reducing to algebraic deductions from them.

This can be shown by means of the following lemma, which holds whatever the terms p_{ik} may be. If p_{ik} $(i, k, = 1, 2, \ldots n)$ is a double skew (or antisymmetrical) system,[1] and if for some fixed suffix a the cyclic relation

$$p_{ik} + p_{ka} + p_{ai} = 0$$

is true for every pair of suffixes i, k, then this relation is also true for any three suffixes i, k, l.

To prove this, take the corresponding relations for the pairs k, l, and l, i,

$$p_{kl} + p_{la} + p_{ak} = 0,$$
$$p_{li} + p_{ia} + p_{al} = 0.$$

Adding, and remembering the condition of skewness

$$p_{ka} + p_{ak} = 0, \text{ \&c.},$$

there remains

$$p_{ik} + p_{kl} + p_{li} = 0. \qquad \text{Q.E.D.}$$

Substituting in equations (22′) the values of the p's given by

[1] I.e. a system of numbers such that a one-to-one correspondence, by a given law, exists between them and the pairs of integers i, k $(=1, 2, \ldots n)$, and such that $p_{ik} = -p_{ki}$ for any pair of indices whatever.

formula (21), and multiplying by $X_i X_j X_k$ so as to clear of fractions, we get the equations of condition

$$\left.\begin{array}{c} X_i\left(\dfrac{\partial X_j}{\partial x_k}-\dfrac{\partial X_k}{\partial x_j}\right)+X_j\left(\dfrac{\partial X_k}{\partial x_i}-\dfrac{\partial X_i}{\partial x_k}\right)+X_k\left(\dfrac{\partial X_i}{\partial x_j}-\dfrac{\partial X_j}{\partial x_i}\right)=0 \\ (i, j, k = 1, 2, \ldots n). \end{array}\right\} \quad (23)$$

We thus find this whole set of equations as a necessary consequence of that group of them—say, the group (20)—in which one of the suffixes is fixed, with the further condition that none of the X's vanish. This last condition was applied at the point where we divided by the product of the X's; it is, however, not essential, and can ultimately be discarded, as we shall now show. In fact, the equations (23) being necessary consequences of the equations (20) for any non-zero values of the X's, however small, and being integral in the X's and their derivatives, it follows that we may pass to the limit when any one of the X's tends to zero. We therefore have, for all values of the X's, that the equations (23)—or a group of them of the type (20)—constitute the necessary and sufficient conditions for the complete integrability of the equations (19), or, in other words, the condition that the n functions $X_i(x_1, x_2, \ldots x_n)$ may be proportional to the derivatives of a single function.

8. **Mixed systems of equations.**

In certain problems we have to deal with *mixed systems*, i.e. those containing some total differential equations and some equations in finite terms:

$$du_\alpha = \sum_1^n{}_i X_{\alpha|i}\, dx_i \qquad (\alpha = 1, 2, \ldots m), \quad . \quad (4)$$

$$F_k(x \mid u) = 0 \qquad (k = 1, 2, \ldots \nu). \quad . \quad . \quad (24)$$

The discussion is essentially the same as in § 3 (p. 16). But we propose to go through it again in order to obtain, in a form suitable for use in concrete cases, the condition of complete integrability of a mixed system of the type (4), (24).

It is obvious in the first place that a necessary condition for the existence of solutions is that the equations (24) (which we shall suppose mutually consistent and independent) are not more in number than m, the number of the unknowns u. If there were exactly m of them, they would completely determine the u's, and we should only have to examine whether the u's satisfy the equations (4). We shall therefore suppose

$$\nu < m,$$

and shall imagine the equations (24) solved with respect to ν of the u's, which will thus be expressed in terms of the x's and the remaining $m - \nu = \mu$ unknowns u.

As on p. 16, we shall call the two groups of u's respectively u''_β $(\beta = 1, 2, \ldots \nu)$ and u'_α $(\alpha = 1, 2, \ldots \mu)$, so that the equations (24) may be written (cf. equations (5'')) in the form

$$u''_\beta = f_\beta(x \mid u') \qquad (\beta = 1, 2, \ldots \nu). \quad . \quad . \quad (24')$$

Corresponding to this division of the u's into two groups it will be convenient to divide the equations (4) into two groups (4a) and (4b) (as was done on p. 16), which we repeat here for the reader's convenience:

$$du'_\alpha = \sum_1^n{}_i X_{\alpha \mid i} \, dx_i \qquad (\alpha = 1, 2, \ldots \mu), \quad . \quad . \quad (4a)$$

$$du''_\beta = \sum_1^n{}_i X_{\mu+\beta \mid i}(x \mid u) \, dx_i \qquad (\beta = 1, 2, \ldots \nu). \quad (4b)$$

We now propose to show that the given mixed system is completely integrable—and it will be called *complete*—if the following conditions are satisfied:

(a) The conditions (5) for the complete integrability of the equations (4) are satisfied when after differentiation the values of u'' given by equations (24') are substituted in them; they need not in general hold when any arbitrary functions are taken for the u'''s;

(b) When the functions u'' are replaced by their values as given by equations (24'), the equations (4b) must be identical with the equations obtained by differentiating the equations (24'); or more concisely, the equations (4b) must reduce to identities on substituting from equations (24').

We shall show that if the mixed system is complete, in the sense now considered, then the equations (4a), when the u'''s in them are expressed in terms of the u's and the x's by means of equations (24′), constitute a completely integrable system of μ total differential equations in μ unknowns; the u''s can therefore be obtained from them, and hence, by equations (24′), the u'''s; by hypothesis (*b*) above, the equations (4*b*) will thus be satisfied. Hence the problem will be solved and its general integral (p. 22) will contain $\mu = m - \nu$ arbitrary constants.

To simplify the formulæ, we shall agree that if

$$\Phi\,(x \mid u', u'')$$

is any function whatever of the x's and the u's, then

$$[\Phi]\,(x \mid u')$$

will denote the same function when the u'''s are replaced by the expressions (24′). We shall obviously have

$$\frac{\partial[\Phi]}{\partial x_j} = \left[\frac{\partial \Phi}{\partial x_j}\right] + \sum_1^\nu{}_\beta \left[\frac{\partial \Phi}{\partial u''_\beta}\right] \frac{\partial f_\beta}{\partial x_j} \qquad (j = 1, 2, \ldots n), \quad (25)$$

$$\frac{\partial[\Phi]}{\partial u'_\gamma} = \left[\frac{\partial \Phi}{\partial u'_\gamma}\right] + \sum_1^\nu{}_\beta \left[\frac{\partial \Phi}{\partial u''_\beta}\right] \frac{\partial f_\beta}{\partial u'_\gamma}, \qquad (\gamma = 1, 2, \ldots \mu). \quad (26)$$

With this convention, we can write hypotheses (*a*) and (*b*) respectively in the forms

$$\left.\begin{aligned} &\left[\frac{\partial X_{\alpha|i}}{\partial x_j}\right] + \sum_1^m{}_\lambda \left[\frac{\partial X_{\alpha|i}}{\partial u_\lambda}\right] \left[X_{\lambda|j}\right] \\ &\qquad = \left[\frac{\partial X_{\alpha|j}}{\partial x_i}\right] + \sum_1^m{}_\lambda \left[\frac{\partial X_{\alpha|j}}{\partial u_\lambda}\right] \left[X_{\lambda|i}\right] \\ &\qquad (i, j = 1, 2, \ldots n) \\ &\qquad (\alpha = 1, 2, \ldots \mu), \end{aligned}\right\} \quad . \quad . \quad (27)$$

$$\left[X_{\mu+\beta|j}\right] = \frac{\partial f_\beta}{\partial x_j} + \sum_1^\mu{}_\gamma \frac{\partial f_\beta}{\partial u'_\gamma}\left[X_{\gamma|j}\right] \qquad (\beta = 1, 2, \ldots \nu). \quad (28)$$

We have therefore to examine the conditions of complete integrability of equations (4*a*), which will be

$$\frac{d[X_{\alpha|i}]}{dx_j} = \frac{d[X_{\alpha|j}]}{dx_i} \qquad (\alpha = 1, 2, \ldots \mu), \quad (29)$$

and we have to show that they are satisfied identically.

Let us transform the left-hand side of (29) by first writing out in full the result of applying the operator $\dfrac{d}{dx_j}$ to a function of the x's and the u''s. We shall get

$$\frac{\partial[X_{\alpha|i}]}{\partial x_j}+\sum_1^\mu{}_\gamma\frac{\partial[X_{\alpha|i}]}{\partial u'_\gamma}[X_{\gamma|j}],$$

or, using formulæ (25) and (26),

$$\left[\frac{\partial X_{\alpha|i}}{\partial x_j}\right]+\sum_1^\nu{}_\beta\left[\frac{\partial X_{\alpha|i}}{\partial u''_\beta}\right]\frac{\partial f_\beta}{\partial x_j}$$

$$+\sum_1^\mu{}_\gamma[X_{\gamma|j}]\left\{\left[\frac{\partial X_{\alpha|i}}{\partial u'_\gamma}\right]+\sum_1^\nu{}_\beta\left[\frac{\partial X_{\alpha|i}}{\partial u''_\beta}\right]\frac{\partial f_\beta}{\partial u'_\gamma}\right\}$$

$$=\left[\frac{\partial X_{\alpha|i}}{\partial x_j}\right]+\sum_1^\mu{}_\gamma\left[\frac{\partial X_{\alpha|i}}{\partial u'_\gamma}\right][X_{\gamma|j}]$$

$$+\sum_1^\nu{}_\beta\left[\frac{\partial X_{\alpha|i}}{\partial u''_\beta}\right]\left\{\frac{\partial f_\beta}{\partial x_j}+\sum_1^\mu{}_\gamma\frac{\partial f_\beta}{\partial u'_\gamma}[X_{\gamma|j}]\right\};$$

and finally, using (28),

$$\left[\frac{\partial X_{\alpha|i}}{\partial x_j}\right]+\sum_1^\mu{}_\gamma\left[\frac{\partial X_{\alpha|i}}{\partial u'_\gamma}\right][X_{\gamma|j}]+\sum_1^\nu{}_\beta\left[\frac{\partial X_{\alpha|i}}{\partial u''_\beta}\right][X_{\mu+\beta|j}].$$

Remembering that the m arguments u consist of the two groups u' and u'', it will at once be seen that this is merely the left-hand side of (27). Interchanging i and j, the right-hand side of (29) similarly is seen to be identical with the right-hand side of (27); equations (27) being supposed to hold, it follows that the equations (29) are satisfied identically.

It follows that *the integration of a complete mixed system of the type* (4), (24), *reduces to that of a complete* (*and therefore integrable*) *total differential system in* μ *unknowns. The general integral therefore contains* $\mu = \mathrm{m} - \nu$ *arbitrary constants.*

If the mixed system is not complete, i.e. if the conditions (a) and (b) are not satisfied without further restrictions, then discussion on the lines of § 3 (p. 16) obviously shows that we must add to the equations (24) so many of the conditions (a) and (b) as do not reduce to identities in virtue of equations (24), since the equations (12) must hold whenever a set of m integrals u_α exists. Repeating the same procedure, we reach either an inconsistency, showing that equations (4), (24), can have no solutions.

or else a complete system with less than μ unknowns. In the latter case the number of constants in the general integral is also less than μ.

A particular result of the foregoing discussion is that if ν independent equations in finite terms are associated with a system of total differential equations in m functions u, the differential system being itself complete, then in the most favourable case (i.e. when the combined system is also complete) the number of constants in the integral is lowered by ν units, from m to $m - \nu$.

In general (i.e. when the mixed system is not complete) the integrals, if they exist, certainly contain less than $m - \nu$ constants.

CHAPTER III

Linear Partial Differential Equations Complete Systems

1. Linear operators.

In this chapter we shall frequently use N to denote the number of independent variables, which will themselves be denoted by the letters $z_1, \ldots z_N$.

Let $f(z_1, \ldots z_N)$ be any function whatever, subject only to the condition of being differentiable to any required order. The term *linear operator* relative to f will be used to denote the operation by means of which an expression of the type

$$\sum_1^N{}_\nu\, a_\nu \frac{\partial f}{\partial z_\nu}$$

is obtained from f, the a_ν's being any functions whatever of the z's. An expression of this kind will sometimes be denoted by a formula of the type Af, in which it is hardly necessary to point out that A is not a quantity, but the symbol of operation just defined.

We have therefore

$$A = \sum_1^N{}_\nu\, a_\nu \frac{\partial}{\partial z_\nu}.$$

It can at once be verified that the linear operator symbol

behaves in exactly the same way as the differentiation symbol when f is a sum, a product, or a composite function (a function of one or more functions); i.e. for two generic functions f_1, f_2, we have identically

$$A(f_1 + f_2) = Af_1 + Af_2, \quad . \quad . \quad . \quad . \quad (1)$$

$$A(f_1 f_2) = f_1 Af_2 + f_2 Af_1, \quad . \quad . \quad . \quad (2)$$

with obvious extensions to any number of terms. Further, if f is given as a function of n arguments $v_1, v_2, \ldots v_n$, which are themselves functions of z, we obviously have

$$Af(v_1, v_2, \ldots v_n) = \frac{\partial f}{\partial v_1} Av_1 + \frac{\partial f}{\partial v_2} Av_2 + \ldots + \frac{\partial f}{\partial v_n} Av_n. \quad (3)$$

Now consider the result of applying successively the two linear operators

$$A = \sum_{1}^{N}{}_{\nu} a_\nu \frac{\partial}{\partial z_\nu},$$

$$B = \sum_{1}^{N}{}_{\nu} b_\nu \frac{\partial}{\partial z_\nu},$$

the b's, like the a's, denoting functions of z which are differentiable to any required order.

The second-order operators

$$A(Bf),\ B(Af)$$

are thus completely defined; they may be written without danger of ambiguity in the form

$$ABf,\ BAf.$$

Writing out the first of these in full, we get by successive stages

$$\begin{aligned} ABf &= \sum_{1}^{N}{}_{\nu} a_\nu \frac{\partial Bf}{\partial z_\nu} = \sum_{1}^{N}{}_{\nu} a_\nu \sum_{1}^{N}{}_{\rho} \frac{\partial}{\partial z_\nu}\left(b_\rho \frac{\partial f}{\partial z_\rho}\right) \\ &= \sum_{1}^{N}{}_{\nu\rho} a_\nu \frac{\partial b_\rho}{\partial z_\nu} \frac{\partial f}{\partial z_\rho} + \sum_{1}^{N}{}_{\nu\rho} a_\nu b_\rho \frac{\partial^2 f}{\partial z_\nu \partial z_\rho} \\ &= \sum_{1}^{N}{}_{\rho} \frac{\partial f}{\partial z_\rho} Ab_\rho + \sum_{1}^{N}{}_{\nu\rho} a_\nu b_\rho \frac{\partial^2 f}{\partial z_\nu \partial z_\rho}. \end{aligned}$$

Similarly, interchanging A and B, and therefore a and b, we get

$$BAf = \sum_{1}^{N}{}_{\rho} \frac{\partial f}{\partial z_{\rho}} Ba_{\rho} + \sum_{1}^{N}{}_{\nu\rho} a_{\rho} b_{\nu} \frac{\partial^2 f}{\partial z_{\nu} \partial z_{\rho}}.$$

It appears from this that the two operators ABf and BAf are not equal; the second-order terms are however the same, as will be seen on interchanging the indices ν and ρ in one of the two double sums. It follows that the difference of the two operators in question is a linear operator of the first order; it is called the *alternate function* or *Poisson's parenthesis* relative to the two operators A and B, and is denoted by the symbol of operation (A, B), so that

$$(A, B)f = ABf - BAf = \sum_{1}^{N}{}_{\nu} (Ab_{\nu} - Ba_{\nu}) \frac{\partial f}{\partial z_{\nu}}. \quad . \quad (4)$$

It follows from the definition of the symbol that

$$(A, B)f = -(B, A)f. \quad . \quad . \quad . \quad . \quad (5)$$

We shall now establish a formal property of linear operators, which we shall use farther on.

Let there be n linear operators

$$A_k f = \sum_{1}^{N}{}_{\nu} a_{k|\nu} \frac{\partial f}{\partial z_{\nu}} \qquad (k = 1, 2, \ldots n),$$

and let any two linear combinations of these (which will also be linear operators),

$$Bf = \sum_{1}^{n}{}_{k} \lambda_k A_k f,$$

$$Cf = \sum_{1}^{n}{}_{h} \mu_h A_h f,$$

be constructed, the λ's and the μ's being any differentiable functions whatever of the independent variables z.

We propose to show that the alternate function $(B, C)f$ is a linear combination of the operators A and of their alternate functions. For the proof, it is sufficient to write out $(B, C)f$ in full; this gives

$$(B, C)f = BCf - CBf = \sum_{1}^{n}{}_{k} \lambda_k A_k (Cf) - \sum_{1}^{n}{}_{h} \mu_h A_h (Bf)$$

$$= \sum_{1}^{n}{}_{kh} [\lambda_k A_k (\mu_h A_h f) - \mu_h A_h (\lambda_k A_k f)].$$

Applying the rule for the differentiation of a product, the last expression becomes

$$\sum_1^n{}_{kh}[(\lambda_k A_k \mu_h) A_h f - (\mu_h A_h \lambda_k) A_k f] + \sum_1^n{}_{kh} \lambda_k \mu_h [A_k A_h f - A_h A_k f],$$

so that finally

$$(B, C)f = \sum_1^n{}_{kh}[(\lambda_k A_k \mu_h) A_h f - (\mu_h A_h \lambda_k) A_k f + \lambda_k \mu_h (A_k, A_h) f].$$

Q.E.D.

2. **Integrals of an ordinary differential system and the partial differential equation which determines them.**

Consider a system of n ordinary differential equations of the first order, in n unknowns x_i. Denoting the independent variable by t, and supposing the equations solved for the derivatives of the unknown functions, we get the equations in what is called the *normal form*:

$$\frac{dx_i}{dt} = X_i(x \mid t) \qquad (i = 1, 2, \ldots n). \quad . \quad (6)$$

Any set of n functions $x_i(t)$ which satisfies the given equations is called a *solution* of the system.

The term *integral* of the system, on the other hand, is used to denote any function $f(x \mid t)$ which reduces to a constant when the x's are replaced by any solution of the equations (6). We can therefore say that f is an integral if the result

$$f(x \mid t) = \text{constant}$$

is a necessary consequence of the differential equations (6).

We shall now show that all the functions f with this property (and no other functions) satisfy a homogeneous linear partial differential equation of the first order; it follows, as we shall see farther on, that the integration of an equation of this form can always be reduced to that of a system of the type (6).

Let $f(x \mid t)$ be an integral of the equations (6); then by definition, when the x's represent functions of t which satisfy equations (6), we have

$$f(x \mid t) = \text{constant},$$

and therefore, differentiating with respect to t,

$$\frac{\partial f}{\partial t} + \sum_1^n {}_i \frac{\partial f}{\partial x_i} \frac{dx_i}{dt} = 0,$$

or, since the functions $x_i(t)$ satisfy equations (6),

$$\frac{\partial f}{\partial t} + \sum_1^n {}_i \frac{\partial f}{\partial x_i} X_i = 0. \qquad (7)$$

This is the partial differential equation referred to. Introducing for shortness the linear operator

$$A = \frac{\partial}{\partial t} + \sum_1^n {}_i X_i \frac{\partial}{\partial x_i}, \qquad (7')$$

we can write it concisely in the form

$$Af = 0.$$

Now by hypothesis equation (7), like the equation $f =$ constant, from which it is derived by differentiation, becomes an identity when the x's in it are replaced by any solutions whatever of the system (6); from this it is easy to deduce that (7) is an identity, that is, that it holds for any values whatever (in a suitable field) which may be assigned to the arguments x, t, of which f is a function. In fact, given $n + 1$ numbers $x_1^0, \ldots x_n^0$, t_0, belonging to a field within which the general existence theorem holds for the system (6), we know that there always exists a solution x_i of the system (6), which takes the values $x_1^0, \ldots x_n^0$ when $t = t_0$. Now equation (7) must hold (whatever t may be) when this particular solution $x_i(t)$ is substituted for the x's. In particular, putting $t = t_0$, the equation is satisfied for the values x_i^0, t_0, arbitrarily chosen in advance. Q.E.D.

It is further evident that any function $f(x \mid t)$ which satisfies equation (7), when the x's and the t's in it are treated as independent variables, constitutes an integral of the system (6). In fact, since equation (7) holds however the x's are chosen, it will be satisfied in particular when we take a solution of the system (6) for the x's; but when this is done the left-hand side of equation (7) becomes identical with $\frac{df}{dt}$. The function f is therefore such that when the x's are solutions of the system (6), $\frac{df}{dt} = 0$, or $f =$ constant.

To sum up, we can state that *the necessary and sufficient condition that a function* $f(x \mid t)$ *may be an integral of the system* (6) *is that it should satisfy the partial differential equation* (7), *in which the* x's *and* t *are* n + 1 *independent variables.*

3. Principal integrals.

Among the integrals f of the system (6) (which, as we have seen in the preceding section, can also be called integrals of equation (7)), there are, for each value t_0 of t, n of special importance which we now proceed to specify.

We take as our starting-point the most general solution of the equations (6), which is known to be a set of n functions of t, containing n arbitrary constants $x_1^0, \ldots x_n^0$:

$$x_i = \phi_i(t \mid x^0) \qquad (i = 1, 2, \ldots n). \quad . \quad . \quad . \quad (8)$$

The constants x^0 are the values of the x's for a given value t_0 of t, so that

$$(\phi_i)_{t = t_0} = x_i^0. \quad . \quad . \quad . \quad . \quad . \quad . \quad (9)$$

We shall show first that the equations (8) are soluble with respect to the x^0's in a region round the point t_0. Write them in the form

$$\phi_i(t \mid x^0) - x_i = 0,$$

and consider the functional determinant of the left-hand side with respect to the x^0's, which is

$$D = \begin{pmatrix} \phi_1 - x_1 & \phi_2 - x_2 & \ldots & \phi_n - x_n \\ x_1^0 & x_2^0 & \ldots & x_n^0 \end{pmatrix},$$

or, since the x^0's are contained only in the ϕ's, and not in the x's,

$$D = \begin{pmatrix} \phi_1 & \phi_2 & \ldots & \phi_n \\ x_1^0 & x_2^0 & \ldots & x_n^0 \end{pmatrix}.$$

Now calculate the value D_0 of this determinant for $t = t_0$. Since the determinant itself contains no derivatives with respect to t, we shall obtain the same result if we differentiate the functions $\phi(t \mid x^0)$ with respect to the x^0's, form their determinant, and finally make $t = t_0$, as if we first make $t = t_0$ in the ϕ's, and then form the determinant of their derivatives. Following the second alternative and remembering the formulæ (9) we see at once that the determinant becomes

$$D_0 = \begin{pmatrix} x_1^0 \, x_2^0 \, . \, . \, . \, x_n^0 \\ x_1^0 \, x_2^0 \, . \, . \, . \, x_n^0 \end{pmatrix} = 1 \neq 0.$$

Now if D, which is a continuous function of t, does not vanish for $t = t_0$, it will have these same properties in some region round t_0, and therefore within this region the equations (8) will be soluble with respect to the x^0's.

Solving the equations, we shall get

$$x_i^0 = w_i(x \mid t) \qquad (i = 1, 2, \ldots n), \quad . \quad . \quad (10)$$

and the w's on the right-hand side constitute n integrals of the system (6). In fact, if we replace the x's in them by any solution of the system (6), (i.e. by a set of n functions obtained from the equations (8) by assigning particular arbitrary values to the constants x^0), then each w necessarily becomes equal to the corresponding x^0, i.e. to a constant.

The integrals of equation (7) obtained in this way are called *principal integrals relative to* $t = t_0$. From the definition it follows that

$$w_i(x^0 \mid t_0) = x_i^0.$$

Writing x instead of x^0, we see that a characteristic property of the principal integrals w_i relative to $t = t_0$ is that each of the functions $w_i(x \mid t)$ reduces to the corresponding variable x when $t = t_0$.

Without undertaking a detailed study of the n principal integrals, we may at least show that none of them can be expressed as a function of the others only; i.e. that considered as n functions of the $n + 1$ variables x and t, they are independent. For this it is necessary and sufficient that the functional matrix (with n rows and $n + 1$ columns) of the w's with respect to the x's and t shall have n for its characteristic; i.e. that the matrix shall contain a determinant of order n which is not zero. Now if we take the determinant

$$\begin{pmatrix} w_1 \, w_2 \, . \, . \, . \, w_n \\ x_1 \, x_2 \, . \, . \, . \, x_n \end{pmatrix}, \quad . \quad . \quad . \quad . \quad . \quad (11)$$

and apply to it the same considerations as we have already used for D, we find that it $= 1$ for $t = t_0$ (since then $w_i = x_i$) and therefore there is a region round t_0 in which it is not zero; hence the characteristic of the matrix is n and the principal integrals are independent.

4. Independent integrals. General integral.

More generally, n integrals $v_1, v_2, \ldots v_n$ of equation (7) are said to be *independent* if the functions $v_i(x \mid t)$ $(i = 1, 2, \ldots n)$ are independent. Of course every function

$$F(v_1, v_2, \ldots v_n) \quad . \quad . \quad . \quad . \quad . \quad (12)$$

of the v's only is also an integral, as follows immediately from formula (3), remembering that for every v_i, with the operator A as defined by (7′), we have

$$Av_i = 0.$$

But the reciprocal theorem is also true, and every integral of equation (7) can be put in the form (12), which therefore represents the *general integral* of equation (7).

To prove this, let f be an integral of equation (7); then the $n + 1$ equations

$$Af = \frac{\partial f}{\partial t} + \sum_1^n{}_j \frac{\partial f}{\partial x_j} X_j = 0,$$

$$Av_i = \frac{\partial v_i}{\partial t} + \sum_1^n{}_j \frac{\partial v_i}{\partial x_j} X_j = 0 \qquad (i = 1, 2, \ldots n),$$

linear and homogeneous in the $n + 1$ quantities $1, X_1, \ldots X_n$, which do not all vanish, will be satisfied. The determinant of their coefficients must therefore vanish, i.e.

$$\begin{pmatrix} f & v_1 & \ldots & v_n \\ t & x_1 & \ldots & x_n \end{pmatrix} = 0.$$

This means that $f, v_1, v_2, \ldots v_n$ are not independent. As the v's are independent, one of the determinants of order n of the functional matrix relative to $v_1, v_2, \ldots v_n$ is certainly not zero. But this is the case considered in Chapter I, pp. 5–8; hence we conclude that f can be expressed in terms of the v's only, without involving t or the x's.

5. Direct study of the most general linear homogeneous partial differential equation.

As a consequence of the relations which we have shown to hold between linear homogeneous partial differential equations of the first order, and systems of ordinary differential equations of the first order, we can in every case reduce the integration of an equation of the former kind to the integration of a system of the latter. In fact, equation (7) differs from the most general possible equation only in having one of its coefficients equal to 1; but it will be at once obvious that every linear homogeneous equation of the first order can be reduced to this form—an elementary remark which we shall examine in more detail.

Consider the equation

$$Af = \sum_{1}^{N} {}_{\nu}\, a_{\nu} \frac{\partial f}{\partial z_{\nu}} = 0 \quad . \quad . \quad . \quad . \quad (13)$$

in N independent variables $z_1, \ldots z_N$.

At least one of the a's, say a_N, will be different from zero. We may therefore divide the equation by a_N. As a result of this step, z_N is in what we may call a privileged position (the coefficient of $\dfrac{\partial f}{\partial z_N}$ being reduced to unity); it is therefore natural to denote it by a special symbol. Calling it t, and introducing the symmetrical notation $x_1, x_2, \ldots x_n$ $(n = N - 1)$ for the remaining variables z, equation (13) becomes

$$\frac{\partial f}{\partial t} + \sum_{1}^{n} {}_{i}\, \frac{a_i}{a_N} \frac{\partial f}{\partial x_i} = 0,$$

which will coincide term for term with equation (7) if we put

$$\frac{a_i}{a_N} = X_i \qquad (i = 1, 2, \ldots n).$$

The corresponding system of equations (6) will thus be

$$\frac{dx_i}{dt} = \frac{a_i}{a_N} \qquad (i = 1, 2, \ldots n). \quad . \quad . \quad . \quad (14)$$

Integrating these n equations, in which t is considered as the independent variable, we get the functions $x_i(t \mid x^0)$, which reduce to x_i^0 when $t = t_0$; solving for each of the x^0's, the resulting expressions

$$w_i(x \mid t) = w_i(z)$$

will be the n principal integrals of equation (13) relative to $t = t_0$, and we may take

$$F(w_1, \ldots w_n)$$

as the form of the general integral, where the symbol F denotes an arbitrary function.

It will frequently be quicker not to find the principal integrals w, but to obtain n *independent* integrals $v_1, v_2, \ldots v_n$ of the equations (14) in any way whatever. The general integral is then at once expressible as an arbitrary function of the v's only, in the form

$$F(v_1, v_2, \ldots v_n).$$

We have seen that we may choose as independent variable (for the system of equations (14)) any one of the variables z which contributes an actual term $\frac{\partial f}{\partial z}$ to equation (13) (the corresponding coefficient a not being zero). The choice may be determined by reasons of convenience for the particular case concerned. In order to avoid prejudging the case before the necessary reasons for our choice appear, we may write the equations (14) in the form

$$\frac{dx_i}{a_i} = \frac{dt}{a_N} \qquad (i = 1, 2, \ldots n),$$

or, returning to the original notation,

$$\frac{dz_1}{a_1} = \frac{dz_2}{a_2} = \ldots = \frac{dz_N}{a_N}. \qquad (15)$$

It will be seen that these can be at once written down from the given partial differential equation; it should be noted that if any of the a's is zero, the differential of the corresponding variable must also be equated to zero.

Examples:

(1) Take the equation in three variables

$$x\frac{\partial f}{\partial x}+y\frac{\partial f}{\partial y}+z\frac{\partial f}{\partial z}=0. \quad . \quad . \quad . \quad . \qquad (16)$$

The corresponding system of two ordinary differential equations is

$$\frac{dx}{x}=\frac{dy}{y}=\frac{dz}{z},$$

$$\text{i.e.} \qquad d\log x = d\log y = d\log z.$$

Writing these in the form

$$d\log y - d\log x = 0,$$

$$d\log z - d\log x = 0,$$

which is the same as

$$d\log\frac{y}{x} = 0,$$

$$d\log\frac{z}{x} = 0,$$

we get

$$\frac{y}{x}=c_1, \quad \frac{z}{x}=c_2 \quad (c_1,\, c_2 \text{ constants}).$$

Hence two independent integrals are

$$\frac{y}{x}, \; \frac{z}{x},$$

and therefore the general integral will be

$$F\left(\frac{y}{x}, \; \frac{z}{x}\right).$$

This is merely the most general homogeneous function of degree zero. In fact, if $\phi(x, y, z)$ denotes the latter, then by definition we must have, for any value of λ whatever,

$$\phi(\lambda x, \lambda y, \lambda z) = \phi(x, y, z);$$

and therefore, putting $\lambda = \frac{1}{x}$,

$$\phi\left(1, \frac{y}{x}, \frac{z}{x}\right) = \phi(x, y, z);$$

ϕ is therefore in effect an arbitrary function F of $\frac{y}{x}, \frac{z}{x}$.

We have thus found for a particular case (which can obviously be generalized) Euler's well-known theorem on homogeneous functions.

(2) Take the equation

$$\begin{vmatrix} \frac{\partial f}{\partial x} & \frac{\partial f}{\partial y} & \frac{\partial f}{\partial z} \\ x & y & z \\ a & b & c \end{vmatrix} = 0,$$

where a, b, c are constants which are not all zero. Putting

$$X = \begin{vmatrix} y & z \\ b & c \end{vmatrix}, \quad Y = \begin{vmatrix} z & x \\ c & a \end{vmatrix}, \quad Z = \begin{vmatrix} x & y \\ a & b \end{vmatrix},$$

the equation may be written in the form

$$X\frac{\partial f}{\partial x} + Y\frac{\partial f}{\partial y} + Z\frac{\partial f}{\partial z} = 0,$$

and the corresponding system of ordinary differential equations is

$$\frac{dx}{X} = \frac{dy}{Y} = \frac{dz}{Z}. \quad . \quad . \quad . \quad . \quad . \quad (17)$$

Taking x as independent variable, we shall have to integrate the two equations

$$\frac{dy}{dx} = \frac{Y}{X}, \ \frac{dz}{dx} = \frac{Z}{X}.$$

But we can find two independent integrals more easily by

another method. In fact, from the form of the equations defining X, Y, Z, we see that the equations

$$\left.\begin{aligned} xX + yY + zZ &= 0 \\ aX + bY + cZ &= 0 \end{aligned}\right\} \quad . \quad . \quad . \quad . \quad . \qquad (18)$$

are satisfied, since the given determinant vanishes if the elements of the first row are replaced by x, y, z, or by a, b, c; expanding, we find that the equations (18) are identically true.

Now if dx, dy, dz satisfy equations (17), i.e. are proportional to X, Y, Z, we can substitute them for X, Y, Z in equations (18), so getting

$$xdx + ydy + zdz = 0,$$

$$adx + bdy + cdz = 0.$$

The left-hand side of each of these equations is an exact differential; hence, integrating, we get as a necessary consequence of equations (17)

$$x^2 + y^2 + z^2 = c_1,$$

$$ax + by + cz = c_2.$$

These are two particular integrals of the system, which are certainly independent, since at least one of the coefficients a, b, c, is not zero. The general integral is thus

$$F(x^2 + y^2 + z^2,\ ax + by + cz).$$

Geometrical interpretation.—When there are three (or two) variables, the foregoing discussion can be given a geometrical interpretation in ordinary space (or in a plane). For this purpose, let any integral $f(x, y, z)$ of an equation

$$(a) \qquad X\frac{\partial f}{\partial x} + Y\frac{\partial f}{\partial y} + Z\frac{\partial f}{\partial z} = 0$$

be considered as the *parameter of a family of surfaces* $f =$ constant. By a suitable choice of the constant on the right-hand side of this equation, we can make one surface of the family pass through a point P arbitrarily chosen in advance; it is plainly only necessary to give this constant the value of f at the chosen point P. The equation (a) which f must satisfy expresses the geometrical fact that at any point P the normal to that surface of the family which passes through P is normal also to the direction of the vector (X, Y, Z), given as a function of position.

The system of equations associated with equation (a), namely

$$(b) \qquad \frac{dx}{X} = \frac{dy}{Y} = \frac{dz}{Z},$$

which expresses the relation of two of the variables to the third, represents on the other hand a property of certain curves. We know from the existence theorem that we can find (and in only one way) two functions $x = x(z)$, $y = y(z)$ which satisfy the system of equations (b), and which take values x_0, y_0, arbitrarily fixed in advance, when z has the value z_0, which is also arbitrary; hence we can state that, given any point P, there passes through it one and only one curve which has the property expressed by the equations (b), i.e. that of being at every point in the direction of the vector (X, Y, Z). An aggregate of curves such that one and only one of them passes through every point of a given field is called a *congruence*.

There is a very simple relation between the family of surfaces which represent the integrals of equation (a) and the congruence of curves which represent the solutions of the system of equations (b), namely, that *each curve of the congruence lies wholly on a surface of the family*. In fact, consider a point $P(x, y, z)$, and let L be the curve of the congruence, and S the surface of the family considered, which pass respectively through P. We shall show that a point which undergoes an infinitely small displacement along L, from P to a point P', does not leave the surface, or in other words that if the equation of the surface S, $f(x, y, z) = C$, is satisfied by the co-ordinates x, y, z of P, it is also satisfied by the co-ordinates $x + dx$, $y + dy$, $z + dz$, of P'. The result is obvious, since for the co-ordinates of P' f becomes

$$f(x, y, z) + \frac{\partial f}{\partial x} dx + \frac{\partial f}{\partial y} dy + \frac{\partial f}{\partial z} dz,$$

and as dx, dy, dz, are by equations (b) proportional to X, Y, Z, the increment of f is proportional to

$$\frac{\partial f}{\partial x} X + \frac{\partial f}{\partial y} Y + \frac{\partial f}{\partial z} Z,$$

which vanishes by equation (a).

The surfaces considered are therefore formed by curves of the congruence; these curves are called their *characteristics*.

We can of course avoid the use of infinitesimal displacements in proving the property that every curve of the congruence lies wholly on a surface $f =$ constant. The argument, which is essentially the same as before, will be as follows. Let the variables x, y, z, in the expression $f(x, y, z)$ be considered as the coordinates of points of a curve L of the congruence, and let λ be any parameter, e.g. the arc of the curve, which fixes definitely the position of a point on L; x, y, z, are thus considered as definite functions of λ. Substituting these functions of λ for x, y, z in the expression $f(x, y, z)$, we shall see that the result is independent of λ, or that $\frac{df}{d\lambda} = 0$. We have in fact

$$\frac{df}{d\lambda} = \frac{\partial f}{\partial x}\frac{dx}{d\lambda} + \frac{\partial f}{\partial y}\frac{dy}{d\lambda} + \frac{\partial f}{\partial z}\frac{dz}{d\lambda};$$

but by equations (b), $\frac{dx}{d\lambda}, \frac{dy}{d\lambda}, \frac{dz}{d\lambda}$, along L, are proportional to X, Y, Z, and f satisfies equation (a); hence $\frac{df}{d\lambda} = 0$. The fact that f remains constant along L is the algebraical equivalent of the geometrical property that the curve L belongs to a surface $f =$ constant.

6. Integrals of a total differential system, and the associated system of partial differential equations which determines them.

Starting from a system of *ordinary* differential equations, we have succeeded in integrating the most general linear homogeneous partial differential equation of the first order. By an analogous procedure, starting from a system of *total differential* equations, we shall succeed in integrating the most general system of linear homogeneous partial differential equations of the first order.

Consider the system of equations

$$du_\alpha = \sum_{1}^{n}{}_i X_{\alpha|i}\, dx_i \qquad (\alpha = 1, 2, \ldots m). \quad (19)$$

We shall apply the term *integral* of this system to every

function $f(x \mid u)$ which is such that it reduces to a constant when the u's are replaced by any *solution* of the equations (19).

Differentiating f we get

$$df = \sum_1^n{}_i \frac{\partial f}{\partial x_i} dx_i + \sum_1^m{}_\alpha \frac{\partial f}{\partial u_\alpha} du_\alpha,$$

and, if the u's are solutions of the equations (19),

$$df = \sum_1^n{}_i \left(\frac{\partial f}{\partial x_i} + \sum_1^m{}_\alpha \frac{\partial f}{\partial u_\alpha} X_{\alpha \mid i} \right) dx_i.$$

The necessary and sufficient condition for the vanishing of this differential, whatever the dx's may be, is evidently that the n equations

$$\frac{\partial f}{\partial x_i} + \sum_1^m{}_\alpha \frac{\partial f}{\partial u_\alpha} X_{\alpha \mid i} = 0 \qquad (i = 1, 2, \ldots n) \quad . \qquad (20)$$

shall be satisfied. They must be satisfied not only when the u's are solutions of the equations (19), as is clear from what has already been said, but also identically, which can be seen in the same way as for the corresponding result in § 2 (p. 36).

Introducing the linear operators

$$\Omega_i = \sum_1^m{}_\alpha X_{\alpha \mid i} \frac{\partial}{\partial u_\alpha}, \quad . \quad . \quad . \quad . \quad . \quad . \quad . \quad (21)$$

$$B_i = \frac{\partial}{\partial x_i} + \Omega_i \qquad (i = 1, 2, \ldots n), \quad . \quad . \qquad (22)$$

the equations (20) may be written in the form

$$B_i f = 0 \qquad (i = 1, 2, \ldots n). \quad . \quad . \qquad (20')$$

The system of equations (20) or (20′) is said to be *associated* with the system of equations (19); the necessary and sufficient condition that f may be an integral of the system of equations (19) is that it should satisfy the associated system of partial differential equations (20), or, in other words, that it should be an integral of the associated system.

7. Principal integrals, as typical cases of independent integrals.

Suppose that the system of equations (19) is completely integrable, and let us assign fixed arbitrary values to the constants x^0 and u^0 in the field in which the X's are regular; then we know that there exist m functions,

$$u_\alpha = \phi_\alpha(x \mid u^0) \qquad (\alpha = 1, 2, \ldots m),$$

regular in the region round the values x^0, u^0, which satisfy the equations (19), and which become respectively equal to the assigned constants u_α^0 when $x_i = x_i^0$. The equations so written are soluble with respect to the quantities u^0 in a region round the point x^0, as can be shown by means of the same arguments as those of § 3 (p. 38). Suppose them solved; we can then express the u^0's in terms of the x's and the u's, and we shall write

$$w_\alpha(x \mid u) = u_\alpha^0.$$

The w's are evidently integrals of the system of total differential equations (19), and are therefore also integrals of the system of partial differential equations (20); we shall now show that they are independent.

Consider the functional matrix of the w's with respect to the x's and the u's; we shall have to show that its characteristic is m, or, which comes to the same thing (since it contains no determinants of order $> m$), that it contains a determinant of order m which is not zero. Now the determinant

$$\begin{pmatrix} w_1 \, w_2 \, \ldots \, w_m \\ u_1 \, u_2 \, \ldots \, u_m \end{pmatrix}$$

becomes $= 1$ when $x_i = x_i^0$, and therefore is different from zero in a region round that point; hence the required result follows.

The m independent integrals w are called *principal integrals* of the total differential system (19), or of the partial differential system (20), corresponding to the values x_i^0 of the independent variables x. We have thus shown that, with the hypothesis that the system of equations (19) is completely integrable, the system of equations (20) (or (20′)) admits of m independent integrals, which can be determined in an infinite number of ways; namely, the principal integrals just considered, which in general vary with the choice of the initial values x^0.

Here too, as on p. 40, we shall say more generally that the m integrals $v_1, v_2, \ldots v_m$ of the system (20) are independent if the functions $v(x \mid u)$ of the $n + m$ variables x and u are independent.

8. The general integral.

By a property already noted of linear operators, if we construct any function whatever of the v's,

$$F(v_1, v_2, \ldots v_m), \quad . \quad . \quad . \quad . \quad (23)$$

we get a new integral of the system of partial differential equations (20). In addition, for the system of equations (20), as before for the single equation (7), the most general function which satisfies the system is included in the expression (23); or this expression, when F is considered as an arbitrary function, constitutes the *general integral* of the system.

To prove this, let $f(x \mid u)$ denote any integral of the system (20), and consider the functional matrix of the $m + 1$ functions (of $m + n$ variables) $v_1, v_2 \ldots v_m, f$:

$$M = \left\| \begin{matrix} \frac{\partial v_1}{\partial x_1} & \cdots & \frac{\partial v_1}{\partial x_n} & \frac{\partial v_1}{\partial u_1} & \cdots & \frac{\partial v_1}{\partial u_m} \\ \cdot & \cdot & \cdot & \cdot & \cdot & \cdot \\ \frac{\partial v_m}{\partial x_1} & \cdots & \frac{\partial v_m}{\partial x_n} & \frac{\partial v_m}{\partial u_1} & \cdots & \frac{\partial v_m}{\partial u_m} \\ \frac{\partial f}{\partial x_1} & \cdots & \frac{\partial f}{\partial x_n} & \frac{\partial f}{\partial u_1} & \cdots & \frac{\partial f}{\partial u_m} \end{matrix} \right\| .$$

If we can show that the characteristic of this matrix is m, it will follow that there exists $(m + 1) - m$, or 1, relation between the $m + 1$ functions which does not involve the x's or the u's (cf. § 7, pp. 9–12). This relation must necessarily contain f explicitly, since there can be no relation connecting the v's alone. We can therefore solve it for f, which will have the form (23), so giving the required result.

We shall first make a slight change in the form of the matrix M, by making it contain only derivatives with respect to the u's. This is easily done, for since

$$B_i v_a = 0, \; B_i f = 0,$$

we get from formula (22)

$$\frac{\partial v_a}{\partial x_i} = -\Omega_i v_a, \frac{\partial f}{\partial x_i} = -\Omega_i f.$$

The matrix thus becomes

$$M = \begin{Vmatrix} -\Omega_1 v_1 & \ldots & -\Omega_n v_1 & \frac{\partial v_1}{\partial u_1} & \ldots & \frac{\partial v_1}{\partial u_m} \\ \cdot & \cdot & \cdot & \cdot & \cdot & \cdot \\ -\Omega_1 v_m & \ldots & -\Omega_n v_m & \frac{\partial v_m}{\partial u_1} & \ldots & \frac{\partial v_m}{\partial u_m} \\ -\Omega_1 f & \ldots & -\Omega_n f & \frac{\partial f}{\partial u_1} & \ldots & \frac{\partial f}{\partial u_m} \end{Vmatrix}.$$

To prove that the characteristic is m, we have to prove:

(1) That every determinant of order $m + 1$ (the highest order possible) vanishes;

(2) That at least one determinant of order m does not vanish.

A generic determinant of order $m + 1$ will be formed by taking all the $m + 1$ rows, and $m + 1$ columns chosen arbitrarily from the $m + n$ of the matrix. These $m + n$ columns are of two types: the first n contain the operators Ω, the remaining m the operators $\frac{\partial}{\partial u}$; let r columns be taken of the first type, and s of the second, with of course $r + s = m + 1$. Now in order to write down in a perfectly general form a row of this determinant, which will contain either the v's or (if it is the last row) f, we shall use the symbol ϕ to denote either one of the v's or f; we can then write the row as follows:

$$\Omega_{h_1}\phi, \Omega_{h_2}\phi, \ldots \Omega_{h_r}\phi, \frac{\partial \phi}{\partial u_{k_1}}, \frac{\partial \phi}{\partial u_{k_2}}, \ldots \frac{\partial \phi}{\partial u_{k_s}},$$

where the suffixes $h_1, h_2, \ldots h_r$ constitute any arrangement of r numbers, chosen from 1 to n, and the suffixes $k_1, k_2, \ldots k_s$ any arrangement of s numbers, chosen from 1 to m. Remembering the definition of Ω given in (21), we see that each of the first r elements of the general row is a linear combination of the other elements; or, as we usually say, that the first r columns of the determinant are linear combinations of the other columns. The determinant can thus be broken up into a linear combination of determinants of order $m + 1$ in which all the columns are of the second type $\left(\text{i.e. are composed of terms } \frac{\partial \phi}{\partial u}\right)$. But there are in all only m columns of the second type; it is therefore impossible to choose $m + 1$ of them without repeating at least one. It follows that in each of these partial determinants there are at least two columns equal, and therefore these determinants all vanish. The general determinant of order $m + 1$, which is a linear combination of them, must therefore also vanish. This proves the first of the required propositions.

The existence of a non-vanishing determinant of order m is a direct consequence of the hypothesis that the integrals v are independent.

We have therefore proved completely that the characteristic of M is m, and therefore that f can be expressed in terms of the independent integrals v, i.e. that f has the form given in (23).

9. Direct study of the most general system of linear homogeneous partial differential equations of the first order. Complete systems. Jacobian systems.

Let us consider a generic system of n linear homogeneous partial differential equations of the first order, in N variables, and with only one unknown function:

$$A_k f = \sum_{1}^{N}{}_{\nu} a_{\nu k} \frac{\partial f}{\partial z_\nu} = 0 \quad (k = 1, 2, \ldots n). \quad . \quad (24)$$

We shall suppose that these n equations are independent, and we can therefore assume $n < N$. In fact, if $n > N$, the equations, which we have supposed independent, considered as algebraic equations in the N quantities $\frac{\partial f}{\partial z_\nu}$, would be mutually inconsistent; and if $n = N$, this would imply that $\frac{\partial f}{\partial z_\nu} = 0$, or $f =$ constant. Further, it is clear that every f which satisfies equations (24) must necessarily also satisfy the following $\frac{1}{2}n(n-1)$ equations (obtained by constructing all possible Poisson's parentheses with the given operators):

$$(A_h, A_k)f = 0 \qquad (h, k = 1, 2, \ldots n). \quad . \quad (25)$$

These are *differential* consequences of the given system. Since derivatives of the second order disappear from equations (25), it may happen that these equations, or some of them, are also *algebraic* consequences of the system, i.e. that they can be obtained algebraically by taking a linear combination of the n given equations.

If *all* the equations (25) are algebraic consequences of the system of equations (24), this system is called *complete.*

In the opposite case, consider the system formed by adding to (24) those of (25) which, together with (24), are linearly independent. The new system will be equivalent to the original one, and will contain one or more additional equations. Repeating the same procedure for the new system, and so on, we shall reach either a complete system or else a system in which the number of equations is equal to or greater than N — the case of mutual inconsistency, as already noted at the beginning of this section.

We need therefore only consider complete systems. The condition of completeness can be written in the following form:

$$(A_h, A_k)f = \sum_1^n{}_l\, p_{hkl} A_l f, \quad . \quad . \quad . \quad (26)$$

where the coefficients p denote functions (*a priori* of any form whatever) of the independent variables z. From the definition, and applying identity (5), it follows that the coefficients p satisfy the relations

$$p_{hkl} = -p_{khl} \qquad (h, k, l = 1, 2, \ldots n).$$

A particular case — of special importance — of a complete system is that in which all Poisson's parentheses are identically zero (i.e. all the coefficients p are zero); when this is so the system is called *Jacobian.*

10. **Equivalence of every complete system to a Jacobian system with the same number of equations. Note on Cramer's rule.**

We propose to show that a complete system can always be replaced by a Jacobian system with the same number of equations; thus the consideration of any complete system can be reduced to that of a Jacobian system.

Starting from the system of equations (24), we shall suppose that it is *complete*; i.e. that the equations (26) are satisfied. We shall adopt the following procedure: we shall construct n distinct linear combinations of the n given equations,

$$B_i f = \sum_{1}^{n}{}_k c_{ik} A_k f = 0, \quad . \quad . \quad . \quad (27)$$

with the condition

$$\left\| c_{ik} \right\| \neq 0 \qquad (i = 1, 2, \ldots n),$$

and we shall choose the coefficients c in such a way that the system (27), which is equivalent to the given system, may be Jacobian.

Before doing this, however, we shall write the given equations in a slightly different form. We know that the matrix of the a's has its characteristic equal to n (since the equations are independent); let us arrange the variables in such an order that the determinant a formed by taking the first n columns of the matrix may be that which does not vanish (or one of those which do not):

$$a = \begin{vmatrix} a_{11} & a_{12} \ldots a_{1n} \\ a_{21} & a_{22} \ldots a_{2n} \\ \cdot \quad \cdot & \cdot \quad \cdot \quad \cdot \quad \cdot \\ a_{n1} & a_{n2} \ldots a_{nn} \end{vmatrix} \neq 0.$$

We shall next divide the variables z into two groups: we shall call the first n of them $x_1, x_2, \ldots x_n$, and the remaining $N - n = m$ we shall call $u_1, u_2, \ldots u_m$. With this notation, the given system can be written in the form

$$\sum_{1}^{n}{}_\nu a_{\nu k} \frac{\partial f}{\partial x_\nu} + U_k f = 0 \qquad (k = 1, 2, \ldots n),$$

where U_k denotes an operator involving only derivatives with respect to the u's, the explicit expression of which does not for the moment concern us. Now solve[1] these n equa-

[1] Cramer's well-known rule may be put in the following form, which we shall frequently use, here and elsewhere.

Let there be given n linear equations

$$\sum_{1}^{n}{}_\nu a_{\nu k} \xi_\nu = \eta_k \qquad (k = 1, 2, \ldots n), \quad . \quad . \quad . \quad . \quad . \quad (a)$$

such that the determinant a of their coefficients is not zero.

We shall denote by a^{rs} the *reciprocal element* of the generic element a_{rs} of the

tions with respect to the terms $\frac{\partial f}{\partial x_\nu}$. Putting them in the form

$$\sum_{1}^{n}{}_{\nu}\, a_{\nu k} \frac{\partial f}{\partial x_\nu} = -U_k f,$$

multiply each equation by the corresponding a^{ik} (the reciprocal element of a_{ik} in the determinant of the a's) and sum with respect to k from 1 to n. We thus get n linear combinations:

$$\frac{\partial f}{\partial x_i} = -\sum_{1}^{n}{}_{k}\, a^{ik} U_k f \qquad (i = 1, 2, \ldots n),$$

which are independent, since by a well-known result the determinant of the coefficients a^{ik} is equal to $\frac{1}{a}$, and therefore is not zero. These equations can be written in the more concise form

$$\frac{\partial f}{\partial x_i} + \Omega_i f = 0 \qquad (i = 1, 2, \ldots n), \quad . \quad (24')$$

where the Ω_i's represent linear operators containing, like the U's,

determinant a; i.e. the algebraic complement (or minor) of a_{rs} divided by a. Then, applying two ordinary theorems on determinants, and indicating by δ_r^s either zero or unity, according as $r \neq s$ or $r = s$, we get

$$\sum_{1}^{n}{}_{k}\, a_{ik} a^{jk} = \delta_i^j, \quad \ldots\ldots\ldots \quad (\beta)$$

$$\sum_{1}^{n}{}_{k}\, a_{ki} a^{kj} = \delta_i^j. \quad \ldots\ldots\ldots \quad (\beta')$$

Applying these properties, the equations (α) can be solved by constructing suitable linear combinations of them. For instance, to find ξ_i, multiply the kth equation by a^{ik}; then giving k all values from 1 to n, and summing, we get

$$\sum_{1}^{n}{}_{k}\, a^{ik} \sum_{1}^{n}{}_{\nu}\, a_{\nu k}\, \xi_\nu = \sum_{1}^{n}{}_{k}\, a^{ik} \eta_k.$$

The left-hand side of this equation can be transformed as follows:

$$\sum_{1}^{n}{}_{k\nu}\, a^{ik} a_{\nu k}\, \xi_\nu = \sum_{1}^{n}{}_{\nu}\, \xi_\nu \sum_{1}^{n}{}_{k}\, a^{ik} a_{\nu k} = \sum_{1}^{n}{}_{\nu}\, \xi_\nu\, \delta_i^\nu = \xi_i;$$

hence the solution is given by the formula:

$$\xi_i = \sum_{1}^{n}{}_{k}\, a^{ik} \eta_k. \quad \ldots\ldots\ldots \quad (\alpha')$$

only derivatives with respect to the u's, and therefore of the form

$$\Omega_i = \overset{m}{\underset{1}{\Sigma_\alpha}} X_{\alpha|i} \frac{\partial}{\partial u_\alpha}.$$

The system of equations (24′) is equivalent to the original system (24). The only formal simplification is the specially simple way in which the terms in $\frac{\partial}{\partial x}$ occur. But we shall show that the system (24′) has the advantage of being both complete and Jacobian; it therefore constitutes precisely the system we are in search of, containing n linear combinations which we have denoted in equations (27) by the operators B_i; the coefficients c_{ik} of (27) will be identical with the coefficients a^{ik}.

We shall first show that the system (24′) is complete. We can write it shortly in the form

$$B_i f = 0, \quad . \quad . \quad . \quad . \quad . \quad (24'')$$

where

$$B_i = \frac{\partial}{\partial x_i} + \Omega_i = \frac{\partial}{\partial x_i} + \overset{m}{\underset{1}{\Sigma_\alpha}} X_{\alpha|i} \frac{\partial}{\partial u_\alpha}. \quad . \quad (28)$$

Since the operators B are linear combinations of the A's, it follows from a theorem proved above on p. 35 that Poisson's parentheses $(B_i, B_j)f$ are linear combinations of the expressions

$$A_k f, \qquad (A_h, A_k)f.$$

Now since the system (24) is complete, it follows that the expressions $(A_h, A_k)f$ are in their turn linear combinations of the expressions Af; so that the operators (B_i, B_j) are seen to be linear combinations of the A's alone. But the A's are linear combinations of the B's (since the B's are independent combinations of the A's); hence ultimately the operators (B_i, B_j) are linear combinations of the B's. In other words, the system (24″) also is itself complete.

We can therefore write

$$(B_i, B_j)f = \overset{n}{\underset{1}{\Sigma_l}} q_{ijl} B_l f, \quad . \quad . \quad . \quad . \quad (29)$$

where the coefficients q are analogous to the p's of formula (26).

To show that the system (24″) is Jacobian, we must prove that all the coefficients q vanish. We note that both sides of equation (29) are linear in the terms $\frac{\partial f}{\partial x}$, and the identity cannot hold unless the coefficients of the same derivative, e.g. $\frac{\partial f}{\partial x_h}$, are the same on both sides. We proceed to find these coefficients.

The left-hand side of (29) can be written in the form

$$B_i\left(\frac{\partial f}{\partial x_j} + \Omega_j f\right) - B_j\left(\frac{\partial f}{\partial x_i} + \Omega_i f\right).$$

We saw on p. 35 that the result contains no second-order derivatives; it is therefore unnecessary to apply the operator B to the derivatives of f, so that the expression in question reduces to

$$B_i\,\Omega_j f - B_j\,\Omega_i f = \sum_1^m{}_a \left[B_i\,X_{a|j} \cdot \frac{\partial f}{\partial u_a} - B_j\,X_{a|i} \cdot \frac{\partial f}{\partial u_a}\right]. \qquad (30)$$

As this contains no terms in $\frac{\partial}{\partial x}$, it follows that the coefficient of every $\frac{\partial f}{\partial x_h}$ is zero. On the right-hand side the coefficient of the corresponding term is q_{ijh} (remembering the definition of B); hence every $q = 0$, and in consequence

$$(B_i, B_j)f = 0,$$

or the system (24′) is Jacobian.

A further remark which will shortly be useful is that from the vanishing identically of each side of equation (29) and from equation (30) it follows that the coefficients of the terms in $\frac{\partial}{\partial u}$ also vanish, or from (30),

$$B_i\,X_{a|j} - B_j\,X_{a|i} = 0. \quad . \quad . \quad . \quad . \qquad (31)$$

11. **Integration by means of the associated system.**

Gathering up the foregoing results, we now see that, given a system of linear homogeneous partial differential equations of the first order, we can find its general integral—if one exists—by means of the integration of a complete system of total differential equations.

We have seen how to transform the given system into a complete system (if it is not so already, and provided it contains no inconsistency). We now note that the Jacobian system (24′) which we reached as a result of transforming the generic complete system (24) for other purposes, is identical with the system (20), which originally arose as the system associated with a generic system of total differential equations. The important point here is that if with the coefficients X belonging to the system (24′) we construct the system of total differential equations (19), this system is completely integrable.

In fact, the condition for this is that

$$\frac{dX_{\alpha|i}}{dx_j} = \frac{dX_{\alpha|j}}{dx_i} \qquad (i, j = 1, 2, \ldots n; \; \alpha = 1, 2, \ldots m),$$

or

$$\frac{\partial X_{\alpha|i}}{\partial x_j} + \sum_1^m{}_\beta \frac{\partial X_{\alpha|i}}{\partial u_\beta} X_{\beta|j} = \frac{\partial X_{\alpha|j}}{\partial x_i} + \sum_1^m{}_\beta \frac{\partial X_{\alpha|j}}{\partial u_\beta} X_{\beta|i};$$

and remembering the definition (28) of the operators B, these can be written shortly in the form

$$B_j X_{\alpha|i} = B_i X_{\alpha|j}.$$

The equations (31) show that the X's obtained from the system (24) satisfy these conditions.

Having transformed the given system into the form (24′), we need therefore only construct the associated system (19) and integrate by the method given in the preceding chapter; the most general solution will be obtained in the form

$$u_\alpha = \phi_\alpha(x \mid u^0) \qquad (\alpha = 1, 2, \ldots m).$$

Solving these m equations with respect to the u^0's, we get $m = N - n$ principal integrals, and constructing any function whatever of these integrals we have the general integral of the given system.

This systematic method of integration is in theory quite general and covers all possible cases, but it is somewhat laborious to apply. In practice it is often shorter to integrate the equations separately, and then to look for the m common integrals which certainly exist, when we have ascertained beforehand that we are dealing with a complete system. The following may be given as an example.

Consider the system

$$\left.\begin{aligned} Af &= \quad x_1\frac{\partial f}{\partial x_1} + x_2\frac{\partial f}{\partial x_2} + x_3\frac{\partial f}{\partial x_3} + x_4\frac{\partial f}{\partial x_4} = 0, \\ Bf &= - x_2\frac{\partial f}{\partial x_1} + x_1\frac{\partial f}{\partial x_2} - x_4\frac{\partial f}{\partial x_3} + x_3\frac{\partial f}{\partial x_4} = 0. \end{aligned}\right\} . \quad (32)$$

We shall first show that it is not only complete, but also Jacobian. To show this as shortly as possible, we put

$$A_1 = x_1\frac{\partial}{\partial x_1} + x_2\frac{\partial}{\partial x_2}, \qquad B_1 = - x_2\frac{\partial}{\partial x_1} + x_1\frac{\partial}{\partial x_2},$$

$$A_2 = x_3\frac{\partial}{\partial x_3} + x_4\frac{\partial}{\partial x_4}, \qquad B_2 = - x_4\frac{\partial}{\partial x_3} + x_3\frac{\partial}{\partial x_4},$$

so that $A = A_1 + A_2$, $B = B_1 + B_2$, and then construct the alternate function of the two given operators. We get by successive transformations

$$\begin{aligned} (A, B)f &= ABf - BAf \\ &= A_1B_1f + A_2B_1f + A_1B_2f + A_2B_2f \\ &\qquad - B_1A_1f - B_1A_2f - B_2A_1f - B_2A_2f \\ &= (A_1, B_1)f + (A_2, B_1)f + (A_1, B_2)f + (A_2, B_2)f. \end{aligned}$$

Now it can be shown directly that

$$(A_1, B_1)f = 0, \qquad (A_2, B_1)f = 0,$$

and interchanging x_1, x_2, and x_3, x_4, it follows that

$$(A_2, B_2)f = 0, \qquad (A_1, B_2)f = 0.$$

Hence

$$(A, B)f = 0,$$

which means that the system is Jacobian. It will therefore have $4 - 2 = 2$ independent integrals, or rather (cf. p. 40) an infinite number of pairs of such integrals.

To find one such pair, note that the first equation (which is of the type considered in the example on p. 43) has as its general integral any homogeneous function of degree zero in the variables

x_1, x_2, x_3, x_4. We need therefore only find two independent integrals of the second equation which are homogeneous of degree zero.

Now the system of ordinary differential equations associated with the second of the equations (32) is

$$-\frac{dx_1}{x_2} = \frac{dx_2}{x_1} = -\frac{dx_3}{x_4} = \frac{dx_4}{x_3}.$$

The equation formed of the first two of these terms can be integrated immediately, and gives

$$x_1^2 + x_2^2 = a^2; \quad . \quad . \quad . \quad . \quad . \quad (33a)$$

similarly the other two terms give

$$x_3^2 + x_4^2 = b^2, \quad . \quad . \quad . \quad . \quad . \quad (33b)$$

where a and b denote constants.

Equating the first and third terms, after substituting in them for x_2 and x_4 the expressions given by equations (33a) and (33b), we get

$$\frac{dx_1}{\sqrt{a^2 - x_1^2}} = \frac{dx_3}{\sqrt{b^2 - x_3^2}},$$

and therefore integrating

$$\sin^{-1}\frac{x_1}{a} - \sin^{-1}\frac{x_3}{b} = c,$$

where c is a third constant.

This last integral can be put in the form

$$\sin^{-1}\frac{x_1}{\sqrt{x_1^2 + x_2^2}} - \sin^{-1}\frac{x_3}{\sqrt{x_3^2 + x_4^2}} = c. \quad . \quad (33c)$$

We also get from (33a) and (33b)

$$\frac{x_1^2 + x_2^2}{x_3^2 + x_4^2} = \frac{a^2}{b^2}. \quad . \quad . \quad . \quad . \quad (33d)$$

Of the four integrals thus found, the last two, (33c) and (33d), are homogeneous of degree zero, and are therefore also integrals of the first equation; and it would be easy to verify that they are independent. Hence the general integral of the system of equations (32) is

$$f\left(\sin^{-1}\frac{x_1}{\sqrt{x_1^2+x_2^2}}-\sin^{-1}\frac{x_3}{\sqrt{x_3^2+x_4^2}},\ \frac{x_1^2+x_2^2}{x_3^2+x_4^2}\right),$$

where f is the symbol of an arbitrary function.

CHAPTER IV

Algebraic Foundations of the Absolute Differential Calculus

1. Effect on some analytical entities of a change of variables.

This chapter is devoted to the study of the effect on some analytical entities of a change of variables. In this first section we propose to give some examples showing the nature of the general considerations which will be subsequently established.

Consider n independent variables $x_1, x_2, \ldots x_n$, which we shall as usual denote collectively by x, and suppose a transformation applied to them which leads to another set of n independent variables $\bar{x}$; it is understood that the transformation used is reversible, i.e. that the transformation formulæ

$$x_i = x_i(\bar{x}) \qquad (i = 1, 2, \ldots n) \quad . \quad . \quad . \quad (1)$$

can be solved for the $\bar{x}$'s in the field considered, so that we have simultaneously the equivalent equations

$$\bar{x}_i = \bar{x}_i(x). \quad . \quad . \quad . \quad . \quad . \quad . \quad (1')$$

The geometrical name for this operation is of course *change of co-ordinates*; to fix the ideas, we may take $n = 3$, so that we are passing from Cartesian orthogonal co-ordinates x, y, z to three generic independent combinations of them (curvilinear co-ordinates) q_1, q_2, q_3.

Now suppose that in dealing with a physical, geometrical, or other question we find that we have to consider not only the variables x, but a certain aggregate of entities connected with them. For instance, in a certain region of physical space referred to Cartesian co-ordinates x, y, z, let the temperature T be defined at every point; then it is a determinate function of x, y, z. Or we may suppose that a field of force exists in the given region, and we shall then have to consider at every point a vector, and hence its components, i.e. three functions X, Y, Z of x, y, z. Now change the variables. We have to find some way of expressing the same quantity or physical phenomenon (temperature, force, &c.); for this purpose we find that we have to introduce certain parameters which in the new system of reference will with advantage take the place of those which were more suitable when we were using Cartesian co-ordinates. These new parameters are naturally called transforms of the original ones; they are obtained from them by a law which cannot be assigned *a priori*, but depends on the nature of the problem, and in part on suitable conventions. For instance, in the new system the temperature T will be a function of q_1, q_2, q_3, such that the same temperature belongs to the same point of space, whether the calculations are made with the original or with the new variables; hence T as a function of the q's will be obtained by substituting for x, y, z in $T(x, y, z)$ their values in terms of q_1, q_2, q_3. This kind of behaviour, which is the simplest we shall have to consider, is called *transformation by invariance*; all functions of position which have a value independent of the system of co-ordinates chosen are transformed in this way.

With the components of a vector, in the other example cited, this does not happen. If in fact, as we may suppose, the vector has a magnitude and a direction which are independent of the system of co-ordinates chosen (we shall think of it as being defined physically as a force), its components, on the contrary, even when the point considered remains unchanged, change their values when the frame of reference is changed. This is obvious in the case of a rotation of Cartesian axes. If, however, the transformation considered is not of this particular kind, we do not know *a priori* what to substitute for the projections X, Y, Z of the vector on the axes of the old system in order to specify the vector in the new system;[1] i.e. we have to determine the law

[1] We shall see in various parts of Chapter V how the introduction of new variables q_1, q_2, q_3 gives rise geometrically to corresponding *co-ordinate surfaces* $q_1 =$ constant, $q_2 =$ constant, $q_3 =$ constant, and *co-ordinate lines* which are their

of transformation which will meet the needs of the case in question. The most suitable criterion to take as a guide in making our choice is found by introducing, alongside the given vector, a scalar quantity with a physical significance which is transformed by invariance. In this case we take two infinitely near points whose co-ordinates differ by dx, dy, dz; then the work of the force whose components are X, Y, Z, in passing from one of these points to the other, will be

$$dW = Xdx + Ydy + Zdz; \quad . \quad . \quad . \quad (2)$$

this scalar quantity has a physical significance which is invariant, and it can therefore be concretely determined. From the mathematical point of view it is an important fact that with any system of orthogonal axes $Oxyz$ the Cartesian components of the force are identical with the coefficients of dx, dy, dz in this expression. Changing to the curvilinear co-ordinates q_1, q_2, q_3, we can find the resulting values of dx, dy, dz by means of the differentials of the new variables, using the formulæ

$$dx = \sum_1^3{}_i \frac{\partial x}{\partial q_i} dq_i, \text{ \&c.}$$

The work dW will take the form

$$dW = \sum_1^3{}_i \left(\frac{\partial x}{\partial q_i} X + \frac{\partial y}{\partial q_i} Y + \frac{\partial z}{\partial q_i} Z\right) dq_i,$$

which is analogous to formula (2).

In fact, putting

$$\frac{\partial x}{\partial q_i} X + \frac{\partial y}{\partial q_i} Y + \frac{\partial z}{\partial q_i} Z = Q_i \qquad (i = 1, 2, 3), \quad . \quad (3)$$

we get $$dW = Q_1 dq_1 + Q_2 dq_2 + Q_3 dq_3.$$

intersections. Bearing this in mind, if we proposed to use geometrical criteria taken from our co-ordinate system in order to specify the elements which determine a vector, we should find ourselves faced by four possibilities, all equally acceptable, and with one or another preferable according to circumstances. At every point, in fact, the tangents to the co-ordinate lines and the normals to the co-ordinate surfaces form two supplementary trihedra, which are in general oblique-angled, and therefore distinct; and a vector may be defined either by its orthogonal projections on, or by its components along, either of these two trihedra.

The quantities Q_1, Q_2, Q_3 here hold the same position as did X, Y, Z in Cartesian co-ordinates; it therefore seems suitable to call them the components of the force in the new system of reference, so that we may say that formula (3) represents the law of transformation of the components of a vector. This law is called *covariance.*

We can also reach this law from a different point of view, which, however, we shall show in a moment to be really a particular case of the preceding argument. Consider an invariant function $u(x, y, z)$; we shall try to find the most convenient law of transformation of its three derivatives $\frac{\partial u}{\partial x}, \frac{\partial u}{\partial y}, \frac{\partial u}{\partial z}$, which are evidently functions of x, y, z. A natural course is to consider the three derivatives $\frac{\partial u}{\partial q_1}, \frac{\partial u}{\partial q_2}, \frac{\partial u}{\partial q_3}$ as being the expressions which correspond to them in the new system of reference; these are of course given by the ordinary formulæ

$$\frac{\partial u}{\partial q_i} = \frac{\partial u}{\partial x}\frac{\partial x}{\partial q_i} + \frac{\partial u}{\partial y}\frac{\partial y}{\partial q_i} + \frac{\partial u}{\partial z}\frac{\partial z}{\partial q_i} \qquad (i = 1, 2, 3). \quad (4)$$

If instead we were to assume transformation by invariance, the three quantities we are considering would represent derivatives of a function only in the original Cartesian system of reference, while in any other they would in general lose this special property.

The formulæ (4) are evidently a particular case of the formulæ (3), in which the derivatives of a single function u have been substituted for the components of the generic vector. The real reason for this is found in the fact that the law of persistence of the derivatives can also be included as a special case of the invariance of a linear differential form. As a particular case, we need merely replace dW (which is not in general an exact differential) by the total differential du, which may be expressed in either of the two forms

$$\frac{\partial u}{\partial x}dx + \frac{\partial u}{\partial y}dy + \frac{\partial u}{\partial z}dz$$

and

$$\frac{\partial u}{\partial q_1}dq_1 + \frac{\partial u}{\partial q_2}dq_2 + \frac{\partial u}{\partial q_3}dq_3.$$

The foregoing remarks will suggest what it is we propose to do, though naturally this will become clearer as we proceed.

Given in a certain system of reference (which may be of any kind whatever) a set of quantities having a certain significance, physical, geometrical, or other, we assign a law of transformation by means of which a set of quantities having the same significance is associated with any other system of reference, and we are thus led to introduce a set of parameters, collectively independent of the system of reference, whether Cartesian or not. This is the basis of the conceptual importance and the fertility of the considerations which we propose to develop.

2. m-fold systems. Forms of degree m and m-ply linear forms.

We shall first define a *system of order* m or m-*fold system.* We apply the term to a system of numbers

$$A_{i_1 i_2 \ldots i_m}$$

which are such that a one-to-one correspondence with a specific law exists between them and the set of m integers $i_1, i_2, \ldots i_m$, where each of the i's can take all integral values from 1 to n. The number of elements of an m-fold system is thus n^m, this being the number of permutations (with repetitions) of n numbers taken m at a time. It is not necessary that these n^m elements should be all different.

A system composed of a single number (which may be represented by a letter without a suffix) may be considered as a system of order zero. A simple (one-fold) system will be the aggregate of n elements which can be represented by the notation

$$A_i \qquad (i = 1, 2, \ldots n);$$

e.g. the set of three components of a vector, for which $m = 1$, $n = 3$.

A double (2-fold) system will be of the type

$$A_{ij} \qquad (i, j = 1, 2, \ldots n),$$

and will consist of n^2 elements; and so on.

A system of order greater than 1 is called *symmetrical* if all the elements in it which differ only as to the order of their suffixes have the same value; e.g. for the case $m = 2$, if $A_{ji} = A_{ij}$. A system is called *antisymmetrical* (*skew*) if when two suffixes are interchanged the element changes its sign but not its value; again for the case $m = 2$, if $A_{ij} = -A_{ji}$. The n coefficients u of a generic linear form[1]

$$\phi = \sum_{1}^{n}{}_i u_i x_i$$

constitute a simple system, which is in fact the most general of its kind, since, given n quantities u_i, it is evidently always possible to consider them as being the coefficients of a linear form ϕ.

Consider next a quadratic form, which we may write as

$$\phi = \sum_{1}^{n}{}_{ij} A_{ij} x_i x_j;$$

as the sum includes all permutations of the suffixes two at a time, the product of x_i and x_j will occur twice, once as $x_i x_j$ and once as $x_j x_i$, so that the coefficient of the product is $A_{ij} + A_{ji}$. This is unchanged if i and j are interchanged; hence we see that the coefficients of a quadratic form constitute a symmetrical double system, which is the most general possible. But if we wish to determine a generic (non-symmetrical) double system by means of the coefficients of a form, a quadric in the independent variables x is no longer sufficient. We shall now require two different n-fold systems of independent variables, e.g. the co-ordinates x and x' of two points between which no *a priori* relation exists, and we must construct the expression (*bilinear form*)

$$F = \sum_{1}^{n}{}_{ij} A_{ij} x_i x_j',$$

which is linear in both the x's and the x''s; the coefficients of this form are the required arbitrary quantities A_{ij}.

More generally, it is easy to see that a generic m-fold system is determined by a multilinear form of m groups of variables, while the coefficients of a form of degree m constitute the most general symmetrical m-fold system.

[1] The term *form* with respect to given arguments (e.g. the independent variables $x_1, x_2, \ldots x_n$) means a polynomial homogeneous in those arguments.

3. **Invariance, covariance, and contravariance of a simple system with respect to linear transformations. Dual variables.**

We now proceed to examine the laws of transformation of systems. We shall at first limit our investigation to a *linear* change of variables and a simple system $u_1, u_2, \ldots u_n$.

We shall suppose that we can pass from the variables x to the new variables $\bar{x}$, and vice versa, by means of the formulæ

$$x_i = \sum_{1}^{n}{}_k c_{ik}\,\bar{x}_k \qquad (i = 1, 2, \ldots n), \quad . \quad . \quad (5)$$

$$\bar{x}_i = \sum_{1}^{n}{}_k c^{ki}\,x_k \qquad (i = 1, 2, \ldots n), \quad . \quad . \quad (5')$$

where the coefficients c are arbitrary constants whose determinant is not zero; the second formula follows from the first by applying Cramer's rule, so that c^{ki} is the reciprocal element of c_{ki} (cf. p. 54, footnote).

The most obvious hypothesis to make is that the u's are functions of position which are transformed by *invariance* (cf. § 1).

We get a slightly less simple, but remarkable, case if we suppose that the u's are transformed by the same law as the co-ordinates, in which case the u's will be called *contravariants.* In particular, the co-ordinates themselves form a contravariant simple system.

Next suppose that the u's are the coefficients of a linear form

$$\phi = \sum_{1}^{n}{}_i u_i\,x_i,$$

and that ϕ is transformed by invariance, i.e. by substituting for the x's the expressions (5), so that ϕ is also a linear form of the new variables $\bar{x}$. We shall take the coefficients of this new form as the transforms $\bar{u}$ of the u's; we shall then say that the u's form a *covariant* system.

Writing out the expressions in full, we have

$$\phi = \sum_{1}^{n}{}_i u_i \sum_{1}^{n}{}_k c_{ik}\,\bar{x}_k = \sum_{1}^{n}{}_{ik} c_{ik}\,u_i\,\bar{x}_k = \sum_{1}^{n}{}_k \bar{x}_k \sum_{1}^{n}{}_i c_{ik}\,u_i.$$

The new coefficients are therefore

$$\bar{u}_k = \sum_1^n {}_i\, c_{ik}\, u_i.$$

Interchanging i and k, so as to get the formulæ in the same shape as (5′), we get

$$\bar{u}_i = \sum_1^n {}_k\, c_{ki}\, u_k \qquad (i = 1, 2, \ldots n),$$

which gives the law of covariance.

Here, too, we naturally add the equivalent formulæ, which are obtained by solving for the original elements u, and are given by the usual formula (Cramer's rule). Writing them first, so that we get them in the order corresponding to that of (5) and (5′), we have finally *the law of covariance expressed by the two groups of equivalent formulæ*

$$u_i = \sum_1^n {}_k\, c^{ik}\, \bar{u}_k, \quad \ldots \ldots \ldots \quad (6)$$

$$\bar{u}_i = \sum_1^n {}_k\, c_{ki}\, u_k \qquad (i = 1, 2, \ldots n). \quad . \quad (6')$$

We shall frequently consider, together with the variables x (which are also called *point* variables), a system of covariant variables u (called *dual* variables); the behaviour of both sets of variables when a linear change of variables is made is shown by formulæ (5) and (6).

To find a geometrical interpretation of dual variables, we may fix our attention on the case $n = 4$, in which x_1, x_2, x_3, x_4 can be considered as homogeneous Cartesian co-ordinates of the points of space. A plane has an equation of the type

$$u_1\, x_1 + u_2\, x_2 + u_3\, x_3 + u_4\, x_4 = 0, \quad . \quad . \quad (7)$$

where with the usual terminology, the coefficients u_1, u_2, u_3, u_4 are Plücker's co-ordinates of the plane. Now, given the geometrical significance of equation (7), its left-hand side must be invariant (except for a non-essential factor, the co-ordinates being homogeneous), and hence the Plucker's co-ordinates u must be transformable by covariance. From the well-known law of duality of projective geometry the u's have been given the name of dual variables. Analogous results hold for any value of n.

4. **Invariance, covariance, and contravariance of an m-fold system with respect to linear transformations. Mixed systems or tensors. Vanishing of a tensor an invariant property.**

We shall now extend the discussion of the preceding section to systems of any order, but still limiting it to the case of linear transformations of the type (5), (5′). We thus define mixed systems, of which covariant and contravariant systems are particular cases.

Consider m sets of n point variables (i.e. m points). Denoting by an upper index the ordinal number of each point, we get the set of arguments

$$\begin{array}{c} x_1^1, \; x_2^1, \; \ldots \; x_n^1; \\ x_1^2, \; x_2^2, \; \ldots \; x_n^2; \\ \cdot \quad \cdot \quad \cdot \quad \cdot \quad \cdot \quad \cdot \quad \cdot \\ x_1^m, \; x_2^m, \; \ldots \; x_n^m. \end{array}$$

Consider also a certain number μ of sets of n dual variables

$$\begin{array}{c} u_1^1, \; u_2^1, \; \ldots \; u_n^1; \\ u_1^2, \; u_2^2, \; \ldots \; u_n^2; \\ \cdot \quad \cdot \quad \cdot \quad \cdot \quad \cdot \quad \cdot \quad \cdot \\ u_1^\mu, \; u_2^\mu, \; \ldots \; u_n^\mu. \end{array}$$

Construct a multilinear form F in all these variables, each term of F containing as factor an element taken from every set of the m x's and the μ u's. The coefficients of these terms, which are *a priori* completely arbitrary, will constitute a generic system of order $m + \mu$. Writing the indices corresponding to the x's below and those corresponding to the u's above, we shall have

$$F = \sum_{1}^{n}{}_{i_1 \ldots i_m j_1 \ldots j_\mu} A_{i_1 \ldots i_m}^{j_1 \ldots j_\mu} x_{i_1}^1 \ldots x_{i_m}^m u_{j_1}^1 \ldots u_{j_\mu}^\mu. \tag{8}$$

Now transforming the x's by the law of contravariance and the u's by the law of covariance, and substituting the expressions so obtained in (8) (i.e. transforming F by invariance), we shall get a multilinear form of the new variables $\bar{x}$, $\bar{u}$; we shall take the coefficients $\overline{A}$ of this new form as being the transforms of the coefficients A. We shall then say that the A's constitute a *tensor* or *mixed system*, *covariant* with respect to the lower indices, *contravariant* with respect to the upper. In particular, m or μ may be zero, leading to the absence from F of the point or dual variables respectively; then the system of coefficients is purely contravariant if the variables in F are all covariant, and vice versa.

The case of the simple system comes at once under this definition. In fact, F in this case becomes the ϕ of the preceding section; if we consider it as linear in the x's, we find that the coefficients u, according to the definition just given, must be called covariants; while if it is considered as linear in the u's, we conclude that the x's form a contravariant system, which agrees with the definitions already assigned.

A covariant, contravariant, or mixed tensor, having $m + \mu$ indices in all, is said to be of *rank* $m + \mu$; a simple system, either covariant or contravariant (i.e. a tensor of rank 1) is also called a *vector*, and its elements are called respectively covariant or contravariant components of the vector.

Following a similar method to that used in the preceding section to find the formulæ (6) and (6′), we could find the general transformation formulæ for mixed systems, and hence, in particular, the formulæ for contravariant and covariant systems. We shall not need these formulæ, as in what follows we shall always go back directly to the definition just given. As an example, however, we propose to find them and give them in full for the simplest case of the mixed system, i.e. the system with a single index each of covariance and of contravariance.

Consider therefore the bilinear form

$$F = \sum_{1}^{n}{}_{ij} A_i^j x_i u_j,$$

and transform it by invariance. Using formulæ (5) and (6), we get

$$F = \sum_{1}^{n}{}_{ij} A_i^j \sum_{1}^{n}{}_k c_{ik} \bar{x}_k \sum_{1}^{n}{}_h c^{jh} \bar{u}_h = \sum_{1}^{n}{}_{ijhk} A_i^j c_{ik} c^{jh} \bar{x}_k \bar{u}_h$$

$$= \sum_{1}^{n}{}_{hk} \bar{x}_k \bar{u}_h \sum_{1}^{n}{}_{ij} A_i^j c_{ik} c^{jh}.$$

The coefficients of this new form are

$$\overline{A}_k^h = \sum_{ij}^{n}{}_1 A_i^j c_{ik} c^{jh}, \quad . \quad . \quad . \quad . \quad . \qquad (9)$$

which gives the law of transformation for mixed systems with two indices. We should get similarly, for the most general mixed system,

$$\overline{A}_{i_1 \dots i_m}^{j_1 \dots j_\mu} = \sum_1^n{}_{k_1 \dots k_m h_1 \dots h_\mu} A_{k_1 \dots k_m}^{h_1 \dots h_\mu} c_{k_1 i_1} \dots c_{k_m i_m} c^{h_1 j_1} \dots c^{h_\mu j_\mu}. \qquad (10)$$

As a *memoria technica*, we may add that the transformation formulæ for the x's and the u's give an easy way of remembering those for a tensor of any kind. The latter are always linear, and the coefficients are composed of the c's in a similar way to those of (5) and (6): to each index of covariance corresponds a c with the indices below, to each index of contravariance a c with the indices above. The opposite holds in the inverse formulæ.

We may sum up the discussion so far in the following definitions.

An m-FOLD COVARIANT *is an* m-*fold system which is transformed in the same way as the coefficients of a multilinear form in point variables; an* m-FOLD CONTRAVARIANT *is one which is transformed in the same way as the coefficients of a multilinear form in dual variables; more generally, a* MIXED SYSTEM *or* TENSOR *is one which is transformed in the same way as the coefficients of a multilinear form in both point and dual variables (including also as particular cases both purely covariant and purely contravariant systems).*

The indices of contravariance are generally written above, those of covariance below; an exception is however made for the variables x, which are as usual denoted by $x_1, x_2, \dots x_n$, with the indices below, even if, as in the present case, we are dealing with a contravariant system and linear transformations.

We shall close this section with a remark which is as obvious as it is fundamental whenever the notion of a tensor occurs. This is the fact that if all the elements of a tensor, with reference to a certain system of variables, vanish, this necessarily also happens for the transformed elements which correspond to any linear change of variables whatever. This is an immediate consequence of the fact that the hypothesis makes the invariant form F vanish identically.

5. Symmetrical double systems.

Since we shall have occasion later on to deal with a remarkable symmetrical covariant double system, we propose to give here some properties of systems of this kind. Let the elements of such a system be

$$a_{ik} = a_{ki}; \quad \ldots \ldots \quad (11)$$

their covariance will be expressed by the fact that the bilinear form

$$F(x \mid x') = \sum_{1}^{n}{}_{ik}\, a_{ik}\, x_i\, x'_k \quad \ldots \quad (12)$$

is invariant in any linear transformation which changes the x's and the x''s into other sets of variables $\bar{x}$, $\bar{x}'$.

We shall first show that such a change of variables leaves the symmetry of the system unchanged; i.e. that

$$\bar{a}_{ik} = \bar{a}_{ki} \quad \ldots \ldots \quad (13)$$

In fact, if we interchange the variables x, x' in the bilinear form (12), we get

$$F(x' \mid x) = \sum_{1}^{n}{}_{ik}\, a_{ik}\, x_k\, x'_i$$

and since the right-hand side of this equation differs from that of equation (12) only by the non-essential interchange of the letters i and k, it follows that

$$F(x' \mid x) = F(x \mid x'). \quad \ldots \ldots \quad (11')$$

Vice versa, if this relation holds, we conclude, by reversing the steps of the argument, that (11) is also true.

Hence the condition of symmetry (11) is completely equivalent to the condition (11′). From this standpoint it is easily seen to be invariant. In fact, changing the variables, and denoting for shortness

$$F\{x(\bar{x}) \mid x'(\bar{x}')\}$$

by

$$\overline{F}(\bar{x} \mid \bar{x}'),$$

equation (11′) changes to the equality

$$\overline{F}(\bar{x}' \mid \bar{x}) = \overline{F}(\bar{x} \mid \bar{x}')$$

which, as we have just seen, is equivalent to (13).

We could show in the same way that if a *contravariant* double system is symmetrical with respect to one system of co-ordinates, it is still symmetrical after any linear change of variables; a mixed system a_i^h, however, has not this property. For an *antisymmetrical* double system, either covariant or contravariant, we could also show similarly that antisymmetry is an invariant property.

We can now use the property just illustrated to establish the covariance of the coefficients of an invariant quadratic form. Let the quadratic form be

$$\phi(x) = \sum_{1}^{n}{}_{ik}\, a_{ik}\, x_i\, x_k. \quad . \quad . \quad . \quad . \quad (14)$$

Changing the variables, $\phi(x)$ evidently becomes a quadratic form in the $\bar{x}$'s, which we shall write

$$\bar{\phi}(\bar{x}) = \sum_{1}^{n}{}_{ik}\, \bar{a}_{ik}\, \bar{x}_i\, \bar{x}_k. \quad . \quad . \quad . \quad . \quad (14')$$

We shall show that the coefficients $\bar{a}_{ik}$ are the transforms by covariance of the coefficients a_{ik}, or in other words are the same as would be obtained by changing the variables in $F(x \mid x')$. In fact, we get $\phi(x)$ from $F(x \mid x')$ by first putting x' equal to x, or

$$\phi(x) = F(x \mid x),$$

and from this, with the usual change of variables, we then get $\bar{\phi}(\bar{x})$, which is thus derived from $F\,(x \mid x')$ by applying successively the two operations

$$x_i' = x_i, \quad . \quad . \quad . \quad . \quad . \quad . \quad (a)$$

$$x_i = x_i(\bar{x}). \quad . \quad . \quad . \quad . \quad . \quad (b)$$

But the same result will obviously be obtained if these two operations are applied in inverse order, i.e. if we pass first from $F(x \mid x')$ to $F(\bar{x} \mid \bar{x}')$ (the coefficients of which are by definition the transforms by covariance of the coefficients a_{ik}), and then, by the operation (a), which implies $\bar{x}' = \bar{x}$ and on account of symmetry does not change the coefficients, to $\bar{\phi}(\bar{x})$; the coefficients of this last expression are therefore the transforms by covariance of the coefficients a_{ik}.

6. Sets of n covariant and contravariant simple systems. Theorem on reciprocal sets.

We now propose to prove a lemma in which we shall have to consider, not a single simple system, but a set of n covariant simple systems. We must therefore distinguish the elements in question by two indices, one showing the ordinal number of the system from which an element is taken, the other (which will be an index of covariance or of contravariance) showing the ordinal number of the element in that system. Consider, therefore, the set of n covariant simple systems

$$\lambda_{\alpha|i} \qquad (\alpha, i = 1, 2, \ldots n), \quad . \quad . \quad . \quad (15)$$

where α represents the ordinal number of the system and is therefore not an index of either covariance or contravariance; and suppose further that the determinant of the λ's does not vanish, or in other words that the n systems are independent. With this hypothesis, to every element $\lambda_{\alpha|i}$ will correspond a reciprocal element (its algebraic complement or minor divided by the value of the determinant), which we shall denote by

$$\lambda_{\alpha}^{i} \qquad (\alpha, i = 1, 2, \ldots n). \quad . \quad . \quad . \quad (15')$$

In a linear change of variables the terms $\lambda_{\alpha|i}$ will be transformed by the law of covariance, and the transforms will be denoted by $\bar{\lambda}_{\alpha|i}$; we shall take the reciprocal elements $\bar{\lambda}_{\alpha}^{i}$ of the terms $\bar{\lambda}_{\alpha|i}$ as representing the transforms of the reciprocal elements λ_{α}^{i}.

We shall now show that this law is identical with contravariance, i.e. that giving α the values $1, 2, \ldots n$, the terms λ_{α}^{i} constitute n contravariant simple systems; or shortly, that the reciprocal set of n covariant systems is a set of n contravariant systems. This is the reason for placing the index i above.

The hypothesis of covariance of the set of n systems (15) means that the n linear forms

$$\tau_{\alpha} = \sum_{1}^{n} {}_{i}\, \lambda_{\alpha|i}\, x_i \qquad (\alpha = 1, 2, \ldots n)$$

are invariant. What we have to prove is that the n linear forms

$$\psi_{\alpha} = \sum_{1}^{n} {}_{i}\, \lambda_{\alpha}^{i}\, u_i \qquad (\alpha = 1, 2, \ldots n),$$

are also invariant, i.e. that

$$\overline{\psi}_\alpha - \psi_\alpha = \sum_1^n {}_i \overline{\lambda}^i_\alpha \overline{u}_i - \sum_1^n {}_i \lambda^i_\alpha u_i = 0 \qquad (\alpha = 1, 2, \ldots n), \quad (16)$$

whatever the u's may be. Now these last expressions are linear in the u's (since the $\overline{u}$'s are merely linear combinations of the u's), so that each of them is of the type

$$\sum_1^n {}_i \xi^i u_i.$$

To show that this vanishes identically (i.e. that all the ξ's are zero) we need only show that it vanishes when we give the u's n *distinct* sets of numerical values, as we shall then have n homogeneous linear equations in the ξ's, whose determinant does not vanish (this condition being implied by the use just now of the adjective *distinct*). We shall give the u's the values

$$\lambda_{\beta|i} \qquad (\beta, i = 1, 2, \ldots n),$$

and hence, from the covariance of these quantities, we shall have to give the $\overline{u}$'s the values $\overline{\lambda}_{\beta|i}$. Using a property of determinants (given as formula (β) in the footnote on p. 55), and substituting in equation (16), we get

$$\overline{\psi}_\alpha - \psi_\alpha = \delta^\beta_\alpha - \delta^\beta_\alpha = 0 \qquad (\alpha, \beta = 1, 2, \ldots n),$$

which proves the result required.

7. Addition of tensors.

Take two tensors (in general mixed) of the same kind, i.e. having the same number of indices of covariance and the same number of indices of contravariance (in particular, two covariant, or two contravariant, systems of the same order):

$$A^{j_1 \ldots j_\mu}_{i_1 \ldots i_m}, \qquad B^{j_1 \ldots j_\mu}_{i_1 \ldots i_m}.$$

Summing corresponding elements (those with the same indices) we get a new system whose general term is

$$A^{j_1 \ldots j_\mu}_{i_1 \ldots i_m} + B^{j_1 \ldots j_\mu}_{i_1 \ldots i_m},$$

depending on the same number of indices. We shall show that

this new system is also a tensor, covariant and contravariant respectively with respect to the indices of covariance and contravariance of the given systems, so that with the notation previously adopted the general term can be written

$$C^{j_1 \ldots j_\mu}_{i_1 \ldots i_m}.$$

To simplify the formulæ we shall prove the result for the case of a single index each of covariance and of contravariance; the reasoning is identical in the general case. Our hypothesis then is that the forms

$$F = \sum_{1}^{n}{}_{ij} A^j_i x_i u_j,$$

$$\Phi = \sum_{1}^{n}{}_{ij} B^j_i x_i u_j,$$

are invariants. The sum

$$F + \Phi = \sum_{1}^{n}{}_{ij} (A^j_i + B^j_i) x_i u_j = \sum_{1}^{n}{}_{ij} C^j_i x_i u_j,$$

will therefore also be an invariant, which is as much as to say that the system

$$C^j_i = A^j_i + B^j_i$$

is covariant with respect to the lower index, and contravariant with respect to the upper.

The tensor C is called the *sum* of the two tensors A and B.

8. **Multiplication of tensors.**

We shall now define the *product* of two tensors. These may be of any kind, in general mixed; we shall suppose that one has m indices of covariance and μ of contravariance, and the other m' and μ' respectively, so that they are represented by

$$A^{j_1 \ldots j_\mu}_{i_1 \ldots i_m}, \qquad B^{j_1 \ldots j_{\mu'}}_{i_1 \ldots i_{m'}}.$$

Construct the system whose general term is the product of any element A by any element B; the element so formed will contain $m + \mu + m' + \mu'$ indices, so that the rank of the *product* system will be the sum of the ranks of the given systems. We shall show that it is a *tensor* which has the $m + m'$ indices of covariance of the given system as indices of covariance, and the $\mu + \mu'$ indices of contravariance as indices of contravariance. To simplify the formulæ we shall as before consider the case of only two indices.

Let the two forms which by hypothesis are invariant be

$$F = \sum_{ih\,1}^{\;n} A_i^h \, x_i \, u_h,$$

$$\Phi = \sum_{jk\,1}^{\;n} B_j^k \, x_j' \, u_k'.$$

Their product will also be an invariant, and is

$$F\Phi = \sum_{ihjk\,1}^{\;n} A_i^h \, B_j^k \, x_i \, x_j' \, u_h \, u_k',$$

or, putting

$$A_i^h \, B_j^k = C_{ij}^{hk},$$

$$F\Phi = \sum_{ihjk\,1}^{\;n} C_{ij}^{hk} \, x_i \, x_j' \, u_h \, u_k'.$$

The invariance of this form means that the indices i and j attached to the letter C are indices of covariance, and h and k are indices of contravariance, which proves the statement just made. The argument is the same in the general case.

9. Contraction of tensors.

We shall now define the operation of contraction, by which we pass from any mixed system to another system having one index of covariance and one of contravariance less than the first.

For convenience of printing, we shall give explicitly only one of the indices of covariance and one of contravariance, replacing the others by points, so that we shall put

$$A_{\cdots r}^{\cdots s}$$

to represent the general term.

Now construct the system

$$B_{\cdots}^{\cdots} = \sum_{r\,1}^{\;n} A_{\cdots r}^{\cdots r},$$

which will contain all the indices, except the two shown on the right; we say then that the tensor has been contracted with respect to these two indices. We shall show that the system so obtained is also a tensor, having the same indices of covariance and of contravariance—except of course the pair used in contracting—as the given tensor. To simplify the formulæ, we shall as usual consider a particular case, but one not differing essentially from the general case. Suppose, therefore, that the form

$$F = \sum_{ihrs\,1}^{n} A_{ir}^{hs}\, x_i\, x'_r\, u_h\, u'_s$$

is invariant whatever may be the variables x, x', u, u', the only restriction being that x, x' are point variables and u, u' dual variables. Their values being arbitrary, we may replace the variables u'_s by n distinct systems of covariant quantities, which we shall denote by $\lambda_{\alpha|s}$, using the notation (15) of § 6; we can then replace the variables x'_r by the quantities λ_α^r, which are the reciprocal elements of the former group, and therefore contravariant (§ 6). We shall thus have the n linear forms

$$F_\alpha = \sum_{ihrs\,1}^{n} A_{ir}^{hs}\, x_i\, u_h\, \lambda_\alpha^r\, \lambda_{\alpha|s} \qquad (\alpha = 1, 2, \ldots n)$$

all invariant. Their sum G will therefore also be invariant. Writing out this sum, and making some slight transformations, we get (remembering the fundamental property of reciprocal elements)

$$G = \sum_{\alpha\,1}^{n} F_\alpha = \sum_{ihrs\,1}^{n} A_{ir}^{hs}\, x_i\, u_h \sum_{\alpha\,1}^{n} \lambda_\alpha^r\, \lambda_{\alpha|s} = \sum_{ihrs\,1}^{n} A_{ir}^{hs}\, x_i\, u_h\, \delta_r^s.$$

Now we know that $\delta_r^s = 0$ if $r \neq s$ and $= 1$ if $r = s$; hence all the terms in the sum for which $r \neq s$ will disappear, and there remains

$$G = \sum_{ihr\,1}^{n} A_{ir}^{hr}\, x_i\, u_h = \sum_{ih\,1}^{n} x_i\, u_h \sum_{r\,1}^{n} A_{ir}^{hr} = \sum_{ih\,1}^{n} x_i\, u_h\, B_i^h.$$

The invariance of this form shows, as was required, that the system

$$B_i^h = \sum_{r\,1}^{n} A_{ir}^{hr}$$

is a tensor covariant with respect to the i's and contravariant with respect to the h's.

The operation of contraction can evidently be repeated several times, contracting successively with respect to various pairs of indices, so that, for example, from the system

$$A^{hks}_{ijr}$$

we can pass, by using two pairs of indices, to the tensor

$$B^h_i = \sum_{1}^{n}{}_{jr} A^{hjr}_{ijr}.$$

If the process is applied to the only pair of indices of a mixed double system, the result is an invariant:

$$A = \sum_{1}^{n}{}_i A^i_i.$$

10. Composition of tensors.

If we combine the operation of *multiplication* of two tensors with that of *contraction*, we get the operation called *composition* (or *inner multiplication*) of two tensors. We shall write the two tensors in the abridged form

$$A^{\cdots}_{\cdots r}, \qquad B^{\cdots s}_{\cdots},$$

where we show only a single index of covariance for one and of contravariance for the other.

The tensor

$$C^{\cdots}_{\cdots} = \sum_{1}^{n}{}_r A^{\cdots}_{\cdots r} B^{\cdots r}_{\cdots}$$

is said to be *compounded* of the first two or is called their *inner product*; its indices of covariance are those of A, except r, and all those of B, and its indices of contravariance are all those of A, and those of B, except s.

It can at once be seen that the system C is a tensor, observing that it is obtained by contraction with respect to the indices r and s from the system

$$\Gamma^{\cdots s}_{\cdots r} = A^{\cdots}_{\cdots r} B^{\cdots s}_{\cdots},$$

which is itself obtained by taking the product of the given systems. Thus, for instance, compounding the systems

$$A^h_{ir}, \qquad B^{ks}_j,$$

with respect to the indices r and s, we get

$$C^{hk}_{ij} = \sum_{1}^{n}{}_r A^h_{ir} B^{kr}_j .$$

11. Change of variables in general. m-fold systems whose elements are functions of position. First general definition of a tensor. Typical tensors of rank 1.

Up to this point we have considered only *linear* changes of variables, and we have defined, with reference to them, covariance, contravariance, and the fundamental operations on systems. We shall now extend these definitions to any change whatever of the variables.

Suppose, therefore, that the formulæ of transformation, instead of equations (5), are

$$x_i = f_i(\bar{x}_1, \bar{x}_2, \ldots \bar{x}_n) \qquad (i = 1, 2, \ldots n), \quad (17)$$

where the f_i's denote arbitrary functions, except for the qualitative restrictions as to differentiability, &c., which will be tacitly imposed whenever necessary, and the condition that the transformation is reversible, i.e. that the equations (17) are soluble for the $\bar{x}$'s and can therefore also be given in the equivalent form

$$\bar{x}_i = g_i(x_1, x_2, \ldots x_n). \quad . \quad . \quad . \quad (17')$$

The *general* transformation (17) involves a *linear* transformation of the differentials. In fact, putting

$$\frac{\partial x_i}{\partial \bar{x}_k} = c_{ik}, \qquad \frac{\partial \bar{x}_i}{\partial x_k} = c^{ki}, \quad . \quad . \quad . \quad (18)$$

we get, differentiating (17) and (17′),

$$d x_i = \sum_{1}^{n}{}_k \frac{\partial x_i}{\partial \bar{x}_k} d\bar{x}_k = \sum_{1}^{n}{}_k c_{ik}\, d\bar{x}_k; \quad . \quad . \quad (19)$$

$$d\bar{x}_i = \sum_{1}^{n}{}_k \frac{\partial \bar{x}_i}{\partial x_k} d x_k = \sum_{1}^{n}{}_k c^{ki}\, d x_k \qquad (i = 1, 2, \ldots n). \quad (19')$$

The second of these groups of formulæ must be identical with the group which would result from solving the first; the quantities c^{ki} must therefore be the reciprocal elements of the quantities

c_{ki}, which justifies the choice of these symbols to represent the derivatives.[1]

From the analogy of formulæ (19), (19′) to (5), (5′), we can at once extend the earlier arguments to m-fold systems whose elements are any functions of position (i.e. of the independent variables $x_1, x_2, \ldots x_n$). We shall say that an m-fold system whose elements are functions of position constitutes a tensor, covariant, contravariant, or mixed, with respect to a generic transformation (17), when it is a tensor of the specified kind (at every point of the field considered) with respect to the linear transformation (19), (19′) between the differentials of the old and the new variables.

In consequence *the differentials of the independent variables provide us with the typical contravariant simple system.* We shall next consider what is the typical covariant simple system.

In § 3 we introduced the dual variables u_i, which were formally defined as the coefficients of a linear form in the variables x. These latter are now to be replaced by their differentials dx, so that we start from a generic Pfaffian

$$\psi = \sum_{1}^{n}{}_i\, u_i\, dx_i,$$

and consider it as invariant for *any change whatever* of the variables x. The coefficients u are considered as functions of position, and hence initially of the x's. When the transformation (17) is made, the dependence on the point co-ordinates is expressed instead in terms of the new variables $\bar{x}$. Substituting in ψ for dx_i from (19), we see in the first place that we still have a Pfaffian

[1] This can also be shown directly, by proving that the terms c^{ki} and c_{ki} have the fundamental property of reciprocal elements. In fact, if in equations (17) we replace the x's by the expressions given by equations (17′), they reduce to identities. Differentiate one of these with respect to x_k, using the rule for a compound function. We shall have

$$\frac{\partial x_i}{\partial x_k} = \sum_{1}^{n}{}_h \frac{\partial x_i}{\partial \bar{x}_h}\frac{\partial \bar{x}_h}{\partial x_k}.$$

Now the left-hand side is 0 or 1 according as $i \neq k$ or $i = k$; on the right-hand side we can introduce the notation (18), so getting

$$\delta^i_k = \sum_{1}^{n}{}_h\, c_{ih}\, c^{kh},$$

which proves the required result.

in the new variables $\bar{x}$; this is obvious, since the original expression is linear in dx. Writing out the result, we get

$$\psi = \sum_1^n {}_i u_i \sum_1^n {}_k \frac{\partial x_i}{\partial \bar{x}_k} d\bar{x}_k = \sum_1^n {}_{ik} u_i \frac{\partial x_i}{\partial \bar{x}_k} d\bar{x}_k = \sum_1^n {}_k d\bar{x}_k \sum_1^n {}_i u_i \frac{\partial x_i}{\partial \bar{x}_k}.$$

The coefficients of the new differentials $d\bar{x}_k$, i.e. the elements $\bar{u}_k$ of the system which is the transform of the coefficients u_k, are therefore

$$\bar{u}_k = \sum_1^n {}_i u_i \frac{\partial x_i}{\partial \bar{x}_k} \qquad (k = 1, 2, \ldots n).$$

Interchanging i and k and adopting the notation (18), we get the law of transformation for the coefficients of a Pfaffian expressed by the formulæ

$$\bar{u}_i = \sum_1^n {}_k c_{ki} u_k,$$

which are identical with the formulæ (6′). Adding the inverse formulæ and replacing the coefficients c_{ki}, c^{ik} by their values as given by (18), we get the *transformation formulæ for the coefficients of a Pfaffian (an invariant) which constitute the typical simple covariant system*, in the explicit form

$$u_i = \sum_1^n {}_k \bar{u}_k \frac{\partial \bar{x}_k}{\partial x_i}, \qquad \ldots \ldots \qquad (20)$$

$$\bar{u}_i = \sum_1^n {}_k u_k \frac{\partial x_k}{\partial \bar{x}_i} \qquad (i = 1, 2, \ldots n). \quad . \quad (20')$$

Suppose in particular that the (invariant) Pfaffian is the exact differential of a function u of position; being invariant, u is such that its expression in terms of the $\bar{x}$'s is obtained from its expression in terms of the x's by substituting $f_i(\bar{x})$ for x_i, and vice versa, so that the formula

$$u(x) = u(\bar{x})$$

is an identity when we substitute in it the expressions given by (17) (or (17′)) for the x's (or the $\bar{x}$'s).

The coefficients u_i, $\bar{u}_i$ of the Pfaffian are respectively $\dfrac{\partial u}{\partial x_i}$ or $\dfrac{\partial u}{\partial \bar{x}_i}$ according as du is considered as expressed in terms of the x's or of the $\bar{x}$'s.

It follows that *the derivatives of an invariant are transformed by covariance, the law being given by formulæ* (20), (20′).

Vice versa, to obtain the formulæ of covariance (20) or (20′) relative to a simple system, without having to go through all the steps from the beginning or to remember them by heart, the easiest *memoria technica* is to consider the elements of the generic system in question as being for the moment the derivatives of a single function, and to apply the rule for differentiating a function of one or more functions. We then automatically get formulæ (20) or (20′) according as we start from an original or a transformed element.

The direct transformation of the differentials further, as we have seen, gives formulæ (19) and (19′), which we can use as the transformation formulæ for a generic contravariant simple system, by substituting the original elements ξ^i for the differentials dx_i and the transformed elements $\bar{\xi}^k$ for the differentials $d\bar{x}_k$.

To sum up, the differentials of the independent variables and the derivatives of a single function give what we may call the pattern of the transformation formulæ for simple contravariant and covariant systems respectively.

12. Second general definition of tensors whose elements are functions of position. Examples.

Take a multilinear form in any number of sets of contravariant variables (i.e. having the same law of transformation as the dx_i) and in any number of sets of covariant variables (i.e. having the same law of transformation as the $u_i = \left.\dfrac{\partial u}{\partial x_i}\right)$. Let the coefficients be considered as functions of position, and the given form as invariant at each separate point. From the definition given in the preceding section it is clear that the coefficients form a *mixed tensor*, whose indices of covariance are those relative to the contravariant variables, and vice versa. Reciprocally, every tensor, in the sense of the first definition, can be identified with the coefficients of a multilinear form of the kind just described. The two definitions are therefore completely equivalent.

From this point everything is analogous to what was said in § 4, and we may therefore dispense with further details, except to repeat once more explicitly the remark made at the end of § 4 as to the vanishing of a tensor (i.e. of all its elements) being an invariant property. The property holds in general for any change of variables of any kind. In other words, if *all* the elements of a generic tensor

$$A^{h_1 h_2 \ldots h_\mu}_{i_1 i_2 \ldots i_m},$$

referred to a particular system of variables, are zero, we may be sure that the equations

$$A^{h_1 h_2 \ldots h_\mu}_{i_1 i_2 \ldots i_m} = 0 \quad (i_1, i_2, \ldots i_m;\ h_1, h_2, \ldots h_\mu = 1, 2, \ldots n)$$

continue to hold however the variables may be changed.

We shall close this section with two examples of tensors which occur fairly often.

Consider first a linear operator A, where

$$Af = \sum_1^n{}_i\, A^i \frac{\partial f}{\partial x_i},$$

whose coefficients A^i are specified functions of position. Let us treat the operator as an *invariant*. Then since the terms $\frac{\partial f}{\partial x_i}$ are covariant, it follows that the A^i's are by definition contravariant, and must therefore have their law of transformation given by the equations (19′), so that we get for the transformed coefficients the expressions

$$\bar{A}^i = \sum_1^n{}_k\, A^k \frac{\partial \bar{x}_i}{\partial x_k},$$

as could easily be verified directly.

Consider next a differential quadratic form

$$\phi = \sum_1^n{}_{ik}\, a_{ik}\, dx_i\, dx_k,$$

which is to be invariant; the coefficients a_{ik} (in general to be considered as functions of position) will then be covariant, and hence their transformation formulæ will be

$$a_{ik} = \sum_1^n{}_{rs}\, \bar{a}_{rs} \frac{\partial \bar{x}_r}{\partial x_i} \frac{\partial \bar{x}_s}{\partial x_k}, \quad . \quad . \quad . \quad . \quad (21)$$

or (solving for the transformed elements)

$$\bar{a}_{ik} = \sum_{1}^{n}{}_{rs}\, a_{rs} \frac{\partial x_r}{\partial \bar{x}_i} \frac{\partial x_s}{\partial \bar{x}_k}. \quad . \quad . \quad . \quad . \quad (21')$$

13. More complex laws of transformation. Scope of the Absolute Differential Calculus.

In a generic change of variables a system, as we have said, is transformed in a way which depends on its definition. The cases so far examined have been the simplest, but others of considerably greater complexity may also occur; we shall now give an example of these.

We have seen that the simple system composed of the first derivatives of an invariant function u is covariant; we now proceed to examine the double system of the first derivatives $\frac{\partial u_i}{\partial x_j}$ of a covariant simple system u_i. As a particular case, if the u_i's are the derivatives $\frac{\partial u}{\partial x_i}$ of a single function u, we cover the case of the transformation of the second derivatives of an invariant function.

To find the transformation formulæ for this system, i.e. the relation between the terms $\frac{\partial u_i}{\partial x_j}$ and the terms $\frac{\partial \bar{u}_i}{\partial \bar{x}_j}$, we start from the transformation formula for the u_i's:

$$\bar{u}_i = \sum_{1}^{n}{}_k \frac{\partial x_k}{\partial \bar{x}_i} u_k.$$

Differentiating it with respect to $\bar{x}_j$, and considering the u_k's on the right as functions of the x's and therefore of the $\bar{x}$'s, we get

$$\frac{\partial \bar{u}_i}{\partial \bar{x}_j} = \sum_{1}^{n}{}_{kh} \frac{\partial x_k}{\partial \bar{x}_i} \frac{\partial x_h}{\partial \bar{x}_j} \frac{\partial u_k}{\partial x_h} + \sum_{1}^{n}{}_k \frac{\partial^2 x_k}{\partial \bar{x}_i \partial \bar{x}_j} u_k. \quad . \quad (22)$$

If the last sum were absent, the law of transformation would be that of covariance. But in fact the presence of the second derivatives of the x's with respect to the $\bar{x}$'s shows that the system we are examining is neither invariant, nor covariant, nor contravariant, nor mixed, and therefore is *not a tensor*; its law of transformation is more complicated than any we have yet examined. A similar result is true more generally for the system composed of the derivatives of any tensor.

It is often necessary to consider the derivatives with respect to the independent variables of the elements of a tensor, covariant, contravariant, or mixed. In order to avoid the complication just observed, it is therefore convenient to replace these derivatives by linear combinations of them with the elements of the tensor, so chosen that those terms which lead to the aforesaid complication disappear in the transformation formulæ. This is the problem which the Absolute Differential Calculus proposes to solve; it does so, as we shall see farther on, by introducing an auxiliary element, namely, an invariant differential quadratic form. We shall therefore devote the next chapter to the study of this important element.

CHAPTER V

Geometrical Introduction to the Theory of Differential Quadratic Forms

(a) *The Line Element on a Surface*

1. **Parametric equations of a surface.**

The meaning of the term " parametric equations of a surface" is known from analytical geometry. We propose, however, here to examine the idea from the beginning, in order to find the formulæ in the shape which is best suited to our purpose.

We shall use the letters y_1, y_2, y_3 throughout this chapter to represent the Cartesian co-ordinates of the points of space referred to three orthogonal axes. Now consider a surface, or more generally a piece of a surface σ, to which alone the following remarks are understood to apply, and suppose that there has been established, in any way whatever, a one-to-one correspondence between the points of σ and the pairs of values which can be assigned to two parameters x_1, x_2 within a certain field C of a plane representative of the arguments x_1, x_2 (cf. the general remarks in Chapter I, § 1).

This implies that the points of σ and with them their Cartesian co-ordinates y_ν are definite (and finite) functions of x_1, x_2 in the field C. We shall accordingly write

$$y_\nu = y_\nu(x_1, x_2) \qquad (\nu = 1, 2, 3), \quad . \quad . \quad . \quad (1)$$

where for subsequent purposes the three functions y_ν must have derivatives, to any order which we may have occasion to consider, which are continuous in the field C.

But this behaviour of the functions is not in itself sufficient to ensure that the equations (1) do effectively define a surface, i.e. that the supposed one-to-one correspondence does in fact exist between C and the points of a portion of a two-dimensional manifold.

It might for instance happen that only the sum $x_1 + x_2$ appeared in the equations (1), in which case the dependence on two parameters would be only apparent, only one of them being essential. In this case the equations (1) would define a piece of a curve. To exclude the possibility of anything of this kind we shall suppose that two of the equations (1) are soluble (within C) for x_1, x_2, so that by solving them, and substituting the values so found in the remaining equation, we can get one (and only one) relation between y_1, y_2, y_3, i.e. the equation of a surface.

This is equivalent to imposing the condition that the characteristic of the functional matrix [1] of the equations (1) is 2. Then the equations (1) will actually represent the parametric equations of a piece of a surface σ; and it could be shown that —with the restriction, if necessary, of the field C to a convenient portion Γ of itself (around an arbitrarily chosen point)—the portion of surface so defined is such that to any point on it there corresponds one and only one set of values of the parameters in the field Γ. Accordingly, with this qualitative restriction as to the field—which we shall always consider as being of the type Γ—in which the parameters x_1, x_2 are made to vary, we are quite justified in calling x_1, x_2 curvilinear co-ordinates on the surface σ defined by equations (1).

Giving x_1 a constant value, and making x_2 vary, we get all the points of a line, which we shall call *the line* x_1 = *constant*, or *the line* x_2, or more shortly, *the line* 2 (since only x_2 varies along

[1] See §§ 6, 7, pp. 8–12.

it); in the same way we can define *the lines* $x_2 =$ *constant*, or *the lines* x_1, or merely *the lines* 1, as those along which only x_1 varies. We can thus think of our surface (or portion of surface) σ as covered by a double network of lines (*co-ordinate lines*) such that two and only two—one line x_1 and one line x_2—pass through every point of it.

2. Expression for ds^2.

We shall now fix two infinitely near points, P, P', on σ; let their curvilinear co-ordinates be

$$x_i,\ x_i + dx_i \qquad (i = 1, 2),$$

and, subject to the equations (1), let

$$y_\nu,\ y_\nu + dy_\nu \qquad (\nu = 1, 2, 3)$$

be their Cartesian co-ordinates.

Note that in order to specify a point P on σ, we may take *arbitrarily* (within Γ) the two co-ordinates x_1, x_2; and so also, in order to reach P', the two increments dx_1, dx_2.

The y's are defined by the equations (1), so that their differentials are connected with the dx's by the equations

$$dy_\nu = \sum_1^2{}_i \frac{\partial y_\nu}{\partial x_i} dx_i \qquad (\nu = 1, 2, 3), \quad . \quad . \quad (2)$$

which are obtained by differentiating the equations (1).

We shall calculate the distance $PP' = ds$, or rather, as being more direct, its square

$$ds^2 = \sum_1^3{}_\nu dy_\nu{}^2 .$$

Substituting the expressions (2) for the dy's, we shall have

$$ds^2 = \sum_1^3{}_\nu \sum_1^2{}_{ik} \frac{\partial y_\nu}{\partial x_i} \frac{\partial y_\nu}{\partial x_k} dx_i\, dx_k = \sum_1^2{}_{ik} dx_i\, dx_k \sum_1^3{}_\nu \frac{\partial y_\nu}{\partial x_i} \frac{\partial y_\nu}{\partial x_k},$$

from which, putting

$$a_{ik} = \sum_1^3{}_\nu \frac{\partial y_\nu}{\partial x_i} \frac{\partial y_\nu}{\partial x_k} \quad . \quad . \quad . \quad . \quad . \quad (3)$$

(by which we define a very important symmetrical double system of regular functions of the x's), we get

$$ds^2 = \sum_{ik}^{2}{}_{1} a_{ik}\, dx_i\, dx_k. \qquad \ldots \quad (4)$$

This quadratic form, which, as we shall see, is fundamental for the study of the metrical properties of our surface, is an obvious generalization of the expression

$$ds^2 = dx^2 + dy^2$$

which in Cartesian co-ordinates gives the distance between two infinitely near points of a plane.

We shall now show that the form (4) is *definite* and *positive* i.e. that it never becomes either zero or negative, whatever values (real and not zero) are assigned to the dx's. That it cannot be negative is at once seen from the fact that it is the sum of the squares of the dy's, which are always real if the dx's are real. It could therefore vanish only if all the dy's vanished, and we shall show that this is impossible for any actual displacement (one in which dx_1 and dx_2 are not both zero).

In fact, let us try to suppose that we can have

$$dy_1 = dy_2 = dy_3 = 0.$$

Using equations (2), these become three linear homogeneous equations in dx_1, dx_2. In order that any two of these may be satisfied by non-zero values of these variables, the corresponding determinant must vanish; since we may choose arbitrarily the pair of equations to be satisfied, we conclude that all three of the functional determinants (of the second order) of the y's with respect to the x's must vanish, which contradicts the hypothesis that the characteristic of the functional matrix is 2.

The general theorem relating to simultaneous linear homogeneous equations could also be applied directly; namely, that the number of independent solutions is the difference between the number of the unknowns and the characteristic of the matrix of the coefficients—in our case $2 - 2$, or 0.

From the proof that the quadratic form under discussion is

definite, it follows, by a known theorem [1] on quadratic forms, that the determinant

$$a = \left\| a_{ik} \right\|$$

composed of the coefficients of ds^2 (called the *discriminant* of the form) is not zero; in particular, when, as in the present case, the form is *positive* as well as definite, we have specifically $a > 0$.

The fundamental form (4) calls for one last remark, almost obvious but important. This is that the system of the coefficients a_{ik} is *covariant with respect to any transformations whatever of the variables* x_1, x_2 (which justifies our having placed the indices i, k below). This covariance follows directly (applying a remark made at the foot of p. 84) from the invariance of the quadratic form ds^2.

3. **Determination of the directions drawn from a generic point.**

In the space y_1, y_2, y_3, a direction drawn from a generic point P may be considered as determined by an infinitesimal segment

[1] The theorem referred to is as follows. Let

$$\phi = \sum_{1}^{n}{}_{ik}\, a_{ik} x_i x_k$$

be a *definite* quadratic form in n variables; we shall show that its discriminant a cannot be zero.

In fact, putting

$$y_i = \sum_{1}^{n}{}_k\, a_{ik} x_k \qquad (i = 1, 2, \ldots n),$$

we get $\qquad \phi = \sum_{1}^{n}{}_i\, y_i x_i.$

Now if $a = 0$, we could make $\phi = 0$ without all the x's being zero (contrary to the hypothesis that the form is definite); we should only have to make all the y's zero, by solving the n linear homogeneous equations

$$\sum_{1}^{n}{}_k\, a_{ik} x_k = 0 \qquad (i = 1, 2, \ldots n),$$

which would be soluble, giving values for the x's which are not all zero, provided $a = 0$.

For definite positive forms a is therefore also necessarily positive. One way of seeing this is to apply one of the infinite number of (real) linear substitutions which reduce ϕ to the canonical form (see e.g. BIANCHI: *Lezioni di geometria analitica*, Appendix, pp. 571–592; Pisa, Spoerri, 1920). It is obvious that a positive form which contains only squares of the variables has its discriminant $\bar{a} > 0$. But the original a and $\bar{a}$ are connected by the relation $a = \bar{a}\Delta^2$, where Δ denotes the determinant of the linear substitution. (See p. 157.) We therefore necessarily also have $a > 0$. Q.E.D.

having one end at P, or if preferred by another point P' infinitely near P, or, which comes to the same thing, by an infinitesimal displacement of P.

Now suppose that P belongs to σ, and consider the directions drawn from P which are *tangent* to the surface. To determine them we have to take points P', infinitely near P and belonging to σ. If therefore we call the surface co-ordinates of P x_1 and x_2, we can determine P' by the surface co-ordinates $x_1 + dx_1$, $x_2 + dx_2$.

Thus *to each pair of infinitesimals* dx_1, dx_2, *there corresponds one and only one tangential direction drawn from* P. To one direction, on the other hand, there correspond an infinite number of pairs of differentials which differ from each other by a (positive) factor, since the length ds of the segment PP' chosen to determine the direction is *a priori* arbitrary, the only condition being that it is infinitesimal.

In order to make the correspondence one-to-one, we shall, in order to determine a direction, replace the differentials by the proportional quantities

$$\lambda^1 = \frac{dx_1}{ds}, \quad \lambda^2 = \frac{dx_2}{ds};$$

these are unchanged if we multiply dx_1 and dx_2 by a positive factor k, since it follows from equation (4) that then ds is also multiplied by k.

These quantities are called *parameters of the direction* and obviously reduce to direction cosines when the surface σ is a plane and x_1, x_2 represent orthogonal Cartesian co-ordinates. The parameters are not independent but are connected by the relation

$$\sum_{1}^{2}{}_{ik}\, a_{ik}\, \lambda^i \lambda^k = 1, \quad . \quad . \quad . \quad . \quad (5)$$

which is obtained by dividing equation (4) by ds^2 and which corresponds to the well-known identity for the Euclidean plane that the sum of the squares of the cosines $= 1$. Since ds is an invariant and the dx's are contravariants, the parameters are also *contravariants*, which justifies our having placed the indices above.

Instead of the parameters two linear combinations of them are sometimes used; these are

$$\lambda_i = \sum_{1}^{2}{}_k\, a_{ik}\, \lambda^k \qquad (i = 1, 2), \quad . \quad . \quad . \quad (6)$$

which are called *moments*. Since the coefficients a_{ik} form a covariant double system (cf. § 2), and the parameters, as we have just shown, form a contravariant simple system, it follows that *the moments are covariants.*[1]

We showed in § 2 that the determinant a is not zero; the equations (6) can therefore be solved, giving the formulæ

$$\lambda^i = \sum_{1}^{2}{}_k\, a^{ik}\, \lambda_k, \quad . \quad . \quad . \quad . \quad . \quad (6')$$

which give the parameters in terms of the moments. The parameters and moments are connected by a particularly simple and remarkable bilinear relation, which follows immediately from (5) and (6). In fact, multiplying the equation (6) for the generic index i by λ^i, and summing for $i = 1$ and $i = 2$, we get from (5)

$$\sum_{1}^{2} \lambda_i\, \lambda^i = 1. \quad . \quad . \quad . \quad . \quad . \quad (5')$$

It follows directly that the moments also are connected by a quadratic relation. We need only substitute in (5′) for λ^i the expression given by formula (6′), which gives at once

$$\sum_{1}^{2}{}_{ik}\, a^{ik}\, \lambda_i\, \lambda_k = 1. \quad . \quad . \quad . \quad . \quad (5'')$$

4. Angle between two directions. Contravariance of the coefficients a^{ik}.

Consider two directions on a surface drawn from a single point P. We shall denote them by $\boldsymbol{\lambda}$ and $\boldsymbol{\mu}$, where these two symbols mean more precisely the two unit vectors which determine the given directions. The parameters and the moments of $\boldsymbol{\lambda}$ will be denoted by λ^i, λ_i, and those of $\boldsymbol{\mu}$ by μ^i, μ_i, respectively. We propose to find the angle ϑ between the two directions as a function of the parameters or of the moments.

Denoting the increments of the co-ordinates y_ν, x_i by dy_ν,

[1] Cf. § 10, p. 79.

dx_i respectively, the direction cosines of $\boldsymbol{\lambda}$, for a displacement ds along $\boldsymbol{\lambda}$, will be

$$\frac{dy_\nu}{ds} = \sum_1^2{}_i \frac{\partial y_\nu}{\partial x_i}\frac{dx_i}{ds} = \sum_1^2{}_i \frac{\partial y_\nu}{\partial x_i}\lambda^i \qquad (\nu = 1, 2, 3). \quad . \quad (7)$$

Similarly, denoting by the symbol δ the increments of the coordinates for a displacement δs along $\boldsymbol{\mu}$, we have for the direction cosines of this direction

$$\frac{\delta y_\nu}{\delta s} = \sum_1^2{}_k \frac{\partial y_\nu}{\partial x_k}\frac{\delta x_k}{\delta s} = \sum_1^2{}_k \frac{\partial y_\nu}{\partial x_k}\mu^k. \quad . \quad . \quad (7')$$

Hence, from the usual formulæ of analytical geometry, we get

$$\cos\vartheta = \sum_1^3{}_\nu \frac{dy_\nu}{ds}\frac{\delta y_\nu}{\delta s} = \sum_1^3{}_\nu \sum_1^2{}_{ik} \frac{\partial y_\nu}{\partial x_i}\frac{\partial y_\nu}{\partial x_k}\lambda^i\mu^k = \sum_1^2{}_{ik}\lambda^i\mu^k \sum_1^3{}_\nu \frac{\partial y_\nu}{\partial x_i}\frac{\partial y_\nu}{\partial x_k},$$

and therefore finally

$$\cos\vartheta = \sum_1^2{}_{ik} a_{ik}\lambda^i\mu^k. \quad . \quad . \quad . \quad . \quad (8)$$

Substituting for μ^k, or λ^i, or both, their expressions in terms of the moments, we get for $\cos\vartheta$ the following equivalent expressions:

$$\cos\vartheta = \sum_1^2{}_i \lambda^i\mu_i, \quad . \quad . \quad . \quad . \quad . \quad (8')$$

$$\cos\vartheta = \sum_1^2{}_i \lambda_i\mu^i, \quad . \quad . \quad . \quad . \quad . \quad (8'')$$

$$\cos\vartheta = \sum_1^2{}_{ik} a^{ik}\lambda_i\mu_k. \quad . \quad . \quad . \quad . \quad (8''')$$

The last of these formulæ enables us to see that the notation a^{ik} is in agreement not only with the convention adopted for reciprocal elements, but also with that of writing the indices of contravariance above. For putting

$$u_i = \lambda_i ds, \quad v_k = \mu_k \delta s \qquad (i, k = 1, 2),$$

where we note that the u's and the v's are *independent* variables (not connected by any relation as are the λ's and the μ's), we can write equation (8‴) in the form

$$ds\delta s \cos\vartheta = \sum_{ik}^{2}{}_{1} a^{ik} u_i v_k;$$

since the left-hand side is invariant, and the right-hand side consists of a bilinear form in two sets of arbitrary covariant variables, it follows that the coefficients a^{ik} are contravariant.

To find $\sin\vartheta$, we can form the product by rows of the two determinants

$$\begin{vmatrix} \lambda^1 & \lambda^2 \\ \mu^1 & \mu^2 \end{vmatrix} \times \begin{vmatrix} \lambda_1 & \lambda_2 \\ \mu_1 & \mu_2 \end{vmatrix}.$$

Applying formulæ (5′), (8′), and (8″), this becomes

$$\begin{vmatrix} 1 & \cos\vartheta \\ \cos\vartheta & 1 \end{vmatrix} = 1 - \cos^2\vartheta = \sin^2\vartheta.$$

We therefore have

$$\sin\vartheta = \sqrt{\begin{vmatrix} \lambda^1 & \lambda^2 \\ \mu^1 & \mu^2 \end{vmatrix} \times \begin{vmatrix} \lambda_1 & \lambda_2 \\ \mu_1 & \mu_2 \end{vmatrix}}, \quad . \quad . \quad (9)$$

where the radical must have the sign $+$, since by definition the angle ϑ between two directions always $\leqslant \pi$, and therefore $\sin\vartheta \geqslant 0$.

The expression (9) can be put in another form. It is easy to verify that

$$\begin{vmatrix} \lambda_1 & \lambda_2 \\ \mu_1 & \mu_2 \end{vmatrix} = \begin{vmatrix} a_{11} & a_{12} \\ a_{21} & a_{22} \end{vmatrix} \cdot \begin{vmatrix} \lambda^1 & \lambda^2 \\ \mu^1 & \mu^2 \end{vmatrix} = a \begin{vmatrix} \lambda^1 & \lambda^2 \\ \mu^1 & \mu^2 \end{vmatrix},$$

and therefore

$$\sin\vartheta = \sqrt{a} \begin{vmatrix} \lambda^1 & \lambda^2 \\ \mu^1 & \mu^2 \end{vmatrix}, \quad . \quad . \quad . \quad . \quad (9')$$

$$\text{or also} \quad \sin\vartheta = \frac{1}{\sqrt{a}} \begin{vmatrix} \lambda_1 & \lambda_2 \\ \mu_1 & \mu_2 \end{vmatrix}, \quad . \quad . \quad . \quad . \quad . \quad (9'')$$

where in each case it is understood that the radical $\sqrt{a}$ is to have its arithmetical value.

5. Associated, and in particular reciprocal, tensors. The typical example of the parameters and moments of a single direction.

Take a generic tensor (with reference to the variables x_1, x_2)

$$A^{h_1 h_2 \ldots h_\mu}_{i_1 i_2 \ldots i_m}$$

of rank $m + \mu$; if we compound it with the coefficients a_{ik} of our expression ds^2, we can transfer any one of the indices h, say h_1, from above to below, so getting

$$B^{h_2 \ldots h_\mu}_{i_1 i_2 \ldots i_m h_1} = \sum_1^2{}_k\, a_{h_1 k}\, A^{k h_2 \ldots h_\mu}_{i_1 i_2 \ldots i_m},$$

which is a tensor of the same rank, but with an index of covariance more and an index of contravariance less, namely h_1. Similarly, compounding with the contravariant double system composed of the reciprocal elements a^{ik} (cf. § 4), we can transfer any one of the indices of covariance, say i_1, from below to above. We need only put

$$C^{h_1 h_2 \ldots h_\mu i_1}_{i_2 \ldots i_m} = \sum_1^2{}_k\, a^{i_1 k}\, A^{h_1 h_2 \ldots h_\mu}_{k i_2 \ldots i_m},$$

in which the system C is also a tensor of the same rank.

These operations can obviously be repeated, so as to transfer not one but several or even all of the indices of the given tensor. All the tensors so obtained are called *associated tensors* of the tensor $A^{h_1 h_2 \ldots h_\mu}_{i_1 i_2 \ldots i_m}$, so that association is a relation which is dependent on a given ds^2. In particular the tensor

$$Z^{i_1 i_2 \ldots i_m}_{h_1 h_2 \ldots h_\mu}$$

$$= \sum_1^2{}_{j_1 j_2 \ldots j_m k_1 k_2 \ldots k_\mu} a^{i_1 j_1}\, a^{i_2 j_2} \ldots a^{i_m j_m}\, a_{h_1 k_1}\, a_{h_2 k_2} \ldots a_{h_\mu k_\mu}\, A^{k_1 k_2 \ldots k_\mu}_{j_1 j_2 \ldots j_m},$$

in which the indices of covariance are the same as the indices of contravariance in A, and vice versa, is said to be *reciprocal* to A, the use of the term being justified by the consideration that the relation is reversible, A being the reciprocal of Z in the same sense as Z is of A. This can be shown explicitly if we suppose the above formulæ defining the system Z solved for the A's.

Equations (6) and (6′) show that the parameters and moments

of a single direction form a particularly simple and striking example of a pair of reciprocal systems.

Remark I.—The definition of the tensors associated with a generic tensor A involves essentially a specific ds^2, whose coefficients a_{ik} and their reciprocals a^{ik} form part of the definition. When it is necessary to emphasize this fact, we can do so by speaking of the tensor or tensors as associated *with respect to the* ds^2 *in question.*

Remark II.—For the symmetrical covariant double system a_{ik} and the contravariant system composed of the reciprocal elements a^{ik}, from which the associated tensors are constructed by composition, we could plainly take the coefficients of any other invariant quadratic ϕ instead of those of ds^2 (provided only that ϕ is irreducible, so that the reciprocal does in fact exist). We should then have associated systems *with respect to the quadric* ϕ.

Remark III.—We may point out at this stage that the idea of associated systems holds good as it stands for any number of variables $x_1, x_2, \ldots x_n$. We need only suppose that the indices take the values $1, 2, \ldots n$, and that the auxiliary element is represented by an irreducible differential quadratic form $\phi = \sum_{ik}{}_1^n a_{ik}\, dx_i\, dx_k$ in n instead of in two variables.

6. **Surface vectors.**

Let **R** be a non-zero vector drawn from a point P of the surface σ, tangentially to the surface; we shall call it a *surface* or *tangential vector*, and we can determine it by its Cartesian components Y_ν ($\nu = 1, 2, 3$) or, in closer agreement with its intimate relationship to the surface, by its magnitude R and its direction, the latter being determined by its parameters λ^i or its moments λ_i. These three quantities are not independent, since the parameters (or moments) are connected by the usual identity; the vector is therefore determined by *two* essential quantities. It will accordingly be convenient to represent it by the two independent quantities

$$R^i = R\lambda^i \qquad (i = 1, 2), \quad . \quad . \quad . \quad (10)$$

or alternatively by the pair

$$R_i = R\lambda_i \qquad (i = 1, 2), \quad . \quad . \quad . \quad (10')$$

which are called respectively the *contravariant* and *covariant components of the vector.*

These obviously form a pair of reciprocal systems, since by the preceding section the parameters and moments are reciprocal, and equations (10) and (10′) show that R^i and R_i differ from the parameters and moments only by a common factor R.

R can be calculated from them by means of the identities

$$\sum_{1}^{2}{}_{ik}\, a_{ik} R^i R^k = R^2, \quad . \quad . \quad . \quad (11)$$

$$\sum_{1}^{2}{}_{ik}\, a^{ik} R_i R_k = R^2, \quad . \quad . \quad . \quad (11')$$

$$\sum_{1}^{2}{}_{i}\, R_i R^i = R^2 \quad . \quad . \quad . \quad . \quad . \quad (11'')$$

(which are merely (5), (5′), and (5″), each multiplied by R^2), and then λ^i and λ_i follow from equations (10) and (10′); thus we see that the vector is completely determined by its contravariant (or by its covariant) components.

To find the relation between the contravariant components and the components Y_ν with respect to Cartesian axes y_1, y_2, y_3, note that the direction cosines of the direction whose parameters are the λ^i's are given by equation (7), and hence the components Y_ν (which are equal to these cosines each multiplied by R) are given by the equation

$$Y_\nu = \sum_{1}^{2}{}_{i} \frac{\partial y_\nu}{\partial x_i} R^i. \quad . \quad . \quad . \quad . \quad (12)$$

It is now obvious that the covariant components can be obtained from the contravariant components, and vice versa, by means of formulæ completely analogous to (6) and (6′), and obtained from these by multiplying them by R.

If we have to deal with *zero* vectors, i.e. having their length R zero and their direction indeterminate, we find that in order to satisfy equations (10) and (10′) in this limiting case we have to take $R^i = 0$, $R_i = 0$. With these values all the other equations ((11), (11′), &c.) are also satisfied, as can at once be seen from the fact that both sides of each equation vanish separately.

By an analogous procedure we can find simple expressions

for the scalar product $\mathbf{R} \times \mathbf{V}$ of two surface vectors $\mathbf{R}$, $\mathbf{V}$, remembering that if ϑ is the angle between the vectors we have

$$\mathbf{R} \times \mathbf{V} = RV \cos\vartheta. \quad . \quad . \quad . \quad . \quad (13)$$

In fact, considering first the general case of two vectors neither of which is zero and whose versors are $\boldsymbol{\lambda}$ and $\boldsymbol{\mu}$ respectively, and multiplying equation (8′) by RV, we get

$$\mathbf{R} \times \mathbf{V} = \sum_{1}^{2}{}_{i} R^i V_i, \quad . \quad . \quad . \quad . \quad (14)$$

while the equations (8), (8″), and (8‴) would give analogous formulæ.

The expression (14) for the scalar product also holds, like formula (13), when one or both vectors are zero, the scalar product (by definition) and the right-hand side being then zero in both formulæ.

7. Parameters and moments of the co-ordinate lines. Element of area.

We shall next obtain the direction parameters of a co-ordinate line, e.g. the line x_1 (i.e. $x_2 =$ constant), considered in the direction of x increasing. For an infinitesimal displacement in this direction, we have

$$dx_2 = 0, \; ds^2 = a_{11}\, dx_1{}^2 + 2a_{12}\, dx_1\, dx_2 + a_{22}\, dx_2{}^2 = a_{11}\, dx_1{}^2.$$

Since ds is essentially positive, and dx_1 is positive by hypothesis, we have, extracting the square root of the last of these formulæ,

$$ds = \sqrt{a_{11}}\, dx_1,$$

where the radical is taken positively. It follows that

$$\lambda^1 = \frac{dx_1}{ds} = \frac{1}{\sqrt{a_{11}}}, \qquad \lambda^2 = \frac{dx_2}{ds} = 0. \quad . \quad (15)$$

Similarly, the parameters of the line 2, in the direction of x_2 increasing, will be

$$\mu^1 = 0, \qquad \mu^2 = \frac{1}{\sqrt{a_{22}}}. \quad . \quad . \quad . \quad (15')$$

Substituting these expressions in the equations (8) and (9′), we can get the angle Ω between the two co-ordinate directions. The resulting formulæ are

$$\cos\Omega = \frac{a_{12}}{\sqrt{a_{11}\,a_{22}}}, \quad . \quad . \quad . \quad . \quad (16)$$

$$\sin\Omega = \frac{\sqrt{a}}{\sqrt{a_{11}\,a_{22}}}. \quad . \quad . \quad . \quad . \quad (16')$$

Equation (16) shows that the necessary and sufficient condition that the co-ordinates x_1, x_2 may be orthogonal is $a_{12} = 0$.

If we take an infinitesimal element of surface, obtained by drawing two infinitesimal segments ds, δs, from a point P along the co-ordinate lines, and completing the parallelogram, the area of this element will be

$$\begin{aligned} d\sigma &= ds\delta s \sin\Omega \\ &= \sqrt{a_{11}}\,dx_1 . \sqrt{a_{22}}\,dx_2 . \frac{\sqrt{a}}{\sqrt{a_{11}\,a_{22}}} = \sqrt{a}\,dx_1\,dx_2. \quad (17) \end{aligned}$$

8. **Fundamental observation (Gauss's) on the intrinsic geometry of a surface.**

We are now in a position to make an observation which will show fully the importance of the quadratic form (4) in the study of the surface. For this purpose we shall first make use of certain intuitive considerations in order to fix the idea of the *intrinsic geometry of a surface.*

Let us give the concept of a surface a material form by thinking of a flexible and inextensible sheet of matter on which figures can be drawn, and such that it can be deformed, bent, and folded in an infinite number of ways, but not torn *or stretched.* When a surface of this kind is deformed the figures drawn on it will take different spatial configurations, but some of their properties will be invariant. For instance, if two lines intersect, they retain this property however the sheet is deformed; the length of a segment of a line remains the same, and hence the distance between two points, *measured along the surface* (i.e. along the shortest line joining them which lies wholly on the surface), is unchanged; the angle between two lines which meet at a point is unchanged; and so on. In short, all those properties which involve no element alien to the surface (or, as it is usually expressed, which can be investigated without leaving the surface) are independent of the deformations of the surface, and constitute its *intrinsic geometry*.

Even in elementary geometry we have examples of this kind. Plane geometry can be, and most of it is, constructed without using points outside the plane, and is therefore intrinsic as regards its plane; it still holds—at least for suitably restricted regions—if the plane is folded, or wrapped round a cone or a cylinder.

Now consider the fact that the fundamental elements for the study of the metrical properties of a figure are: (*a*) the distance between two infinitely near points, and (*b*) the angle between two directions. In fact, the length of any line whatever is found by integration from the length of its infinitesimal elements, the area of a figure can be calculated by breaking it up into elementary parallelograms, and so on. Now the formulæ (4) and (8) (or (8′), &c.) provide us with precisely these two fundamental elements for the study of the intrinsic geometry of a surface, whenever the coefficients of ds^2 are known as functions of the x's; these coefficients therefore determine the metrical (intrinsic) properties of the surface, and are invariant for any deformation whatever of the surface which does not involve stretching. Hence the particular interest of all those theorems which can be expressed analytically in terms only of the surface co-ordinates x and the coefficients a_{ik} of the fundamental form; namely, the fact that they express properties belonging to the intrinsic geometry of the surface. The introduction into mathematics of this idea, and the fundamental observation relating to it, are due to Karl Friedrich Gauss.

9. Note on developable surfaces.

A *developable* surface is one which is flexible and inextensible and can be made to coincide with a region of a plane, without tearing or overlapping. Examples are the cylinder and the cone, and any surface formed of several portions of a plane. The intrinsic geometry of surfaces of this kind, as we have seen in the preceding section, is identical with that of the plane, and their line element can take the same forms as that of the plane; e.g.

we can choose a system of surface co-ordinates x_1, x_2, such that $ds^2 = dx_1^2 + dx_2^2$.

Consider a simple infinity of planes, which we may think of as represented by a linear equation in the Cartesian co-ordinates y_1, y_2, y_3, whose coefficients are continuous functions of a parameter u. The *envelope* of this family of planes is a developable surface to which they are tangent planes. This proposition is rendered intuitive by the following argument based on infinitesimals.

Let $\varpi_1, \varpi_2, \varpi_3, \ldots$ be planes of the family corresponding to successive infinitesimal increments of the parameter u; and let g_1 be the intersection of ϖ_1 and ϖ_2, g_2 the intersection of ϖ_2 and ϖ_3, and so on. By definition, the geometrical locus of all these lines is the *envelope* surface. The lines $g_1, g_2, \ldots$ are called its *characteristics* or *generators*; each of the planes ϖ contains two of them, forming an infinitesimal angle (cf. fig. 1), and the envelope may be considered as made up of an infinite number of these infinitesimal plane regions. It is thus clear that the envelope surface can be developed into a plane by successive rotations about the generators $g_1, g_2 \ldots$.

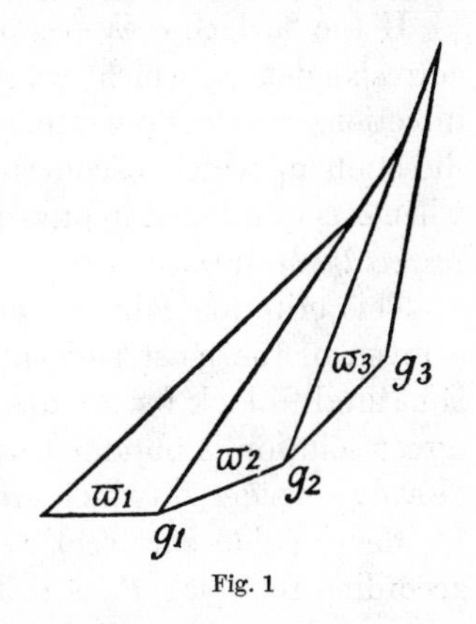

Fig. 1

We shall shortly have occasion to consider the envelope of a particular family of planes (depending on a single parameter), namely, the tangent planes to any surface whatever σ, at all points of a specified line T lying on the surface. The envelope of these planes is a developable surface σ_T, which is called the *developable circumscribed to σ along T*; since the tangent planes to σ at points on T are also tangent planes to σ_T, it follows that the circumscribed developable touches σ along the line T.

(b) *Parallelism with respect to a Surface*

10. **Geometrical definition.**

In Euclidean plane geometry, when two points P, P_1, are fixed, then to every direction drawn from P there corresponds one and only one direction drawn from P_1 and *parallel* to the first. We now propose to extend this idea from the intrinsic geometry of the plane to that of any surface σ whatever.

For this purpose consider a point P of σ, the corresponding tangent plane ϖ, and a generic direction drawn from P tangentially to σ and therefore lying in ϖ. We shall consider the direction as determined by the corresponding versor (unit vector) $\mathbf{u}$, and shall accordingly refer merely to the direction $\mathbf{u}$ instead of the direction whose versor is $\mathbf{u}$. Let P_1 be any other point on σ, and ϖ_1 the tangent plane at P_1.

If the surface σ is *developable*, we can obviously establish a correspondence, which we shall call *parallelism*, between the directions drawn tangentially from P and those from P_1. The direction $\mathbf{u}_1$ which becomes parallel to $\mathbf{u}$ in the ordinary sense when σ is developed upon a plane will be called *parallel to* $\mathbf{u}$ *with respect to the surface*.

This criterion fails in the case of a non-developable surface σ (even of the most elementary type, such as a sphere), and it is natural to look for an adequate generalization of it. The most direct solution is obtained by adding to the elements of position already considered (which are sufficient without further definition for developable surfaces) a connecting law, *a priori* arbitrary, according to which P_1 is to be considered as reached from P by moving along a *specified* curve T lying on σ (the curve of displacement).

We can now, with reference to this curve T, define parallel displacement from P to P_1 as follows. Consider the developable circumscribed to σ along T; this surface, which we shall call σ_T, is, as we know, tangential to σ along the given curve, and in particular at P and P_1. Hence the directions tangential to σ at these two points are also tangential to σ_T. We can now take for our definition of surface parallelism on σ along T the parallelism which we have associated with the developable σ_T, and we shall agree to say that the parallel at P_1 along the line T to a generic direction (in the surface) $\mathbf{u}$ at P is the direction (in the surface) $\mathbf{u}_1$ which, on the developable σ_T, is parallel to $\mathbf{u}$ in the sense just defined.[1]

[1] A simple and so to speak automatic way of constructing parallel directions is to roll the surface σ along a plane. Cf. PERSICO: "Realizzazione cinematica del parallelismo superficiale", in *Rend. della R. Acc. dei Lincei*, Vol. XXX (2nd half-year, 1921), pp. 127–128.

11. First consequences. Equipollence of vectors with respect to a surface.

A necessary consequence of the foregoing definition is that —contrary to what happens for developables—the direction $\mathbf{u}_1$ which is parallel with respect to the surface to the direction $\mathbf{u}$ at P, is *not* uniquely determined by P, $\mathbf{u}$, P_1, alone, but in general *depends also on the curve of displacement.* From this point of view the geometrical concept of parallelism can be compared with the physical concept of work, which involves the integral of an expression of the form $X_1 dx_1 + X_2 dx_2$ (where x_1, x_2 are co-ordinates, of any kind, of the points of σ). This integral in general depends on the line T of integration; only in the particular case when $X_1 dx_1 + X_2 dx_2$ is a perfect differential is there no such dependence.

Returning to parallelism along T, we must first point out that *angles are unchanged by parallel displacement.* That is to say, if $\mathbf{a}$, $\mathbf{b}$ are two generic directions (in the surface) at P, their parallels at P_1 with respect to the surface, $\mathbf{a}_1$, $\mathbf{b}_1$, contain the same angle. This is obvious if we notice that we have parallelism in the ordinary sense in the plane upon which σ_T is developed, and that further the operation of development does not change angles.

Up to this point of the discussion we have referred solely to directions, with their corresponding versors. It is clear that the same construction as that used to pass from $\mathbf{u}$ to $\mathbf{u}_1$ can be applied to a tangential vector $\mathbf{R}$ of any (non-unit) length R. If $\mathbf{u}$ is the corresponding versor, we have $\mathbf{R} = R\mathbf{u}$, from which we get a vector $\mathbf{R}_1 = R\mathbf{u}_1$, i.e. a vector localized at P_1, having the same length as $\mathbf{R}$ and the same direction as the versor $\mathbf{u}_1$ which is parallel to $\mathbf{u}$ with respect to the surface. We shall naturally say that the vectors $\mathbf{R}$ and $\mathbf{R}_1$ are *equipollent* with respect to the surface, with reference to the path T. In substance this concept of equipollence with respect to a surface reduces at once to parallelism, two tangential vectors being equipollent when they are parallel and have the same length.

The case where the curve of displacement T is a geodesic [1]

[1] I.e., with the usual definition, any line on σ such that at every point its osculating plane is perpendicular to the tangent plane to σ. The lines which give the shortest path lying on the surface between two given points always have this property. Further, the reciprocal theorem is also true (under certain restrictions); hence to define geodesics we can use sometimes one and sometimes the other criterion. We shall return to the question farther on (cf. p. 130).

on σ calls for a special note in relation to parallelism. T is then also a geodesic on σ_T. In order to see that this is so, note that σ and σ_T have the same tangent planes at all points of T; hence, if the various osculating planes of T are normal to one of the surfaces, they are also normal to the other. When σ_T is developed on a plane, the geodesic T becomes a straight line (an immediate consequence of its characteristic property of giving the shortest path between any two points on it), and the directions $\mathbf{u}$ and $\mathbf{u}_1$, which become parallel in the plane as a result of the development, will make equal angles with this line. Since development does not change angles, we deduce that *parallel directions on σ at points of a geodesic make equal angles with this geodesic.*[1] In particular, if $\mathbf{u}$ coincides with the direction of T at P, then $\mathbf{u}_1$ will coincide with the direction of T at P_1, or in other words, the directions of a geodesic at its various points are all parallel (along the geodesic itself); more shortly, *the geodesics are auto-parallel curves.* It follows from these arguments that auto-parallelism is a characteristic property of geodesics and can be used to define them.[2]

12. **Infinitesimal displacement. Infinitesimal form of the law of parallelism.**

Suppose in particular that P_1 is infinitely near to P, so that the path T is reduced to the elementary arc PP_1, which is uniquely determined (except for infinitesimals of order higher than the first) by its extremities. For the development in this case we need only give the plane ϖ_1 an elementary rotation round the straight line r in which it intersects ϖ. Incidentally we may note that the direction of this line is said to be *conjugate* to the direction PP_1, at P or at P_1 (both points giving the same result, except for infinitesimals). We shall denote by $-\boldsymbol{\omega}$ the infinitesimal vector, parallel to r, which in magnitude, direction, and

[1] Taking this property as defining parallelism with respect to a surface we can deduce from it for the sphere an elegant geometrico-kinematical construction from which various other properties follow easily. Cf. G. CORBELLINI: "Genesi cinematica intrinseca del parallelismo di Levi-Civita", in *Rend. della R. Acc. dei Lincei*, Vol. XXXII (1st half-year, 1923), pp. 72–76.

[2] This statement will be recognized as an obvious extension to surfaces of any kind whatever of the primary intuition of the nature of the straight line, expressed by Euclid in the words εὐθεῖα γραμμή ἐστιν, ἥτις ἐξ ἴσου τοῖς ἐφ' ἑαυτῆς σημείοις κεῖται (a straight line is that which lies equally with respect to all its points).

sense represents the elementary rotation by means of which ϖ_1 is brought into coincidence with ϖ. Then $\boldsymbol{\omega}$ will be the elementary rotation which will bring ϖ_1 back from the plane of development ϖ to its original position as tangent plane at P_1. Let $\mathbf{R}$ be a generic tangential vector drawn from P; in order to find the equipollent vector $\mathbf{R}_1$ at P_1, we draw from P_1, when in the plane of development, the vector equipollent in the ordinary sense to $\mathbf{R}$, and then bring the plane ϖ_1 back to its original position, carrying with it the vector so constructed. Thus the vector $\mathbf{R}_1$ is merely $\mathbf{R}$, after having undergone a displacement (of no interest if we consider the vector independently of its point of application) and also the rotation $\boldsymbol{\omega}$. From the elementary principles of rigid dynamics we find that the difference between the vectors $\mathbf{R}_1$ and $\mathbf{R}$, i.e. the vectorial increment $d\mathbf{R}$ of the vector $\mathbf{R}$ during the parallel displacement from P to P_1, is given by

$$d\mathbf{R} = \boldsymbol{\omega} \wedge \mathbf{R},$$

i.e. the vector product $\boldsymbol{\omega}\mathbf{R}$.

As both $\boldsymbol{\omega}$ and $\mathbf{R}$ are vectors in the plane ϖ it follows that the increment $d\mathbf{R}$ is perpendicular to this plane, or, in particular, is zero.[1]

We shall now show that this condition, combined with the condition that $\mathbf{R}_1$ is a tangential vector (i.e. belongs to ϖ_1), completely determines the vector $\mathbf{R}_1$, so that we may take as the differential definition of parallelism with respect to a surface the following geometrical relations, in which $\mathbf{n}$ denotes the normal to ϖ:

$$d\mathbf{R} \parallel \mathbf{n}, \quad . \quad . \quad . \quad . \quad . \quad (18)$$
$$\mathbf{R}_1 \parallel \varpi_1.$$

To prove this, note that the equation

$$\mathbf{R} = \mathbf{R}_1 - d\mathbf{R}$$

must be satisfied; i.e. that it must be possible to resolve the vector $\mathbf{R}$ into one component $\mathbf{R}_1$ parallel to a given plane and

[1] The last-mentioned case will occur if $\mathbf{R}$ has the direction of $\boldsymbol{\omega}$, i.e. the conjugate of PP_1; in this, and only in this case, the parallel $\mathbf{R}_1$ with respect to the surface coincides with the Euclidean parallel. This remark is due to PROFESSOR E. BOMPIANI, who has made use of it to generalize the theory of systems conjugate to surfaces belonging to non-Euclidean spaces; cf. *Atti del R. Ist. Veneto*, Vol. LXXX, 1921, p. 1120.

another, $-d\mathbf{R}$, parallel to a given direction not contained in the plane; it is known that this can be done in only one way.[1]

13. **The intrinsic character of parallelism.**

Returning to the question of parallel displacement along an arc T of finite length, we see at once that if T is a segment of a geodesic, parallelism depends solely on the intrinsic properties of the surface σ; i.e. it depends on the nature of the linear element ds, and not on the configuration of the surface in space, as might *a priori* have been supposed from the geometrical construction (which uses the surrounding space) or the equivalent formulæ (18) and $\mathbf{R}_1 \parallel \varpi_1$.

In fact, we need only recall the two general properties of the conservation of angles and the autoparallelism of geodesics. The parallel $\mathbf{u}_1$ at P_1 to a generic direction $\mathbf{u}$ drawn from P is determined by the conditions of (a) belonging to the surface σ, and (b) of making the same angle at P_1 with the geodesic of displacement as $\mathbf{u}$ does at P. It will be seen that we are dealing with angular properties which depend solely on the metric of σ.

This argument for a geodesic T can easily be extended to the general case, if we suppose T divided up into elementary displacements, from a generic point P to a very near point P_1. In a displacement of this kind the elementary change in the direction $\mathbf{u}$ is determined, as we have seen, by the extremities PP_1; the nature of the line joining these extremities has no effect, and we may therefore think of it as a displacement along an infinitesimal segment of a geodesic. But a displacement of this kind depends only on the intrinsic properties of the surface; hence we see that in general this is true also for the change in $\mathbf{u}$, and therefore for parallelism, whatever may be the line of displacement.

The same result holds good for equipollence, i.e. for the displacement of vectors of any (non-unit) length whatever. In fact (§ 11), this length by definition remains unchanged.

[1] Some interesting geometrical consequences, especially for the case of ruled surfaces, have been pointed out by A. Myller in some notes in *Comptes Rendus*; cf. Vol. 174 (1922), pp. 997–998; Vol. 175 (1922), pp. 939–941; Vol. 176 (1923), pp. 483–485. Cf. also a recent note by O. Mayer: "Une interprétation géométrique de la seconde forme quadratique fondamentale d'une surface en relation avec la théorie du parallélisme", *ibid.*, Vol. 178 (1924), pp. 954–956.

14. The symbolic equation of parallelism.

The condition (18) can be put in a more expressive form if we note that it is equivalent to saying that the vector $d\mathbf{R}$ is perpendicular to every direction which is tangential to σ at P, or in other words, if we think of such a direction as being determined by an infinitesimal displacement of the point P along the surface, that $d\mathbf{R}$ is perpendicular to all these displacements. In symbols, if δP denotes the infinitesimal vector representing the displacement, we shall have

$$d\mathbf{R} \times \delta P = 0 \quad . \quad . \quad . \quad . \quad . \quad (19)$$

for *any* δP whatever which is tangential to σ—an equation similar in form to the equation of virtual work. If dY_ν $(\nu = 1, 2, 3)$ denotes the components of $d\mathbf{R}$, and δy_ν $(\nu = 1, 2, 3)$ the components of δP (in both cases referred to the orthogonal Cartesian co-ordinates y_1, y_2, y_3), we have identically

$$d\mathbf{R} \times \delta P = \sum_{1}^{3}{}_{\nu}\, dY_\nu\, \delta y_\nu, \quad . \quad . \quad . \quad (20)$$

and the vectorial relation (19) is thus transformed into the scalar relation

$$\sum_{1}^{3}{}_{\nu}\, dY_\nu\, \delta y_\nu = 0; \quad . \quad . \quad . \quad . \quad (19')$$

this, or the original equation (19), may be called the *symbolic equation of parallelism.*

15. Intrinsic equations of parallelism.

As the symbolic equation involves geometrical elements which do not belong to the surface, it does not show directly that parallelism is a concept depending only on the intrinsic properties of the surface. But we can deduce from it without much difficulty other equations which have this important characteristic.

In order to do so, we shall naturally try to find the values in terms of intrinsic elements of the quantities dY_ν and δy_ν which occur in equation (19'). Take first the displacements δy_ν. The only condition imposed on them—other than that of being infinitesimal—is that they represent a displacement along the surface of σ; they can therefore be expressed in terms of the corresponding (arbitrary) variations δx_1, δx_2 of the surface coordinates, by differentiating the equations (1). We accordingly have

$$\delta y_\nu = \sum_{1}^{2}{}_k \frac{\partial y_\nu}{\partial x_k} \delta x_k.$$

As the vector $\mathbf{R}$ is tangential, we can define it intrinsically by means of its contravariant components, and substitute for the Y_ν's the expressions (12).

Putting

$$\tau_k = \sum_{1}^{2}{}_j \sum_{1}^{3}{}_\nu \frac{\partial y_\nu}{\partial x_k} d\left(\frac{\partial y_\nu}{\partial x_j} R^j\right) \qquad (k = 1, 2) \quad . \quad (21)$$

for the sake of shortness, the identity (20) can therefore finally be written in the form

$$d\mathbf{R} \times \delta P = \sum_{1}^{2}{}_k \tau_k \delta x_k; \quad . \quad . \quad . \quad (20')$$

since δx_1, δx_2 are completely arbitrary, it follows that the symbolic equation of parallelism is equivalent to the two following equations:

$$\tau_k = 0 \qquad (k = 1, 2). \quad . \quad . \quad . \quad (22)$$

These are the two equations which define the increments dR^1, dR^2 to be assigned to the components of a generic vector $\mathbf{R}$ when it undergoes a parallel displacement along the elementary path dx_1, dx_2; that they are really intrinsic equations will be clear when the expressions τ_k are written out in full, as will now be done.

Differentiating the product on the right-hand side of the equations (21), and using the expression for the coefficients a_{ik} given in formula (3), the expression for τ_k becomes

$$\tau_k = \sum_{1}^{2}{}_j a_{kj} dR^j + \sum_{1}^{3}{}_\nu \sum_{1}^{2}{}_{jl} R^j \frac{\partial y_\nu}{\partial x_k} \frac{\partial^2 y_\nu}{\partial x_j \partial x_l} dx_l,$$

or $$\tau_k = \sum_{1}^{2}{}_j a_{kj} dR^j + \sum_{1}^{2}{}_{jl} R^j dx_l \sum_{1}^{3}{}_\nu \frac{\partial y_\nu}{\partial x_k} \frac{\partial^2 y_\nu}{\partial x_j \partial x_l}. \qquad (21')$$

We have now to show that the result of the summation with

respect to ν can be expressed in terms of intrinsic elements; more precisely, that it is a linear combination of the derivatives of the coefficients a_{ik}. Consider its general term, and note that we can write

$$\frac{\partial y_\nu}{\partial x_k}\frac{\partial^2 y_\nu}{\partial x_j \partial x_l} = \frac{\partial}{\partial x_j}\left(\frac{\partial y_\nu}{\partial x_k}\frac{\partial y_\nu}{\partial x_l}\right) - \frac{\partial^2 y_\nu}{\partial x_j \partial x_k}\frac{\partial y_\nu}{\partial x_l},$$

or analogously

$$\frac{\partial y_\nu}{\partial x_k}\frac{\partial^2 y_\nu}{\partial x_j \partial x_l} = \frac{\partial}{\partial x_l}\left(\frac{\partial y_\nu}{\partial x_k}\frac{\partial y_\nu}{\partial x_j}\right) - \frac{d^2 y_\nu}{\partial x_l \partial x_k}\frac{\partial y_\nu}{\partial x_j}.$$

In order to maintain symmetry in the indices j and l, we shall take half the sum of the expressions on the right of these equations to represent the value of the term in question. Noting that the sum of the two terms preceded by the minus sign is exactly the derivative with respect to x_k of the product $\frac{\partial y_\nu}{\partial x_j}\frac{\partial y_\nu}{\partial x_l}$, we get

$$\frac{\partial y_\nu}{\partial x_k}\frac{\partial^2 y_\nu}{\partial x_j \partial x_l} = \frac{1}{2}\left[\frac{\partial}{\partial x_j}\left(\frac{\partial y_\nu}{\partial x_k}\frac{\partial y_\nu}{\partial x_l}\right) + \frac{\partial}{\partial x_l}\left(\frac{\partial y_\nu}{\partial x_k}\frac{\partial y_\nu}{\partial x_j}\right) - \frac{\partial}{\partial x_k}\left(\frac{\partial y_\nu}{\partial x_j}\frac{\partial y_\nu}{\partial x_l}\right)\right].$$

Now sum with respect to ν. Remembering the values of the coefficients a_{ik}, we get

$$\sum_{1}^{3}{}_\nu \frac{\partial y_\nu}{\partial x_k}\frac{\partial^2 y_\nu}{\partial x_j \partial x_l} = \frac{1}{2}\left[\frac{\partial a_{kl}}{\partial x_j} + \frac{\partial a_{jk}}{\partial x_l} - \frac{\partial a_{jl}}{\partial x_k}\right].$$

Hence this sum has been put in the required form. The right-hand side of this equation is represented shortly by the symbol

$$[jl, k]$$

(*Christoffel's symbol of the first kind*); which is easy to remember, the arrangement of the indices corresponding to that of the negative term of the linear combination above, while the two positive terms have the same indices but differently arranged. We shall investigate presently some properties of these symbols; for the moment we need only remark that they represent certain functions of the surface co-ordinates x_1, x_2 which depend only on the fundamental quadratic form.

Returning to the expression (21′) for the quantities τ_k, we can now write it in the form

$$\tau_k = \sum_{1}^{2}{}_j\, a_{kj}\, dR^j + \sum_{1}^{2}{}_{jl}\, [jl, k]\, R^j\, dx_l \qquad (k = 1, 2). \quad (21'')$$

Before continuing the argument, it is important to note that the quantities τ_k, which (as shown by equation (21′)) depend on two vectors (**R** and the displacement dx_1, dx_2) as well as on the coefficients of ds^2 and their first derivatives, are *covariant*. This follows from the invariance of the linear form $\sum_{1}^{2}{}_k\, \tau_k\, dx_k$, which is itself shown by the identity (20′).

The system which is the reciprocal of the τ_k's, namely,

$$\tau^i = \sum_{1}^{2}{}_k\, a^{ik}\, \tau_k \qquad (i = 1, 2),$$

is accordingly *contravariant*; using equation (21″), it can be put in the form

$$\tau^i = dR^i + \sum_{1}^{2}{}_{jl}\, R^j\, dx_l \sum_{1}^{2}{}_k\, a^{ik}\, [jl, k];$$

or, putting

$$\sum_{1}^{2}{}_k\, a^{ik}\, [jl, k] = \{jl, i\}$$

(*Christoffel's symbol of the second kind*), in the form

$$\tau^i = dR^i + \sum_{1}^{2}{}_{jl}\, \{jl, i\}\, R^j\, dx_l. \quad . \quad . \quad (21''')$$

The *equations of parallelism* (22), as is *a priori* to be expected from their geometrical significance, are *invariant* whatever system of curvilinear co-ordinates x_1, x_2 is chosen. This is evident from the fact that they express the vanishing of the covariant system τ_k (cf. remarks on pp. 71, 84). The equations of parallelism can of course also be put in the equivalent form

$$\tau^i = 0 \qquad (i = 1, 2), \quad . \quad . \quad . \quad (22')$$

which also shows that they are invariant.

Solving them for the differentials dR^i, we get

$$dR^i = -\sum_{1}^{2}{}_{jl}\, \{jl, i\}\, R^j\, dx_l \qquad (i = 1, 2). \qquad (23)$$

This is the final form of the differential equations of parallelism. It gives the increments of the contravariant components of a surface vector in an equipollent displacement along the elementary path dx_i, expressed in terms of the dx_i's, the components of the vector, and certain functions of position (to be considered as given) depending only on the coefficients of ds^2 and therefore on the intrinsic nature of the surface.

16. **Christoffel's symbols.**

We have introduced the symbols

$$[jl, k] = \frac{1}{2}\left[\frac{\partial a_{jk}}{\partial x_l} + \frac{\partial a_{kl}}{\partial x_j} - \frac{\partial a_{jl}}{\partial x_k}\right], \quad . \quad . \quad (24)$$

$$\{jl, i\} = \sum_1^2{}_k\, a^{ik}\,[jl, k], \quad . \quad . \quad . \quad (25)$$

which can also be formally extended to quadratic forms in n variables; we now propose to examine their more elementary properties.

First, it is obvious that both symbols are symmetrical with respect to the coupled indices, i.e. that

$$[jl, k] = [lj, k], \qquad \{jl, i\} = \{lj, i\}.$$

Consequently for a form in n variables there are n of each kind corresponding to each pair of indices. Hence there are in all $\frac{1}{2}n^2(n+1)$ of each kind (the number of first derivatives of the coefficients a_{ik}).

It is easy to express the derivatives of the a_{ik}'s in terms of Christoffel's symbols. Writing down equation (24) and the corresponding equation obtained by interchanging l and k, and adding them, we get the following formula, which frequently occurs:

$$\frac{\partial a_{lk}}{\partial x_j} = [jl, k] + [jk, l]. \quad . \quad . \quad . \quad (24')$$

From equation (25), applying Cramer's rule in the usual way, we can get the symbols of the first kind in terms of those of the second kind. Multiplying by a_{im} and summing with respect to i, we get in fact

$$[jl, m] = \sum_1^n{}_i\, a_{im}\,\{jl, i\}. \quad . \quad . \quad . \quad (25')$$

Lastly, we shall prove a formula which is frequently used and gives the derivatives of the determinant a (or more precisely of its logarithm) in terms of Christoffel's symbols.

Applying the usual rule for differentiating a determinant of order n, we see that the derivative of a with respect to any one of the x's (say x_i) is the sum of n determinants, any one of which (say the kth) is obtained from a by replacing the elements of the kth row by their derivatives. A determinant of this kind, expanded from the kth row, can be written in the form

$$\sum_{1}^{n}{}_j \frac{\partial a_{jk}}{\partial x} a^{jk} a$$

(the co-factor of a_{jk} being a^{jk} multiplied by a); hence

$$\frac{\partial a}{\partial x_i} = \sum_{1}^{n}{}_{kj} \frac{\partial a_{jk}}{\partial x_i} a^{jk} a,$$

or dividing by a, and using formula (24′),

$$\frac{\partial \log a}{\partial x_i} = \sum_{1}^{n}{}_{kj} ([ji, k] + [ki, j]) a^{jk}.$$

Finally, by formula (25), we get

$$\frac{\partial \log a}{\partial x_i} = \sum_{1}^{n}{}_j \{ji, j\} + \sum_{1}^{n}{}_k \{ki, k\}.$$

The two sums in this formula differ only in the letter chosen to denote the index of summation; hence we have

$$\frac{\partial \log a}{\partial x_i} = 2\sum_{1}^{n}{}_k \{ki, k\}.$$

This formula is more frequently written in the form obtained by dividing by 2, i.e.

$$\frac{\partial \log \sqrt{a}}{\partial x_i} = \sum_{1}^{n}{}_k \{ki, k\} \qquad (i = 1, 2, \ldots n). \quad (26)$$

17. Equations of parallelism in terms of covariant components.

It is easy to find equations analogous to (23) for the differentials of the *covariant* components of the vector **R**. These components in fact are obtained from the contravariant components by means of the relations (cf. §§ 3, 6, pp. 90, 96)

$$R_i = \sum_{1}^{2}{}_j a_{ij} R^j;$$

hence, differentiating and changing j into k in the second sum,

$$dR_i = \sum_{1}^{2}{}_{jl} \frac{\partial a_{ij}}{\partial x_l} dx_l R^j + \sum_{1}^{2}{}_k a_{ik} dR^k.$$

Now substitute for dR^k the expression given for it by formula (23), and we shall have

$$dR_i = \sum_{1}^{2}{}_{jl} \frac{\partial a_{ij}}{\partial x_l} dx_l R^j - \sum_{1}^{2}{}_{jlk} a_{ik} \{jl, k\} R^j dx_l.$$

In the first sum, we can express the derivatives of the coefficients a_{ij} in terms of the symbols of the first kind, so getting

$$\sum_{1}^{2}{}_{jl} ([jl, i] + [il, j]) dx_l R^j;$$

in the second, we can sum with respect to k (cf. formula (25′)), so getting

$$- \sum_{1}^{2}{}_{jl} [jl, i] R^j dx_l.$$

We thus have

$$dR_i = \sum_{1}^{2}{}_{jl} [il, j] dx_l R^j.$$

In order to make the contravariant components disappear altogether, we substitute for R^j from the formula

$$R^j = \sum_{1}^{2}{}_k a^{jk} R_k;$$

summing with respect to j (which, by formula (25), changes the symbol of the first kind to one of the second kind), we get

$$dR_i = \sum_{1}^{2}{}_{kl} \{il, k\} R_k dx_l.$$

Finally, changing k into j in order to show more clearly the analogy with the equations (23), we have the equations

$$dR_i = \sum_{1}^{2}{}_{jl} \{il, j\} R_j dx_l \qquad (i = 1, 2), \quad . \quad (27)$$

which are equivalent to (23). They are in fact the result of combining certain formulæ and identities with the equations (23); and reciprocally, starting from (27), an analogous process will give (23), as can easily be verified.

18. **Some analytical verifications.**

We are now in a position to give an analytical proof of some properties of parallelism which have already been obtained as immediate consequences of the geometrical definition.

Consider first the parallel displacement of a vector **R** along a *finite* segment T of a curve, from P to P_1. Let the curve be defined by the parametric equations

$$x_i = x_i(s), \qquad . \quad . \quad . \quad . \quad . \quad (28)$$

where s represents any parameter (which may, if we wish, be the length of the arc measured from an arbitrary origin P_0). The quantities R^i are to be considered as functions of s with arbitrarily assigned values at P. The equations (23), divided by ds, become

$$\dot{R}^i = -\sum_{1}^{2}{}_{jl} \{jl, i\} R^j \dot{x}_l \qquad (i = 1, 2),$$

where the dot indicates differentiation with respect to s, and the quantities $\dot{x}_l$ are of course obtained by differentiating equations (28), and are therefore to be considered as given functions. These are two linear differential equations of the first order, in the normal form with respect to the derivatives of the two unknown functions R^1, R^2; hence, as is known from the calculus, they uniquely determine these two functions when the (arbitrary) initial values are given. We have thus a confirmation of the geometrically obvious fact of the *possibility of displacing* an arbitrarily assigned surface vector, and of the *uniqueness of the result.*

Using the differential equations already found, we shall now prove that the *length of a vector* and the *angle between two vectors are unchanged* by a parallel displacement. These two results can be proved simultaneously, as follows. Let **R**, **V**, be two vectors. Give them a parallel displacement along an infinitesimal path, and calculate the change in their scalar product due to this displacement. We shall have (cf. formula (14))

$$d(\mathbf{R} \times \mathbf{V}) = \sum_{1}^{2}{}_i R^i \, dV_i + \sum_{1}^{2}{}_i V_i \, dR^i;$$

substituting for dR^i and dV_i from (23) and (27), this becomes

$$d(\mathbf{R} \times \mathbf{V}) = \sum_{1}^{2}{}_{ijl} R^i \{il, j\} V_j \, dx_l - \sum_{1}^{2}{}_{ijl} V_i \{jl, i\} R^j \, dx_l.$$

Interchanging i and j in one of the two sums, we see that the sums are equal, and therefore

$$d(\mathbf{R} \times \mathbf{V}) = 0;$$

i.e. the scalar product is unchanged by an infinitesimal (and therefore also by a finite) displacement. Now let $\mathbf{V}$ coincide with $\mathbf{R}$, so that $\mathbf{R} \times \mathbf{V} = \mathbf{R}^2$, and we at once obtain the result that the length of a vector is unchanged by a parallel displacement. Combining this result with equation (13), we see that as the scalar product of two vectors and their respective lengths are all unchanged, the angle between them (provided neither vector is of zero length) must also remain the same.

19. **Permutability.**

While a tangential vector is intrinsically defined by two numbers, the geometrical notion corresponding to it, as we have already said, is a segment of a tangent line at a point P of the surface σ—an entity which does not belong wholly to σ, at least in general. If, however, we are dealing with an *infinitesimal* vector, the element of the tangent plane in which it lies coincides with the element of the surface σ around P, and we may say that we are using only points lying in σ. Hence, for a generic infinitesimal tangential vector we can use the ordinary notion of a displacement from the origin P to the final point P_1, where P_1 also lies on σ. As the length R reduces in this case to a linear element ds, it follows from the definition of direction parameters that the quantities R^i, which are equal to $\lambda^i ds$, are identical with the increments dx_i of the curvilinear co-ordinates in passing from P to P_1.

Next, consider two systems of differentials dx_i, δx_i, and the corresponding infinitesimal vectors (or displacements) $dP = PP_1$, $\delta P = PP_2$ (assumed to lie in σ). We shall use the symbol df to denote the increment of f (where f is a generic vector or any scalar or vector quantity derived from it) corresponding to a parallel displacement from P to P_1; the symbol δf will be defined in the same way for the displacement from P to P_2.

With this convention $d\delta P$ will represent the vectorial increment of δP for a displacement from P to P_1, and $d\delta x_i$ the increment of the associated contravariant system δx_i. For the latter, equation (23) gives

$$d\delta x_i = -\sum_{1}^{2}{}_{jl} \{jl, i\} \delta x_j \, dx_l \qquad (i = 1, 2). \quad . \quad (29)$$

Similarly, the displacement of dP from P to P_2 gives the increments δdx_i, for which we have

$$\delta dx_i = -\sum_{1}^{2}{}_{jl} \{jl, i\} \, dx_j \, \delta x_l. \quad . \quad . \quad . \quad (29')$$

Interchanging j and l in one of these two sums, and using the property of symmetry of Christoffel's symbols, we see that

$$d\delta x_i = \delta dx_i, \quad . \quad . \quad . \quad . \quad . \quad . \quad (30)$$

which proves that the two operators d and δ, as just defined, are *permutable.*

The geometrical meaning of this result is particularly simple. Note first of all that for infinitesimal vectors—the only kind considered here—the elements of the contravariant system are merely the differences of corresponding co-ordinates. Hence, if the co-ordinates of P are the x_i's, we shall have in the first place $x_i + dx_i$ as the co-ordinates of P_1, and $x_i + \delta x_i$ as the co-ordinates of P_2. Let Q be the point on σ reached by constructing at P_1 the vector equipollent to δP; as the contravariant system for this vector is $\delta x_i + d\delta x_i$, we get finally

$$x_i + dx_i + \delta x_i + d\delta x_i$$

as the co-ordinates of Q.

Similarly let Q^* be the point on σ reached by constructing at P_2 the vector equipollent to dP; we get the co-ordinates of

Q^* by interchanging the operators d and δ in the co-ordinates of Q, which gives

$$x_i + \delta x_i + dx_i + \delta dx_i.$$

Applying equation (30), we see that Q *coincides with* Q^*. A more illuminating way of expressing the same result is to say that *the parallelogram rule holds for infinitesimal vectors which are equipollent with respect to a surface.*[1]

It may be noted that in the foregoing argument second-order quantities of the type $d\delta x$ have been taken into account, but $(dx_i)^2$, $(\delta x_i)^2$ have been neglected. If the latter were to be taken into account, by considering the vectors δP, dP and the equipollent vectors at P_2 and P_1 as vectors in space, we should no longer have a parallelogram, nor even a closed quadrilateral. In fact, referring to the space construction already given (cf. p. 105) for vectors equipollent with respect to a surface, we see that while $d\delta P$ and δdP are both in the direction of the normal to σ at P, yet their lengths are in general different, since the three points P, P_1, P_2 and their respective tangent planes have *a priori* no relation between them except that of being infinitely near one another.

The formulæ (29) or (29′) provide a definition of the second differentials which is invariant with respect to any change of variables. In order to grasp the significance and value of this fact, we must recall the conventions as to second differentials which are adopted in the elementary theory of the calculus.

To fix the ideas, consider the simpler case of a single independent variable. Ordinarily the convention $d^2x = 0$ is adopted; i.e. the increments dx are considered independent of x, as is quite legitimate. But this simplification does not hold if we change the independent variable by putting $x = f(\xi)$, from which, on the hypothesis that we have a reversible transformation, we can reciprocally find ξ as a function $F(x)$ of x. In fact, differentiating twice the formula $\xi = F(x)$, we get

$$d\xi = F'(x)\,dx,$$
$$d^2\xi = F''(x)\,(dx)^2 + F'(x)\,d^2x,$$

[1] This property might be taken as the starting point for an intrinsic proof of the properties of parallelism, depending only on the metric of σ, and making no use of the surrounding space. The method can be applied directly to manifolds V_n of any number of dimensions. Cf. H. Weyl, *Raum, Zeit, Materie*, § 14 (Berlin, Springer, 1923).

which shows that even if we make $d^2x = 0$, $d^2\xi$ will not in general be zero.

If then there are n variables, it is usual to consider only systems of differentials which are completely independent of the variables x_i, so that we have not only $d^2x_i = 0$, but also, for any two systems dx_i, δx_i of these differentials whatever,

$$d\delta x_i = \delta dx_i = 0 \qquad (i = 1, 2, \ldots n).$$

Now change the variables, by putting $x_i = f_i(\bar{x})$, and therefore $\bar{x}_i = F_i(x)$. Using the condition $\delta dx_i = 0$, we get

$$\delta d\bar{x}_i = \sum_1^n{}_{jl} \frac{\partial^2 F_i}{\partial x_j \partial x_l} dx_j \, \delta x_l,$$

so that the property

$$\delta d\bar{x}_i = d\delta\bar{x}_i$$

also holds, but these differentials will not in general be zero. The usual convention is therefore legitimate, and is suggested by obvious reasons of simplicity, when in a given question we are dealing always with the same variables; but it is not invariant for a change of variables.

If instead we adopt the formulæ (29) and (29′), and suppose that

$$d\delta x_i = \delta dx_i = -\sum_1^2{}_{jl} \{jl, i\} \, \delta x_j \, dx_l, \quad . \quad (31)$$

we get, for the same geometrical interpretation of this formula,

$$d\delta\bar{x}_i = \delta d\bar{x}_i = -\sum_1^2{}_{jl} \{\overline{jl, i}\} \, \delta\bar{x}_j \, d\bar{x}_l,$$

where the line above the letters denotes that Christoffel's symbols refer to the variables $\bar{x}$, i.e. to the transformed quadratic form

$$ds^2 = \sum_1^2{}_{ik} \bar{a}_{ik} \, d\bar{x}_i \, d\bar{x}_k.$$

We could of course verify by direct substitution that the form of the expressions (31) is unchanged by the change of variables. We are in fact dealing with an immediate corollary of the invariance of the equations $\tau^i = 0$ (cf. § 15), which follows at once by putting $R^i = \delta x_i$ in these equations.

On account of this invariant property the second differentials, defined as in (31), are called *contravariant*, although strictly speaking the term applies not to them but to the expressions

$$d\delta x_i + \sum_{jl}^{2}{}_{1} \{jl, i\} \delta x_j \, dx_l,$$

which in any case (see § 15) constitute a simple contravariant system.

(c) *Extension of the Foregoing Notions to* n-*dimensional Manifolds of any Metric*

20. n-dimensional manifolds.

Alongside the extension of the use of geometrical terms which was developed in Chapter I, we shall now introduce, on the lines of the discussion in subdivision (*a*) of this chapter, the fundamental notion of an n-dimensional metric manifold, where n is any integer.

If there are n variables $x_1, x_2, \ldots x_n$, we know that the aggregate of values which can be assigned to them is called an n-dimensional manifold. Now suppose that together with these variables and their field of variation there is also given *a priori* a differential quadratic form

$$ds^2 = \sum_{ik}^{n}{}_{1} a_{ik} \, dx_i \, dx_k, \quad . \quad . \quad . \quad . \quad (32)$$

in which the coefficients a_{ik} are given functions of the x's, and $a_{ik} = a_{ki}$. We shall agree to consider ds as the *distance* between the two infinitely near points whose co-ordinates are $x_1, x_2, \ldots x_n$ and $x_1 + dx_1, x_2 + dx_2, \ldots x_n + dx_n$; we shall in consequence agree that ds is to be *invariant* for any change of co-ordinates. Having thus introduced into the manifold the notion of an elementary distance, we get from it at once by integration the notion of the *length* of a line, and also deduce from it, as we shall see, the most direct criteria for defining all the properties of extension (angles, areas, volumes, &c.).

A manifold with which has been associated a quadratic form of the type (32), or in other words, a manifold whose metric is given, is called a *metric manifold*, and will be here denoted briefly by V_n. Since ds^2 is invariant, the coefficients a_{ik} obviously form a symmetrical covariant double system; we shall throughout the discussion suppose that they and their first and second derivatives are finite and continuous functions, and so chosen as to make the quadratic form definite and positive.[1] Thus the distance between two real points will always be real; the determinant a of the coefficients a_{ik} will always be positive. With the usual notation the reciprocal elements will be denoted by a^{ik}, &c.

We shall now extend the concept of direction to a generic V_n. We shall consider direction as determined by two infinitely near points, i.e. by a system of dx's. As before, we shall apply the term *parameters* to the n contravariant quantities

$$\lambda^i = \frac{dx_i}{ds} \qquad (i = 1, 2, \ldots n)$$

which define a direction (and are uniquely determined by it), and we shall apply the term *moments* to the covariant quantities

$$\lambda_i = \sum_{1}^{n}{}_k\, a_{ik}\, \lambda^k \qquad (i = 1, 2, \ldots n).$$

Thus for any value of n we have again two simple systems, reciprocal with respect to ds^2, or to the form (32) (cf. p. 96, Remark III).

The parameters are connected by a relation completely analogous to (5), and the formulæ (5′), (5″), and (6′) can be extended without difficulty, the summations being now from 1 to n instead of from 1 to 2. The aggregate consisting of a direction and a positive number R will be called a *vector* **R** in a V_n (R being the magnitude of the vector); the products of R by the parameters of the direction will be called the *contravariant components* R^i, and the products of R by the moments the *covariant components* R_i. We shall then have a set of formulæ analogous to (11), (11′), (11″).

Suppose the x's expressed as regular functions (i.e. finite and continuous, together with all their derivatives which enter the

[1] At the end of the chapter (p. 141) we shall also consider shortly the case of an indefinite quadratic form. This case was at first neglected as offering little likelihood of useful application, but the theory of relativity has now invested it with very great importance.

discussion, in the field considered) of p parameters $u_1, u_2, \ldots u_p$, where p is a positive integer less than n:

$$x_i = f_i(u_1, u_2, \ldots u_p) \qquad (i = 1, 2, \ldots n). \quad . \quad (33)$$

We shall make the hypothesis that at least one set of p functions f is independent, i.e. that p is the characteristic of the functional matrix of the f's with respect to the u's. Hence the x's are connected by $n - p$ relations and no more, namely, those which we should get by eliminating the u's from equations (33). In this way we define a subordinate p-dimensional manifold W_p, whose co-ordinates are the u's. W_p is said to be contained or *immersed* in V_n, since to every system of p values assigned to the u's there corresponds, by (33), a system of n values assigned to the x's (i.e. every point of W_p belongs to V_n), while not all the systems of values which can be assigned to the x's satisfy the equations (33) (i.e. not all the points of V_n belong to W_p). Now, remembering the analogy with the case $n = 3$, $p = 2$ (cf. p. 87), we naturally assign to the distance between two points of the subordinate variety the same value (32) as that of the distance between the same two points when they are considered as belonging to V_n; i.e. we construct ds^2 for the subordinate manifold by substituting in (32) for the dx's their values obtained by differentiating the equations (33). In this way we can easily find the coefficients of the fundamental quadratic form in the du's, and the metric of the p-dimensional manifold W_p, immersed in V_n, will be completely defined. For $p = 1$ the definition coincides with that given in Chapter I, § 1, for a line, of which the equations (33) are the parametric equations.

If $p = n - 1$, the W_p is often called a *surface*, or more properly a *hypersurface.*

21. **Euclidean manifolds. Any V_n can always be considered as immersed in a Euclidean space.**

If ds^2 reduces to the sum of the squares of the differentials, as in the case of orthogonal Cartesian co-ordinates, the quadratic form is said to be *Euclidean*, and the co-ordinates, by an obvious analogy with the elementary cases $n = 2$ and $n = 3$, are called *orthogonal Cartesian co-ordinates*. When this is so, all Christoffel's symbols obviously vanish identically, since the coefficients a_{ik}

are constants. Given a generic V_n, and therefore a generic ds^2, it is not in general possible to bring about a change of variables such that ds^2 takes the Euclidean form, or in other words to establish a system of Cartesian co-ordinates in V_n; if it is possible V_n is called a *Euclidean manifold*, and we shall denote it by S_n. We shall find later on the conditions to be satisfied by the a_{ik}'s in order that V_n may be Euclidean. V_n, however, can always be considered as immersed in an N-dimensional Euclidean variety, where $N \geqslant n$, as we shall now show.

We propose to determine N functions of the x's,

$$y_1(x),\ y_2(x), \ldots\ y_N(x), \qquad . \quad . \quad . \quad (34)$$

such that when we differentiate them and take the sum of the squares of the differentials we get a form, quadratic in the dx's, which is identical with the given ds^2, so that we have identically

$$\sum_1^N{}_\nu\, dy_\nu{}^2 = \sum_1^n{}_{ik}\, a_{ik}\, dx_i\, dx_k.$$

Expressing the dy's in terms of the dx's, we have

$$\sum_1^N{}_\nu \sum_1^n{}_{ik} \frac{\partial y_\nu}{\partial x_i}\frac{\partial y_\nu}{\partial x_k}\, dx_i\, dx_k = \sum_1^n{}_{ik}\, a_{ik}\, dx_i\, dx_k,$$

or, equating the coefficients of $dx_i\, dx_k$,

$$\sum_1^N{}_\nu \frac{\partial y_\nu}{\partial x_i}\frac{\partial y_\nu}{\partial x_k} = a_{ik} \qquad (i, k = 1, 2, \ldots n). \qquad (35)$$

We have thus obtained $\frac{1}{2}n(n+1)$ partial differential equations of the first order in the N unknowns y; unless any of these are mutually inconsistent (and a more detailed discussion would show that this is not so) we deduce that the problem is soluble for $N = \frac{1}{2}n(n+1)$, and *a fortiori* for $N > \frac{1}{2}n(n+1)$. The y's can evidently be considered as Cartesian co-ordinates in a *Euclidean* manifold (space) S_N in which the given V_n is immersed, V_n being parametrically represented by the values of the y's in (34) (cf. formula (33)). It is therefore possible to immerse a generic V_n in a Euclidean space S_N provided $N \geqslant \frac{1}{2}n(n+1)$. For particular V_n's, however, a smaller number of dimensions may suffice; e.g. for a Euclidean V_n, n dimensions are sufficient; in this case the y's are Cartesian co-ordinates of V_n itself.

If N has the smallest possible value, the difference $N - n$ is called the *class* of the V_n. Since the minimum N is not greater than $\frac{1}{2}n(n + 1)$, the class cannot be greater than $\frac{1}{2}n(n + 1) - n$, or $\frac{1}{2}n(n - 1)$. Further, N cannot be less than n,[1] and therefore the minimum value of the class is 0. For $n = 2$, the class is 1, which shows that every binary ds^2 may be considered as belonging to an ordinary surface. In other words, the parametric expressions which were our starting point (p. 86) impose no restrictions on the study of the intrinsic properties of a ds^2 in two variables.

22. **Angular metric.**

We shall now extend to the generic V_n the notion of the *angle between two directions.* The most direct method is by the formal extension of formula (8) (and its equivalents) by summing from 1 to n instead of from 1 to 2; this however will be legitimate, if we wish to avoid imaginary values of ϑ, only when we have shown that the expression on the right < 1.

In order to do this, we shall examine some algebraic properties of quadratic forms.

Let

$$\phi_{zz} = \sum_{1}^{n}{}_{ik}\, a_{ik}\, z_i\, z_k$$

be a definite positive quadratic form. Suppose that the z's are linear combinations of two different systems of non-proportional variables, so that we may put

$$z_i = \lambda x_i + \mu y_i;$$

we therefore have

$$\begin{aligned}\phi_{zz} &= \sum_{1}^{n}{}_{ik}\, a_{ik}\, (\lambda x_i + \mu y_i)(\lambda x_k + \mu y_k) \\ &= \sum_{1}^{n}{}_{ik}\, a_{ik}\, [\lambda^2\, x_i\, x_k + \lambda\mu\, (x_i\, y_k + y_i\, x_k) + \mu^2\, y_i\, y_k].\end{aligned}$$

[1] A quadratic form $\phi = \sum_{1}^{n}{}_{ik}\, a_{ik}\, \xi_i\, \xi_k$ is called *irreducible* when the number of independent variables cannot be reduced by substituting for the ξ's linear combinations of them. This is always so when the form is definite, as in this case the determinant a of the coefficients is certainly not zero (p. 90). N cannot therefore be less than n.

Splitting up the right-hand side into three sums, and putting

$$\sum_{1}^{n}{}_{ik}\, a_{ik}\, x_i\, x_k = \phi_{xx},$$

$$\sum_{1}^{n}{}_{ik}\, a_{ik}\, x_i\, y_k = \sum_{1}^{n}{}_{ik}\, a_{ik}\, y_i\, x_k = \phi_{xy} = \phi_{yx},$$

$$\sum_{1}^{n}{}_{ik}\, a_{ik}\, y_i\, y_k = \phi_{yy},$$

we have finally

$$\phi_{zz} = \lambda^2 \phi_{xx} + 2\, \lambda\, \mu\, \phi_{xy} + \mu^2 \phi_{yy}. \quad . \quad . \quad (36)$$

This may be considered as a quadratic form in λ and μ; it is easy to show that it is definite and positive, i.e. that it is always greater than 0 when λ and μ are not zero. In fact, ϕ_{zz}, considered as a quadratic form in the z's, is always positive, provided at least one of the z's is not zero; and this condition is equivalent to our hypothesis that the x's and y's are not proportional.

From (36) we therefore get

$$\lambda^2 \phi_{xx} + 2\, \lambda\, \mu\, \phi_{xy} + \mu^2 \phi_{yy} > 0,$$

whatever λ and μ may be.

Hence, from an ordinary property of quadratic inequalities, we get

$$\phi_{xx}\, \phi_{yy} - \phi_{xy}^2 > 0, \quad . \quad . \quad . \quad . \quad (37)$$

which is the formula we wished to prove.

We now return to the proposed formal extension of formula (8). What we have to prove is that

$$\left(\sum_{1}^{n}{}_{ik}\, a_{ik}\, \lambda^i \mu^k\right)^2 < 1,$$

i.e. that

$$1 - \left(\sum_{1}^{n}{}_{ik}\, a_{ik}\, \lambda^i \mu^k\right)^2 > 0,$$

whatever λ^i and μ^i may be, provided they are not proportional (since we exclude the obvious case where the directions coincide or are opposite).

This inequality can now be proved at once. Introducing the

quadratic relations between the parameters, we can write it in the form

$$\left(\sum_{1}^{n}{}_{ik}\, a_{ik}\, \lambda^i \lambda^k\right)\left(\sum_{1}^{n}{}_{ik}\, a_{ik}\, \mu^i \mu^k\right) - \left(\sum_{1}^{n}{}_{ik}\, a_{ik}\, \lambda^i \mu^k\right)^2 > 0;$$

and this is merely (37), with the x's and y's replaced by the λ^i's and μ^i's.

We may therefore assume

$$\cos\vartheta = \sum_{1}^{n}{}_{ik}\, a_{ik}\, \lambda^i \mu^k, \quad . \quad . \quad . \quad . \quad (38)$$

and the other expressions equivalent to it will also hold good, namely,

$$\cos\vartheta = \sum_{1}^{n}{}_{i}\, \lambda^i \mu_i, \quad . \quad . \quad . \quad . \quad . \quad (38')$$

$$\cos\vartheta = \sum_{1}^{n}{}_{i}\, \lambda_i \mu^i, \quad . \quad . \quad . \quad . \quad . \quad (38'')$$

$$\cos\vartheta = \sum_{1}^{n}{}_{ik}\, a^{ik}\, \lambda_i \mu_k, \quad . \quad . \quad . \quad (38''')$$

in which the *moments* (cf. § 20) of one or both directions take the place of the corresponding parameters.

In the provisionally excluded case of two coincident or opposite directions ($\lambda^i = \pm \mu^i$), we must naturally agree that $\cos\vartheta = \pm 1$. With this convention the four formulæ just given still hold good, the right-hand side in each case also reducing to ± 1 in virtue of the fundamental relations

$$\sum_{1}^{n}{}_{ik}\, a_{ik}\, \lambda^i \lambda^k = \sum_{1}^{n}{}_{i}\, \lambda_i \lambda^i = \sum_{1}^{n}{}_{ik}\, a^{ik}\, \lambda_i \lambda_k = 1$$

between the parameters and moments of a single direction (cf. § 20).

Now consider our V_n, immersed in a Euclidean space S_N. Given two directions $\boldsymbol{\lambda}$, $\boldsymbol{\mu}$ belonging to V_n, and drawn from the same point, the angle between them is defined in two ways, since the directions $\boldsymbol{\lambda}$, $\boldsymbol{\mu}$ may be specified either by their parameters λ^i, μ^i relative to V_n, or by their parameters λ'^ν, μ'^ν relative to S_N,[1] and formula (38) may be applied to either set. We

[1] We may note in passing that in a Euclidean space, referred to rectangular Cartesian co-ordinates, the parameters of a direction coincide with its moments; also (as follows directly from the properties of linear orthogonal substitutions) the formulæ of covariance are identical with those of contravariance (cf. § 3, p. 67).

shall call the angle calculated in the first way ϑ and in the second ϑ', and we shall show that $\cos\vartheta = \cos\vartheta'$.

Remembering that for the Cartesian co-ordinates y of S_N, ds^2 has the form $\sum_{\nu=1}^{N} dy_\nu^2$, we have

$$\cos\vartheta = \sum_1^n {}_{ik}\, a_{ik}\, \lambda^i \mu^k,$$

$$\cos\vartheta' = \sum_1^N {}_\nu\, \lambda'^\nu \mu'^\nu.$$

Now the parameters λ'^ν, μ'^ν are given by the formulæ analogous to (7), (7')

$$\lambda'^\nu = \frac{dy_\nu}{ds} = \sum_1^n {}_i \frac{\partial y_\nu}{\partial x_i} \frac{dx_i}{ds} = \sum_1^n {}_i \frac{\partial y_\nu}{\partial x_i} \lambda^i,$$

$$\mu'^\nu = \frac{\delta y_\nu}{\delta s} = \sum_1^n {}_k \frac{\partial y_\nu}{\partial x_k} \frac{\delta x_k}{\delta s} = \sum_1^n {}_k \frac{\partial y_\nu}{\partial x_k} \mu^k.$$

We have therefore

$$\cos\vartheta' = \sum_1^N {}_\nu \sum_1^n {}_{ik} \frac{\partial y_\nu}{\partial x_i} \frac{\partial y_\nu}{\partial x_k} \lambda^i \mu^k = \sum_1^n {}_{ik}\, \lambda^i \mu^k \sum_1^N {}_\nu \frac{\partial y_\nu}{\partial x_i} \frac{\partial y_\nu}{\partial x_k},$$

and therefore by (35)

$$\cos\vartheta' = \sum_1^n {}_{ik}\, \lambda^i \mu^k a_{ik} = \cos\vartheta.$$

Q.E.D.

Now consider two vectors **R**, **V**, whose directions are $\boldsymbol{\lambda}$, $\boldsymbol{\mu}$ respectively. We can extend the definition of the scalar product by giving this name to the invariant

$$\mathbf{R} \times \mathbf{V} = RV \cos\vartheta,$$

where ϑ is the angle between the two directions. For each of the various expressions for $\cos\vartheta$ we shall get a corresponding expression for the scalar product by multiplying (38), (38'), (38''), (38''') in turn by RV. The resulting formulæ are:

$$\mathbf{R} \times \mathbf{V} = \sum_1^n {}_{ik}\, a_{ik} R^i V^k = \sum_1^n {}_i R^i V_i = \sum_1^n {}_i R_i V^i = \sum_1^n {}_{ik}\, a^{ik} R_i V_k.$$

If one of the vectors, say **V**, is of unit length, we shall call the product $\mathbf{R} \times \mathbf{V}$ the *projection* of the vector $\mathbf{R}$ on the direction determined by the unit vector $\mathbf{V}$, or its *component* in this direction.

The orthogonality of two non-zero vectors is evidently expressed by the vanishing of their scalar product.

We can now make some useful remarks relating to certain particular directions. Let $\mathbf{s}_i$ denote[1] the direction of the coordinate line i (i.e. the unit vector in that direction, in the sense of the x_i's increasing); remembering that for a displacement along the line i we have $dx_r = 0$ for $r \neq i$ and $ds = \sqrt{a_{ii}}\,dx_i$, we see that the parameters s_i^r of the direction $\mathbf{s}_i$ are all zero except the ith, so that we have

$$s_i^r = 0 \ (r \neq i), \quad s_i^i = \frac{1}{\sqrt{a_{ii}}}.$$

On the other hand, the direction $\mathbf{n}_j$ of the normal to the coordinate hypersurface $x_j =$ constant (the normal meaning the direction perpendicular to any direction drawn on the hypersurface) has its *moments* $n_{j|i}$ all zero except the jth. For $\mathbf{n}_j$ must be perpendicular to each of the $n - 1$ directions $\mathbf{s}_i$ $(i \neq j)$, so that applying formula (38′) to the values just found for the parameters of $\mathbf{s}_i$ we can write

$$n_{j|i} \frac{1}{\sqrt{a_{ii}}} = 0 \qquad (i \neq j),$$

whence $n_{j|i} = 0$ for $i \neq j$. The value of $n_{j|j}$ is therefore determined by the quadratic identity between the moments, which gives

$$n_{j|j} = \frac{1}{\sqrt{a^{jj}}},$$

if we suppose that the sense of $\mathbf{n}_j$ is that of the x_j's increasing; for the opposite sense the radical must have the minus sign.

That the direction $\mathbf{n}_j$ so defined (at a generic point) is actually perpendicular to any direction $\boldsymbol{\lambda}$ (through the same point) on the hypersurface $x_j =$ constant, follows from the fact that for every such direction the parameter λ^j is zero, and therefore $\sum_1^n{}_i\, n_{j|i} \lambda^i = 0.$

[1] The suffix i is not of course an index of covariance.

Applying the above remarks, it can at once be seen that the angle ω_{ik} between the co-ordinate lines i and k is given by the formula

$$\cos\omega_{ik} = \frac{a_{ik}}{\sqrt{a_{ii}\, a_{kk}}},$$

while the angle σ_{ik} between the co-ordinate hypersurfaces $x_i =$ constant and $x_k =$ constant (i.e. the angle between their normals $\mathbf{n}_i$ and $\mathbf{n}_k$) is given by

$$\cos\sigma_{ik} = \frac{a^{ik}}{\sqrt{a^{ii}\, a^{kk}}}.$$

These formulæ show the real meaning of the coefficients of ds^2, and the geometrical interpretation to be given to their vanishing.

We shall now try to find the geometrical meaning of the covariant and contravariant components of a vector $\mathbf{R}$. For this purpose we shall calculate the orthogonal projections of $\mathbf{R}$ on the directions $\mathbf{s}_i$ and $\mathbf{n}_i$. We get for these

$$\mathbf{R} \times \mathbf{s}_i = \sum_1^n{}_r R_r\, s_i^r = \frac{R_i}{\sqrt{a_{ii}}},$$

$$\mathbf{R} \times \mathbf{n}_i = \sum_1^n{}_r R^r\, n_{i|r} = \frac{R^i}{\sqrt{a^{ii}}}.$$

These results show that R_i and R^i represent the projections of the vector $\mathbf{R}$ on the co-ordinate direction i and on the normal to the co-ordinate hypersurface $x_i =$ constant, multiplied by $\sqrt{a_{ii}}$ and $\sqrt{a^{ii}}$ respectively.

23. Definition of geodesics.

We shall fix any two points A, B in a generic V_n, and we shall try to find the shortest of the lines which join A and B. In a certain sense this problem is analogous to that of finding the points at which a function is a maximum or minimum, the solution of which is of course an important application of the calculus. Here, however, we are not trying to find *points*, and hence the values of one (or in general of n) variables which satisfy the required condition; we are trying to determine a *line*, and hence, analytically, to determine the form of n functions (the parametric equations of the line). The problem is therefore of a higher order of difficulty: while the former led to equations in finite terms, the latter leads to differential equations. The solution of this problem and of others related to it is the principal object of the calculus of variations. We shall recall shortly the fundamental idea of this calculus, which does not differ in principle from the idea which leads to the solution of the other more elementary problem of the maxima and minima of a given function.

To fix the ideas, we shall suppose that there is only one

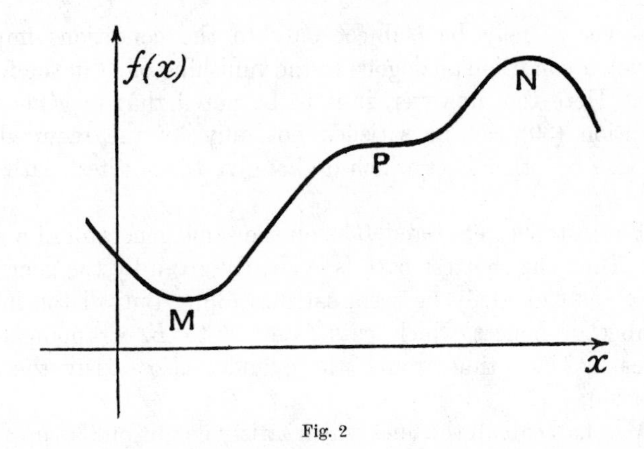

Fig. 2

variable. We know that if a function $f(x)$ has a maximum or minimum at x_0, its differential $df = f'(x_0)dx$ is zero at that point (and therefore $f'(x_0) = 0$), whatever dx may be; in other words, for an infinitely small displacement to left or right from the point x_0, f remains constant (except for infinitesimals of higher order). This can also be seen intuitively from the graphical representation of the function (cf. the points M and N in the diagram). The converse, however, is not true, i.e. when $df = 0$ it does not necessarily follow that there is a maximum or minimum (cf. for instance the point P in the diagram). The maxima and minima must be looked for *among* the points where $df = 0$.

Let us now see how we can apply this method to the determination of the shortest line joining A and B, without going outside a given V_n (we may think of a line drawn on a surface, i.e. the case $n = 2$). Let g be such a line; draw a line g', having the same extremities as g, and infinitely near g, but otherwise completely arbitrary. We can consider g' as derived from g by an infinitesimal deformation, i.e. by displacing each point $(x_1, x_2, \ldots x_n)$ of g to $(x_1 + \delta x_1, \ldots x_n + \delta x_n)$. If g is the shortest of these lines, its length is not changed[1] by this deformation (except for infinitesimals of higher order); hence if l is the length of g and $l + \delta l$ that of g', we have

$$\delta l = 0, \quad . \quad . \quad . \quad . \quad . \quad . \quad (39)$$

whatever g' may be (subject only to the conditions imposed above), a condition analogous to the vanishing of df in the former case. Here too, however, it is to be noted that in general the condition (39) can be satisfied not only by the required line but also by other lines which do *not* give the shortest path from A to B.

For instance, let A and B be on the same generator of a cylinder. Then the shortest path is evidently given by the generator, which, as can easily be seen, satisfies (39). But all the infinite number of helices which pass from A to B, wrapping themselves 1, 2, . . . times round the cylinder also satisfy the same equation.

We shall call all the lines which satisfy condition (39) *geodesics*. They possess important characteristic properties, which can be deduced from (39); e.g. the osculating plane at any point of a geodesic on a surface is normal to the tangent plane to the surface—the property adopted on p. 103 as the definition of a geodesic. The *lines of minimum length* between two given points must be looked for among the geodesics through the two points.

This is the definition which we shall use below; but it is to be noted that some writers in defining geodesics start from another property. We could in fact show that when a point A is fixed on a geodesic g, then for all points B (on g) sufficiently near A, g is the *only* geodesic joining A and B, and is therefore the shortest line joining them. Hence we can also say that a geodesic is a line such that it forms the shortest path

[1] For a more rigorous and complete discussion the reader is referred to treatises on analysis.

between *any* two of its points, *provided they are sufficiently close together*.

With this restriction the two definitions are equivalent.

24. **Differential equations of geodesics.**

We shall now examine the property, concisely expressed by the equation (39), that the length remains unchanged in an infinitesimal displacement which does not move the extremities, and see how to express it by means of n differential equations which the n functions

$$x_i = x_i(s) \qquad (i = 1, 2, \ldots n)$$

defining the curve g must satisfy.

Let the equations of g' be

$$x_i = x_i(s) + \delta x_i \qquad (i = 1, 2, \ldots n),$$

where the δx's are to be considered as infinitesimal functions of s, vanishing for $s = 0$ and $s = l$, and having finite and continuous first and second derivatives, but otherwise arbitrary.

Take an infinitesimal segment PP_1 of g, of length ds; we have to calculate δds, i.e. the increment (or, as it is called, the variation) of ds in the deformation which displaces P to P' and P_1 to P_1'. If dx_i $(i = 1, 2, \ldots n)$ is the difference between the co-ordinates of P and of P_1, the corresponding difference after the deformation—which we shall denote by $d(x_i + \delta x_i) = dx_i + d\delta x_i$, where $d\delta x_i$ is of course the differential of the function δx—is calculated as follows:

The co-ordinates of P' are $x_i + \delta x_i$;
those of P_1' are $(x_i + dx_i) + \delta(x_i + dx_i)$;
therefore the required difference is $dx_i + \delta dx_i$.

It follows that

$$\delta dx_i = d\delta x_i, \qquad (40)$$

a result which we shall at once make use of.

We shall now take the expression for ds^2, and calculate its variation, differentiating with the operator δ. We have

$$2ds \, . \, \delta ds = \sum_1^n{}_{jk} \delta a_{jk} \, dx_j \, dx_k + \sum_1^n{}_{jk} a_{jk} \, dx_k \, \delta dx_j + \sum_1^n{}_{jk} a_{jk} \, dx_j \, \delta dx_k.$$

Since the last two sums are identical, we can write this in the form (using equation (40))

$$2ds\,.\,\delta ds = \sum_1^n{}_{jk}\,\delta a_{jk}\,dx_j\,dx_k + 2\sum_1^n{}_{jk}\,a_{jk}\,dx_j\,d\delta x_k.$$

Dividing by $2ds$, and denoting differentiation with respect to s by a dot, this gives

$$\delta ds = \tfrac{1}{2}\sum_1^n{}_{jl}\,\delta a_{jl}\,\dot{x}_j\,\dot{x}_l\,ds + \sum_1^n{}_{jk}\,a_{jk}\,\dot{x}_j\,d\delta x_k;$$

and from this, since

$$\delta a_{jl} = \sum_1^n{}_k\frac{\partial a_{jl}}{\partial x_k}\,\delta x_k,$$

we get δds in the form

$$\delta ds = \tfrac{1}{2}\sum_1^n{}_{jlk}\frac{\partial a_{jl}}{\partial x_k}\,\dot{x}_j\,\dot{x}_l\,\delta x_k\,ds + \sum_1^n{}_{jk}\,a_{jk}\,\dot{x}_j\,d\delta x_k.$$

Now since the length of g is

$$l = \int_A^B ds,$$

the variation which is to be equated to 0 in (39) is

$$\delta l = \int_A^B \delta ds,$$

or
$$\delta l = \int_A^B \tfrac{1}{2}\left(\sum_1^n{}_{jlk}\frac{\partial a_{jl}}{\partial x_k}\,\dot{x}_j\,\dot{x}_l\,\delta x_k\right)ds + I, \quad . \quad . \quad (41)$$

where we have put

$$I = \int_A^B \sum_1^n{}_{jk}\,a_{jk}\,\dot{x}_j\,d\delta x_k. \quad . \quad . \quad . \quad . \quad (42)$$

We shall leave (41) aside for a moment and examine the possibility of transforming the integral in (42) also into a form which explicitly contains the arbitrary variations δx_k. Integrating by parts we get

$$I = \left[\sum_1^n{}_{jk}\,a_{jk}\,\dot{x}_j\,\delta x_k\right]_A^B - \int_A^B \sum_1^n{}_{jk}\,d\,(a_{jk}\,\dot{x}_j)\,\delta x_k.$$

The integrated part vanishes, since $\delta x_k = 0$ at the extremities; differentiating the product, the other part gives

$$I = -\int_A^B \sum_{1}^{n}{}_{jk}\, a_{jk}\, \ddot{x}_j\, ds\, \delta x_k - \int_A^B \sum_{1}^{n}{}_{jk}\, \dot{x}_j\, da_{jk}\, \delta x_k. \qquad (42')$$

Expanding da_{jk}, the sum under the second integral sign may be written

$$\sum_{1}^{n}{}_{jkl} \frac{\partial a_{jk}}{\partial x_l}\, \dot{x}_j\, \dot{x}_l\, \delta x_k\, ds,$$

or, interchanging j and l,

$$\sum_{1}^{n}{}_{jkl} \frac{\partial a_{lk}}{\partial x_j}\, \dot{x}_j\, \dot{x}_l\, \delta x_k\, ds.$$

We shall take half the sum of these two expressions to represent the value of either. Substituting in (42′), we get

$$I = -\int_A^B \sum_{1}^{n}{}_{jk}\, a_{jk}\, \ddot{x}_j\, \delta x_k\, ds - \int_A^B \tfrac{1}{2}\left(\sum_{1}^{n}{}_{jkl}\left[\frac{\partial a_{jk}}{\partial x_l} + \frac{\partial a_{lk}}{\partial x_j}\right] \dot{x}_j\, \dot{x}_l\, \delta x_k\right) ds.$$

We now return to (41) and insert in it this expression for I. Putting all the terms under a single integral sign, and taking out the common factor $\delta x_k ds$, we get

$$\delta l = -\int_A^B \sum_{1}^{n}{}_{k} \left\{ -\tfrac{1}{2}\sum_{1}^{n}{}_{jl} \frac{\partial a_{jl}}{\partial x_k}\, \dot{x}_j\, \dot{x}_l + \sum_{1}^{n}{}_{j}\, a_{jk}\, \ddot{x}_j + \tfrac{1}{2}\sum_{1}^{n}{}_{jl}\left[\frac{\partial a_{jk}}{\partial x_l} + \frac{\partial a_{lk}}{\partial x_j}\right] \dot{x}_j\, \dot{x}_l \right\} \delta x_k\, ds,$$

or, remembering the definition of Christoffel's symbols,

$$\delta l = -\int_A^B \sum_{1}^{n}{}_{k} \left\{ \sum_{1}^{n}{}_{j}\, a_{jk}\, \ddot{x}_j + \sum_{1}^{n}{}_{jl}\, [jl, k]\, \dot{x}_j\, \dot{x}_l \right\} \delta x_k\, ds.$$

Putting

$$p_k = \sum_{1}^{n}{}_{j}\, a_{jk}\, \ddot{x}_j + \sum_{1}^{n}{}_{jl}\, [jl, k]\, \dot{x}_j\, \dot{x}_l, \quad . \quad . \qquad (43)$$

the formula appears in the concise form

$$\delta l = -\int_A^B \sum_{1}^{n}{}_{k}\, p_k\, \delta x_k\, ds. \quad . \quad . \quad . \qquad (44)$$

The result of all these calculations is that (39) can be written in the form

$$\int_A^B \sum_1^n {}_k\, p_k\, \delta x_k\, ds = 0. \quad . \quad . \quad . \quad . \quad (39')$$

Now since (39′) must hold however the arbitrary functions δx_k are chosen (subject to the qualitative conditions stated above), we must necessarily have

$$p_k = 0 \qquad (k = 1, 2, \ldots n); \quad . \quad . \quad (45)$$

for if not, we need only take each δx_k with the same sign as the corresponding p_k (which can be done without going outside the conditions imposed); the sum would then certainly be positive at all points of the arc of integration and therefore the integral would not vanish. This is the fundamental argument in the calculus of variations; by means of it we get from (39′) (which is only (39) expanded) the n differential equations (45) which written at length are

$$\sum_1^n {}_j\, a_{jk}\, \ddot{x}_j + \sum_1^n {}_{jl}\, [jl, k]\, \dot{x}_j\, \dot{x}_l = 0 \qquad (k = 1, 2, \ldots n). \quad (45')$$

It is convenient to write these equations in the form obtained by solving for the $\ddot{x}$'s. To do this we introduce the quantities

$$p^i = \sum_1^n {}_k\, a^{ik}\, p_k, \quad . \quad . \quad . \quad . \quad (46)$$

and replace the equations (45) by the equivalent system of linear combinations

$$p^i = 0,$$

or

$$\ddot{x}_i + \sum_1^n {}_{jl}\, \{jl, i\}\, \dot{x}_j\, \dot{x}_l = 0. \quad . \quad . \quad . \quad (47)$$

These n differential equations of the second order in the n unknown functions $x_i(s)$ are equivalent to equation (39) and are therefore the *characteristic equations of a geodesic*; when integrated, they give the parametric equations of the curve. By the ordinary theory of such equations, the integrals will contain $2n$ arbitrary constants, which can be determined by the condition that the geodesic passes through two specified points, or that it starts from a given point and has a specified direction.

It may be noted that the equations (47) contain only intrinsic elements, as the definition of a geodesic would lead us to expect.

25. Geodesic curvature.

The discussion in the preceding section provides us with an opening for the introduction of an illuminating and fertile geometrical notion relating to any curve $x_i = x_i(s)$ in our V_n.

We must first show that the quantities p_k defined by (43), corresponding to a generic curve $x_i = x_i(s)$, are *covariant* (and in consequence the p^i's are contravariant), so that we shall naturally associate with every point of the curve l (which is *a priori* any curve whatever) the vector $\mathbf{p}$ of which they are the components. Suppose then that we pass by any transformation from the variables x to new variables $\bar{x}$, and let $\bar{p}_k$ represent the values of the p_k's calculated in the new system. We get from (44), through the invariance of δl,

$$\delta l = -\int_A^B \sum_1^n {}_k \bar{p}_k \, \delta \bar{x}_k \, ds,$$

and therefore

$$0 = \int_A^B \left[\sum_1^n {}_k \bar{p}_k \, \delta \bar{x}_k - \sum_1^n {}_k p_k \, \delta x_k \right] ds.$$

By a similar argument to that used in passing from (39′) to (45) we deduce from this that at every point of l we must have

$$\sum_1^n {}_k \bar{p}_k \, \delta \bar{x}_k - \sum_1^n {}_k p_k \, \delta x_k = 0,$$

which expresses the invariance of

$$\sum_1^n {}_k p_k \, \delta x_k$$

(a linear form in the arbitrary contravariant variables δx_k) and therefore the covariance of the p_k's. It follows from (46) that the reciprocal contravariant system consists of the p^i's, i.e. of the left-hand side of equations (47).

We shall use the term *geodesic curvature* of the curve l at any point on l to denote the vector $\mathbf{p}$ whose covariant components are defined by (43), its contravariant components being in consequence defined by (46), or by

$$p^i = \ddot{x}_i + \sum_{1}^{n}{}_{jl} \{jl,\ i\}\, \dot{x}_j\, \dot{x}_l \qquad (i = 1, 2, \ldots n). \quad (43')$$

An immediate corollary is that *the geodesics are the lines whose geodesic curvature is everywhere zero.*

More generally we have for the length of the vector $\mathbf{p}$ an important property, pointed out by Lipka,[1] which we shall merely state without proof: *The absolute value of the geodesic curvature is represented*, as in ordinary space, *by the ratio between the angle of contingence and the elementary arc*, where the angle of contingence is defined as the angle, at one extremity of the arc, between the tangent at that point and the parallel to the tangential direction at the other extremity.

Another important property is that *the geodesic curvature is normal to the curve*, which is equivalent to saying that

$$\sum_{1}^{n}{}_{k}\, p_k\, \dot{x}_k = 0, \quad . \quad . \quad . \quad . \quad . \quad (48)$$

since the parameters of the tangent to the curve are the $\dot{x}_i$'s.

To prove this, take the identity

$$\sum_{1}^{n}{}_{jk}\, a_{kj}\, \dot{x}_k\, \dot{x}_j = 1$$

(obtained by dividing (32) by ds^2) and differentiate it with respect to s. We get

$$2\sum_{1}^{n}{}_{jk}\, a_{kj}\, \dot{x}_k\, \ddot{x}_j + \sum_{1}^{n}{}_{jk}\, \dot{a}_{kj}\, \dot{x}_k\, \dot{x}_j = 0,$$

$$\text{or} \qquad 2\sum_{1}^{n}{}_{jk}\, a_{kj}\, \dot{x}_k\, \ddot{x}_j + \sum_{1}^{n}{}_{jkl} \frac{\partial a_{kj}}{\partial x_l}\, \dot{x}_k\, \dot{x}_j\, \dot{x}_l = 0,$$

and therefore, by (24′),

$$2\sum_{1}^{n}{}_{jk}\, a_{kj}\, \dot{x}_k\, \ddot{x}_j + \sum_{1}^{n}{}_{jkl}\, [jl,\, k]\, \dot{x}_k\, \dot{x}_j\, \dot{x}_l + \sum_{1}^{n}{}_{jkl}\, [kl,\, j]\, \dot{x}_k\, \dot{x}_j\, \dot{x}_l = 0.$$

[1] "Sulla curvatura geodetica delle linee appartenenti ad una varietà qualunque" in *Rend. della R. Acc. dei Lincei*, Vol. XXXI (1st half-year, 1922), pp. 353–356.

Interchanging k and j, we see that the third sum is the same as the second; hence, taking out the factor 2, we have

$$\sum_{1}^{n}{}_{jk}\, a_{kj}\, \dot{x}_k\, \ddot{x}_j + \sum_{1}^{n}{}_{jkl}\, [jl, k]\, \dot{x}_k\, \dot{x}_j\, \dot{x}_l = 0.$$

This is merely (48), with p_k replaced by its value as given by (43); hence the assertion made above is proved.

In ordinary space, as will at once be seen, the geodesic curvature coincides in direction with the principal normal, and in magnitude with the flexion or principal curvature of the curve.

26. **Extension of the notion of parallelism. Bianchi's derived vectors.**

We propose next to extend to a V_n the notion of parallelism or, more generally, of equipollence defined above for a V_2.

In this case we have no criterion analogous to that used for the V_2, as in general the circumscribed developable which formed the starting-point of the former argument does not exist.

The differential law of parallelism, however, expressed by the symbolic equation (19), can be immediately adapted to the case of a V_n. To do so, consider a vector $\mathbf{R}$ drawn from a point P of V_n, and let $\mathbf{R} + d\mathbf{R}$ denote the equipollent vector drawn from a point P_1 of V_n, very near to P. We can think of the V_n, and therefore of the vectors $\mathbf{R}$, $\mathbf{R} + d\mathbf{R}$, as immersed in a Euclidean space S_N, where N is a sufficiently large integer; we can therefore define the vectors $\mathbf{R}$, $\mathbf{R} + d\mathbf{R}$, not only by their (covariant or contravariant) components with respect to V_n, but also by their components Y_ν, $Y_\nu + dY_\nu$ ($\nu = 1, 2, \ldots N$) with respect to a system of Cartesian co-ordinates $y_1, y_2, \ldots y_N$ in S_N. Now consider an arbitrary infinitesimal displacement δP, contained in V_n and drawn from P; it can be specified either intrinsically, by means of the δx_i's ($i = 1, 2, \ldots n$), or with reference to Cartesian co-ordinates by means of the δy_ν's ($\nu = 1, 2, \ldots N$); but it is to be noted that while the first set are arbitrary the second are not, on account of the equations (§ 21) which define the y's as functions of the x's. We can also say, in geometrical language, that the displacement must satisfy the condition of being tangential, i.e. of belonging to V_n. We shall define the vector $d\mathbf{R}$, and therefore the parallel displacement, by means of the symbolic equation (19), which can be expanded (p. 107) into the form, analogous to (19′):

$$\sum_{1}^{N}{}_{\nu}\, dY_{\nu}\, \delta y_{\nu} = 0, \qquad \ldots \quad (49)$$

which holds for all displacements satisfying the given condition.

The only difference between formula (49) and (19′), which defines parallelism with respect to a surface, is that the summation for ν is from 1 to N, instead of from 1 to 3. All successive steps in the calculation follow automatically as in § 15, p. 107.

We shall first write the equation in terms of the δx's by putting

$$d\mathbf{R} \times \delta P = \sum_{1}^{N}{}_{\nu}\, dY_{\nu}\, \delta y_{\nu} = \sum_{1}^{n}{}_{k}\, \tau_k\, \delta x_k; \quad . \quad (50)$$

after transformations analogous to those formerly used, we find for the τ's the expressions

$$\tau_k = \sum_{1}^{n}{}_{j}\, a_{kj}\, dR^j + \sum_{1}^{n}{}_{jl}\, [jl,\, k]\, R^j\, dx_l \quad (k = 1, 2, \ldots n), \quad (51)$$

an obvious generalization of formula (21″). Evidently, in view of the identity (50), we are here dealing with *covariant* expressions (with respect to any transformations of the x's). The reciprocal system

$$\tau^i = \sum_{1}^{n}{}_{k}\, a^{ik}\, \tau_k$$

can also be expressed in the form

$$\tau^i = dR^i + \sum_{1}^{n}{}_{jl}\, \{jl,\, i\}\, R^j\, dx_l \qquad (i = 1, 2, \ldots n), \quad (51')$$

in complete analogy to (21‴).

From (49) and (50) we finally reach the *intrinsic equations of parallelism*:

$$\tau_k = 0 \qquad (k = 1, 2, \ldots n);$$

these are equivalent to $\tau^i = 0$, or to

$$dR^i + \sum_{1}^{n}{}_{jl}\, \{jl,\, i\}\, R^j\, dx_l = 0 \qquad (i = 1, 2, \ldots n), \quad (52)$$

which define the increments dR^i of the contravariant components of a vector **R** for a displacement parallel with respect to V_n from P (of co-ordinates x) to P_1 (of co-ordinates $x + dx$).

For the covariant components we find, as on p. 113, the equivalent equations

$$dR_i = \sum_1^n{}_{jl} \{il, j\} R_j \, dx_l \qquad (i = 1, 2, \ldots n). \qquad (52')$$

This equation and (52) alike show that parallel displacement is an intrinsic operation with respect to the metric of V_n. This was not *a priori* evident from the geometrical definition we adopted, which is expressed in formula (49), involving the use of a surrounding space S_N.

The equations (52) and (52′) are, so to speak, identical with the equations (23) and (27) which hold for a V_2, the only difference being in the number of dimensions. It is of course understood that $\{jl, i\}$ and $\{il, j\}$ denote Christoffel's symbols of the second kind constructed with the ds^2 of V_n.

All the properties deduced from the equations of parallelism with respect to a surface can be extended without difficulty to parallelism in V_n; in particular, the properties that parallel displacement along any finite curve whatever is always possible, and in only one way, and that parallel displacement leaves unchanged the scalar product of two vectors, and therefore lengths and angles. We shall show in the following section that we can also extend the property of autoparallelism of geodesics, which we proved geometrically in the case of surfaces.

We may also call attention here to the notion due to Bianchi [1] of the *derivative of a generic vector* **R** along a curve T, **R** being a function of the points of T. If the vectors **R**(s) at various points of T are not parallel along T, the contravariant simple system τ^i, defined by (51′), is not identically zero. Accordingly the quantities

$$(D\mathbf{R})^i = \frac{\tau^i}{ds} = \frac{dR^i}{ds} + \sum_1^n{}_{jl} \{jl, i\} R^j \frac{dx_l}{ds} \qquad (i = 1, 2, \ldots n)$$

may be considered as *contravariant components* of a non-zero vector D**R** which is also a function of the points of T. The

[1] Cf. "Sul parallelismo vincolato di Levi-Civita nella metrica degli spazi curvi", in *Rend. della R. Acc. di Napoli*, Vol. XXVIII, 1922, pp. 150–171.

vector $D\mathbf{R}$ has been called by Bianchi *associated*, and its direction and length the *direction* and *curvature associated* point by point *with the vector* $\mathbf{R}(s)$. We prefer, however, the qualification *derived*, because in Euclidean spaces $D\mathbf{R}$ is precisely the vector commonly known as the derivative of $\mathbf{R}$ with respect to the arc s of T. In fact, if we assume Cartesian co-ordinates, the Christoffel's symbols vanish, and the preceding expressions for the (ordinary) components of $(D\mathbf{R})^i$ reduce to $\frac{dR_i}{ds}$.

Returning to a general manifold V_n, if $\mathbf{R}(s)$ reduces to the versor which is tangential to the curve T, i.e. in particular if $R^i = \frac{dx_i}{ds} = \dot{x}_i$, we find that we are again dealing with the vector $\mathbf{p}$ of geodesic curvature considered in the preceding section.

It can be shown that in every case $D\mathbf{R}$ (if not zero) is perpendicular to $\mathbf{R}$, and that it has other interesting properties demonstrated by Bianchi. For further details the reader is referred to the paper just cited in the footnote, or to the Appendix to Vol. II of the same writer's *Lezioni di geometria differenziale*.[1]

27. Autoparallelism of geodesics.

Analytically we may derive this property from the equations of parallelism by using the differential equations found above for geodesics.

Let $\boldsymbol{\lambda}$ denote a unit vector defined at all points of the geodesic under consideration and having everywhere the same direction as the geodesic. We shall show that $\boldsymbol{\lambda}$ may be considered as undergoing a parallel displacement along the geodesic.

Let its parameters be λ^i. From the definition of these parameters, and using the parametric equations $x_i = x_i(s)$ of the geodesic in question, we plainly have

$$\lambda^i = \frac{dx^i}{ds} = \dot{x}_i,$$

and therefore

$$\frac{d\lambda^i}{ds} = \ddot{x}_i.$$

[1] Second edition. Bologna, Zanichelli, 1923.

Now the $\ddot{x}$'s and the $\dot{x}$'s are connected by the equations (47) (the characteristic equations of the geodesics). Substituting λ^i and $\frac{d\lambda^i}{ds}$ for $\dot{x}_i$ and $\ddot{x}_i$, these become

$$p^i = \frac{d\lambda^i}{ds} + \sum_1^n{}_{jl} \{jl, i\} \lambda^j \dot{x}_l = 0, \quad . \quad . \quad (53)$$

or multiplying by ds,

$$p^i\, ds = d\lambda^i + \sum_1^n{}_{jl} \{jl, i\} \lambda^j dx_l = 0,$$

which are the equations expressing the parallel displacement of the vector $\boldsymbol{\lambda}$.

It is worth noting that a comparison of (51′) and (53) shows that the quantities $p^i\, ds$ are a particular case of the τ^i's, the generic vector $\mathbf{R}$ being replaced by the unit vector $\boldsymbol{\lambda}$ of contravariant components $\dot{x}_i$. There follows immediately the contravariance of the quantities p^i, or, which comes to the same thing, the covariance of the quantities p_i, which we proved directly in § 25.

28. Remarks on the case of an indefinite ds^2.

We agreed (§ 20) to say that an n-dimensional V_n is metrically defined when there is associated with it a differential quadratic form, with real coefficients a_{ik},

$$\phi = \sum_1^n{}_{ik}\, a_{ik}\, dx_i\, dx_k.$$

We then introduced the hypothesis that ϕ is definite and positive, and this is the only case we have considered in the foregoing sections. We now propose to make some remarks on the case in which ϕ is still supposed irreducible (or such that its discriminant a is not zero), but is no longer definite, being capable of taking positive values for certain sets of differentials dx_i and negative values for certain others.

In this case also, fixing a generic point P of co-ordinates x_i and an infinitely near point P' of co-ordinates $x_i + dx_i$, we put

$$ds^2 = \phi = \sum_1^n{}_{ik}\, a_{ik}\, dx_i\, dx_k, \quad . \quad . \quad . \quad (54)$$

and we shall call ds^2 (which can now be positive, negative, or zero) the square of the line element (the distance), or better the *interval* between the two points P and P'.

Among the ∞^n (real) systems of differentials dx_i, or, as we shall say, considering only ratios, among the ∞^{n-1} *directions* drawn from P, there are ∞^{n-2} which satisfy the quadratic equation

$$ds^2 = 0. \quad . \quad . \quad . \quad . \quad . \quad . \quad . \qquad (55)$$

For a moment we shall interpret the differentials dx_i as referring to Cartesian co-ordinates with origin P. Then the directions just defined, which are said to be of *zero interval*, constitute a quadric cone of vertex P. This cone divides the sheaf of directions drawn from P into two regions, in one of which

$$ds^2 > 0, \quad . \quad . \quad . \quad . \quad . \quad . \quad . \qquad (56)$$

and in the other

$$ds^2 < 0. \quad . \quad . \quad . \quad . \quad . \quad . \quad . \qquad (57)$$

All the directions in the first region are said to be *of the first kind* or *timelike* (the term being suggested by the interpretation given to these symbols in the theory of relativity); those in the second region are said to be *of the second kind* or *spacelike*. The parameters of a direction of either kind are defined by the formulæ

$$\lambda^i = \frac{dx_i}{|\,ds\,|} \qquad (i = 1, 2, \ldots n); \quad . \quad . \quad . \qquad (58)$$

there is no analogous result for the directions of zero interval corresponding to which $ds^2 = 0$.

For timelike directions $ds^2 > 0$; hence, if ds denotes the arithmetic value of the square root of ds^2, we have $|\,ds\,| = ds$, and the argument is exactly as it was for the definite quadric.

For spacelike directions, on the contrary, we have

$$|\,ds^2\,| = -\,ds^2 = -\sum_1^n{}_{ik}\, a_{ik}\, dx_i\, dx_k,$$

so that the quadratic identity satisfied by the parameters λ^i is

$$\sum_1^n{}_{ik}\, a_{ik}\, \lambda^i \lambda^k = -\,1, \quad . \quad . \quad . \quad . \qquad (59)$$

with -1 on the right, instead of $+1$ as for the timelike directions.

Granted these results, the systematic extension of the preceding sections to the indefinite case would certainly not seem likely to be difficult. As we are not aware that this has yet been done exhaustively, the reader's attention may be called to it. We propose merely to point out an essential fact, almost evident *a priori* and often used in the theory of relativity; namely, that the definitions, geometrical representations, and formulæ in the foregoing sections can certainly be carried over and applied to the indefinite case, provided (*a*) that we take account of the exceptional behaviour of the directions of zero interval, and (*b*) that we make the obvious formal modifications necessitated by (59) when we have to deal with spacelike directions.

We leave the matter here,[1] with two examples to conclude the discussion:

(1) The condition that two directions, whether timelike or spacelike, of parameters λ^i, μ^i, may be orthogonal is in every case expressed by the equation

$$\sum_{1}^{n}{}_{ik} a_{ik} \lambda^i \mu^k = 0.$$

(2) If we consider only lines wholly composed of timelike elements ($ds^2 > 0$), the discussion in § 24 holds without modification, and we reach the same equations (47) of the geodesics.

[1] See Chapter XI, p. 287.

PART II

The Fundamental Quadratic Form and the Absolute Differential Calculus

CHAPTER VI

Covariant Differentiation; Invariants and Differential Parameters; Locally Geodesic Co-ordinates

1. Covariant differentiation.

Returning to the remarks made on p. 86 of Chapter IV, we now propose to generalize the operation of differentiation by substituting for the ordinary derivatives of the elements of a tensor certain linear combinations of these derivatives and of the elements of the given system, which will in their turn constitute a mixed (or in particular, covariant) system with one index of covariance more than the given system. Explicitly, if $A^{h_1 \dots h_\mu}_{i_1 \dots i_m}$ is the given generic system whose elements are functions of the x's or, in geometrical terms, functions of position, we shall deduce from it another system $A^{h_1 \dots h_\mu}_{i_1 \dots i_m l}$, where l is a new index of covariance, which reduces to the system $\dfrac{\partial A^{h_1 \dots h_\mu}_{i_1 \dots i_m}}{\partial x_l}$ in the particular case when the co-ordinates are Cartesian.

To simplify the formulæ, we shall consider first a mixed system A^h_i with a single index i of covariance and a single index h of contravariance.

Fixing our attention on a specific point of V_n (i.e. ignoring the fact that the A's are defined as functions of position), we know that the law of transformation of the functions A^h_i for a change of variables is defined by the invariance of the form

$$F = \sum_1^n{}_{ih} A_i^h \, \xi^i \, u_h, \quad . \quad . \quad . \quad . \qquad (1)$$

in which the ξ^i's constitute a generic contravariant system, or, in other words, are the contravariant components of a generic vector $\boldsymbol{\xi}$; similarly the u_h's can be considered as the covariant components of a generic vector $\mathbf{u}$.

Now, since a set of values of the A_i^h's is associated with every point of V_n, we can at every point choose two arbitrary vectors $\boldsymbol{\xi}$, $\mathbf{u}$, and construct an invariant form with them and the A's.

Suppose this choice made at an arbitrary but determined point P, and consider also a generic point P_1 infinitely near to P. We shall agree to take for $\boldsymbol{\xi}$ and $\mathbf{u}$ at P_1 the vectors *parallel* to those chosen at P; as the displacement is infinitesimal, the curve of displacement is immaterial. We shall use the operator δ to denote in general the increment of a quantity in passing from P to P_1, and we propose to calculate δF. Differentiating (1) with the operator δ, we have

$$\delta F = \sum_1^n{}_{ih} \left\{ \delta A_i^h \, \xi^i \, u_h + A_i^h \, \delta \xi^i \, u_h + A_i^h \, \xi^i \, \delta u_h \right\}.$$

Now, by the convention just adopted as to the vectors $\boldsymbol{\xi}$ and $\mathbf{u}$, the differentials $\delta \xi^i$ and δu_h must be calculated by the formulæ of parallelism ((52) and (52'), pp. 138, 139), while δA_i^h is given by the usual rule of differentiation

$$\delta A_i^h = \sum_1^n{}_l \frac{\partial A_i^h}{\partial x_l} \, \delta x_l,$$

the A's being by hypothesis functions of position. Using these results, we have

$$\delta F = \sum_1^n{}_{ihl} \frac{\partial A_i^h}{\partial x_l} \, \xi^i \, u_h \, \delta x_l - \sum_1^n{}_{ihjl} A_i^h \, \{jl, i\} \, u_h \, \xi^j \, \delta x_l$$

$$+ \sum_1^n{}_{ihjl} A_i^h \, \{hl, j\} \, \xi^i \, u_j \, \delta x_l.$$

Interchanging i and j in the second sum and h and j in the third, so as to get the factor $\xi^i \, u_h \, \delta x_l$ in all three sums, and collecting all the terms under a single summation sign, we have

$$\delta F = \sum_1^n {}_{ihl} \left[\frac{\partial A_i^h}{\partial x_l} - \sum_1^n {}_j A_j^h \{il, j\} + \sum_1^n {}_j A_i^j \{jl, h\} \right] \xi^i u_h \delta x_l. \quad (2)$$

Now the left-hand side of this equation is invariant on account of its meaning, while ξ^i, δx_l, u_h are arbitrary contravariant or covariant systems; hence the coefficients of this form (the expression in square brackets) constitute by definition a system which is covariant with respect to i and l and contravariant with respect to h. We can therefore put

$$(A_i^h)_l = \frac{\partial A_i^h}{\partial x_l} - \sum_1^n {}_j A_j^h \{il, j\} + \sum_1^n {}_j A_i^j \{jl, h\}. \quad . \quad (3)$$

This system is called the *covariant derivative* of the system A_i^h. It is sometimes denoted by the symbol $A_{i|l}^h$, and also, when no ambiguity is possible, simply by A_{il}^h.

It is obvious that in Cartesian co-ordinates (which exist when we are dealing with Euclidean forms; cf. § 21 of Chapter V, p. 121) the system reduces to that of ordinary derivatives.

The method used above can be applied, *mutatis mutandis*, to a generic mixed system. We shall always get for δF (as follows at once on carrying out the necessary operations) a multilinear form whose coefficients we shall define as elements of the covariant derived system. These coefficients consist of a first term which is the ordinary derivative, followed by as many terms preceded by the minus sign as there are indices of covariance of the given system, and as many terms preceded by the plus sign as there are indices of contravariance. If we denote by (i) the aggregate of indices $i_1 \ldots i_m$ and by (h) the aggregate $h_1 \ldots h_\mu$, the general formula is [1]

$$\left.\begin{aligned} A_{(i)|l}^{(h)} = \frac{\partial A_{(i)}^{(h)}}{\partial x_l} &- \sum_1^m {}_r \sum_1^n {}_j A_{i_1 \ldots i_{r-1} j\, i_{r+1} \ldots i_m}^{(h)} \{i_r l, j\} \\ &+ \sum_1^\mu {}_\rho \sum_1^n {}_j A_{(i)}^{h_1 \ldots h_{\rho-1} j\, h_{\rho+1} \ldots h_\mu} \{jl, h_\rho\}. \end{aligned}\right\} \quad (4)$$

[1] Cf. A. PALATINI: "Sui fondamenti del Calcolo Differenziale assoluto", in *Rend. del Circolo Mat. di Palermo*, Vol. XLIII, 1919, pp. 192–202. Another vectorial illustration of covariant differentiation was given by the late Prof. HESSENBERG in his paper "Vektorielle Begründung der Differentialgeometrie", in *Math. Ann.*, Vol. 78, 1917, pp. 187–217.

2. **Particular cases.**

Consider first a covariant simple system A_i, which we can always interpret as consisting of the covariant components (moments) of a vector **A**. In this case the terms contributed by the indices of contravariance are absent, and (4) (or (3)) gives

$$A_{i|l} = \frac{\partial A_i}{\partial x_l} - \sum_1^n {}_j \{il, j\} A_j. \quad . \quad . \quad . \quad (5)$$

It is easy to see that this double system is not in general symmetrical; from (5) however we get at once the important relation

$$A_{i|l} - A_{l|i} = \frac{\partial A_i}{\partial x_l} - \frac{\partial A_l}{\partial x_i}. \quad . \quad . \quad . \quad (6)$$

The vanishing of the covariant derivative $A_{i|l}$ has a simple geometrical significance. In this case, multiplying (5) by dx_l, we have

$$\frac{\partial A_i}{\partial x_l} dx_l = \sum_1^n {}_j \{il, j\} A_j \, dx_l;$$

comparing this with equation (52′) of the preceding chapter, in which we suppose all the dx's to vanish except the lth, we see that it expresses the fact that the vector **A** undergoes a parallel displacement along the line l.

Analogously, for the derivatives of a contravariant simple system A^i, we have

$$A^i_{|l} = \frac{\partial A^i}{\partial x_l} + \sum_1^n {}_j A^j \{jl, i\}. \quad . \quad . \quad . \quad (5')$$

Next, consider a system of order zero, i.e. an invariant f. In this case (4) becomes

$$f_l = \frac{\partial f}{\partial x_l}, \quad . \quad . \quad . \quad . \quad . \quad . \quad (7)$$

or *the covariant and the ordinary derivatives are identical.* If we construct the system of covariant second derivatives, applying formulæ (5) to (7), we shall have

$$f_{lk} = \frac{\partial^2 f}{\partial x_l \partial x_k} - \sum_1^n {}_j \{lk, j\} \frac{\partial f}{\partial x_j}; \quad . \quad . \quad . \quad (8)$$

these are not the same as the ordinary second derivatives but, like them, are symmetrical.

For a covariant double tensor (4) becomes

$$A_{ik|l} = \frac{\partial A_{ik}}{\partial x_l} - \sum_1^n {}_j \{il, j\} A_{jk} - \sum_1^n {}_j \{kl, j\} A_{ij}; \qquad (9)$$

and for a contravariant double tensor it becomes

$$A^{ik}_{|l} = \frac{\partial A^{ik}}{\partial x_l} + \sum_1^n {}_j \{jl, i\} A^{jk} + \sum_1^n {}_j \{jl, k\} A^{ij}. \qquad (9')$$

3. **Ricci's lemma.**

If formula (9) is applied to the system of the coefficients of ds^2, we get, remembering the expression for the derivatives of these coefficients in terms of Christoffel's symbols (Chap. V, § 16, p. 111),

$$a_{ik|l} = 0 \qquad (i, k, l = 1, 2, \ldots n). \quad . \quad (10)$$

This important theorem, that *the covariant derivatives of the coefficients* a_{ik} *are zero*, can be proved directly from the definition of covariant differentiation. To do so, we must choose two arbitrary vectors $\boldsymbol{\xi}$, $\boldsymbol{\eta}$, and construct the expression

$$F = \sum_1^n {}_{ik}\, a_{ik}\, \xi^i \eta^k;$$

we then calculate δF corresponding to a parallel displacement of the vectors $\boldsymbol{\xi}$, $\boldsymbol{\eta}$, and we shall get a trilinear form in ξ^i, η^k, δx_l, whose coefficients, by definition, will give the required derived system.

Now F is merely the scalar product of the vectors $\boldsymbol{\xi}$ and $\boldsymbol{\eta}$, which, as we know, is not changed by a parallel displacement; hence we shall have $\delta F = 0$ for any values whatever of $\boldsymbol{\xi}$, $\boldsymbol{\eta}$, and δx's, which means that all the coefficients of this form vanish identically.

Similarly we can show that the covariant derivatives of the reciprocals a^{ik} vanish; in this case we have to use the expression

$$F = \sum_1^n {}_{ik}\, a^{ik} u_i v_k,$$

which is again the scalar product of the (arbitrary) vectors **u** and **v**.

4. Contravariant differentiation.

There is in the absolute differential calculus a kind of law of reciprocity or duality in accordance with which we can deduce from every theorem or formula a reciprocal theorem or formula, by interchanging the words *covariant* and *contravariant*, and lowering or raising the indices. We have already had several examples of this; we shall now make some brief remarks on the operation of contravariant differentiation, which corresponds to that of covariant differentiation just described.

The shortest way to deduce from a system $A^{(h)}_{(i)}$ the system $A^{(h)k}_{(i)}$ which has the properties reciprocal to those of the covariant derivatives, is to find the covariant derivative of the given system and then compound it with the system of the a^{kl}'s; i.e. to make

$$A^{(h)k}_{(i)} = \sum_{1}^{n}{}_l \, a^{kl} A^{(h)}_{(i)l}.$$

We could find for this system an expression analogous to (4) and properties corresponding exactly to those of the covariant derivatives; or we could find these properties directly from those of the covariant derivatives, by using the foregoing formula of definition. We shall therefore not pursue the argument in detail, and shall instead resume our discussion of the fundamental properties of covariant differentiation.

5. Conservation of the rules of the ordinary differential calculus.

First, consider a tensor, in general mixed, which is the sum of two others of the same rank and species, i.e.

$$A^{(h)}_{(i)} = B^{(h)}_{(i)} + C^{(h)}_{(i)}.$$

It will at once be seen that the covariant derivative of the system A is obtained, like an ordinary derivative, by adding together that of B and that of C, or

$$A^{(h)}_{(i)l} = B^{(h)}_{(i)l} + C^{(h)}_{(i)l}. \quad . \quad . \quad . \quad . \quad (11)$$

This formula follows either from the linearity of (4), or from the consideration that the form F relative to A is the sum of a form relative to B and a form relative to C, so that a similar result holds for δF; the coefficients of the latter expression (which are by definition the derivatives $A^{(h)}_{(i)l}$) will therefore be the sums of the corresponding coefficients of the other two (which are by definition the derivatives $B^{(h)}_{(i)l}$ and $C^{(h)}_{(i)l}$). The reasoning can be extended without difficulty to a sum of any number of terms.

Next, consider the derivative of a product. If $B^{(h')}_{(i')}$, $C^{(h'')}_{(i'')}$ are two generic tensors, we shall denote their product by

$$A^{(h)}_{(i)} = B^{(h')}_{(i')} \cdot C^{(h'')}_{(i'')},$$

where the symbol (i) stands for the aggregate of the indices (i') and (i'') together, and similarly for (h). We shall show that

$$A^{(h)}_{(i)l} = B^{(h')}_{(i')l} \cdot C^{(h'')}_{(i'')} + B^{(h')}_{(i')} \cdot C^{(h'')}_{(i'')l}. \quad . \quad . \quad (12)$$

To simplify the formulæ we shall suppose that the systems A and B have each only one index of covariance and one of contravariance. We know (Chapter IV, § 8, p. 76) that if

$$\phi = \Sigma B^{h'}_{i'} \xi^{i'} u_{h'},$$

$$\psi = \Sigma C^{h''}_{i''} \eta^{i''} v_{h''}$$

are the invariant forms for the systems B and C, that for the system A is

$$F = \phi\psi.$$

We shall therefore have

$$\delta F = \psi \delta \phi + \phi \delta \psi,$$

and equating the coefficients of $\xi^{i'} \eta^{i''} u_{h'} v_{h''} \delta x_l$ on both sides of this equation we get equation (12) (for the particular case considered).

Now consider the derivative of a compounded mixed system (Chapter IV, p. 79)

$$A^{(h)}_{(i)} = \sum_{1}^{n}{}_{(r)(s)} B^{(h')(s)}_{(i')(r)} C^{(h'')(r)}_{(i'')(s)}, \quad . \quad . \quad . \quad (13)$$

where (i) and (h) have the meanings already explained, and (r) and (s) denote the aggregate of all the indices affected by the process of contraction. We shall show that

$$A^{(h)}_{(i)l} = \sum_{1}^{n}{}_{(r)(s)} \left[B^{(h')(s)}_{(i')(r)l}\, C^{(h'')(r)}_{(i'')(s)} + B^{(h')(s)}_{(i')(r)}\, C^{(h'')(r)}_{(i'')(s)l} \right]. \qquad (14)$$

In particular, if each aggregate reduces to a single index and if the process of contraction is applied only to one index, (13) becomes

$$A^{hk}_{ij} = \sum_{1}^{n}{}_{r}\, B^{h}_{ir}\, C^{kr}_{j}, \quad . \quad . \quad . \quad . \quad . \quad (13')$$

and (14) becomes

$$A^{hk}_{ijl} = \sum_{1}^{n}{}_{r}\, [B^{h}_{irl}\, C^{kr}_{j} + B^{h}_{ir}\, C^{kr}_{jl}]. \quad . \quad . \quad (14')$$

We shall give the proof for this simpler case, merely pointing out that it can be immediately extended to the general case.

We start from the invariant forms relative to the systems B and C

$$\phi_{a} = \sum_{1}^{n}{}_{ihr}\, B^{h}_{ir}\, \xi^{i}\, u_{h}\, \lambda^{r}_{a},$$

$$\psi_{a} = \sum_{1}^{n}{}_{jks}\, C^{ks}_{j}\, \eta^{j}\, v_{k}\, \lambda_{a|s},$$

where we have followed the same procedure as in Chapter IV, p. 78, and introduced a set of n contravariant systems λ^{r}_{a} $(a = 1, 2, \ldots n)$ and the associated reciprocal set. The invariant form

$$F = \sum_{1}^{n}{}_{a}\, \phi_{a}\, \psi_{a}$$

has the A's as coefficients, as we saw in Chapter IV.

Applying the symbol of operation δ to this we get

$$\delta F = \sum_{1}^{n}{}_{a}\, [\psi_{a}\, \delta\phi_{a} + \phi_{a}\, \delta\psi_{a}],$$

and equating the coefficients of $\xi^{i}\, \eta^{j}\, u_{h}\, v_{k}\, \delta x_{l}$ on both sides of this equation, we get (14′).

To sum up, we have shown that the fundamental rules of ordinary differentiation hold good for covariant differentiation.

6. Applications.

We note first of all that if we start from a generic simple system (a function of position), say a covariant system V_i, and consider its reciprocal V^i, we have by definition

$$V^i = \sum_1^n{}_k \, a^{ik} \, V_k;$$

hence, taking the covariant derivative and using Ricci's lemma,

$$V^i_{|l} = \sum_1^n{}_k \, a^{ik} \, V_{k|l}. \quad . \quad . \quad . \quad . \quad (15)$$

We shall next calculate the covariant derivative of the scalar product X of two vectors, which, as we know already, is identical with the ordinary derivative.

Let $\mathbf{U}$, $\mathbf{V}$, be two generic vectors, and put

$$X = \mathbf{U} \times \mathbf{V} = \sum_1^n{}_i \, U_i \, V^i.$$

Taking the covariant derivative, we have

$$X_l = \sum_1^n{}_i \, [U_{i|l} \, V^i + U_i \, V^i_{|l}].$$

In the second term on the right we can replace $V^i_{|l}$ by the expression for it in (15), so that

$$\sum_1^n{}_i \, U_i \, V^i_{|l} = \sum_1^n{}_{ik} \, a^{ik} \, U_i \, V_{k|l} = \sum_1^n{}_k \, U^k \, V_{k|l}.$$

Changing k into i, and substituting in X_l, we get the formula

$$X_l = \sum_1^n{}_i \, [U_{i|l} \, V^i + U^i \, V_{i|l}], \quad . \quad . \quad . \quad (16)$$

which is often used.

In particular, if $\mathbf{V} = \mathbf{U}$, we have $X = \mathbf{U}^2$, and therefore

$$X_l = 2\mathbf{U} \frac{\partial \mathbf{U}}{\partial x_l} = 2\sum_1^n{}_i \, U^i \, U_{i|l}. \quad . \quad . \quad (16')$$

7. Divergence of a vector and of a double tensor. Δ_2 of an invariant.

Take a covariant simple system X_i, which we can always think of as the aggregate of the components of a vector **X**, and construct the invariant

$$\Theta = \sum_{1}^{n}{}_{il} a^{il} X_{i|l}, \quad . \quad . \quad . \quad . \quad . \quad (17)$$

where the terms $X_{i|l}$ denote covariant derivatives. In the particular case of the fundamental form being Euclidean, we have $a^{il} = \delta_i^l$, and also the covariant and ordinary derivatives are identical; hence in this case (17) becomes

$$\Theta = \sum_{1}^{n}{}_i \frac{\partial X_i}{\partial x_i}.$$

In three dimensions this expression is called the *divergence* of the vector **X**. We shall extend the use of this term to the general case (17).

We can transform (17) by means of (15). Writing X instead of V, (15) becomes, for $l = i$,

$$X^i_{|i} = \sum_{1}^{n}{}_k a^{ik} X_{k|i}.$$

Summing with respect to i, the right-hand side gives Θ, as can be seen at once from (17) by putting l instead of k and then interchanging l and i. Hence we have

$$\Theta = \sum_{1}^{n}{}_i X^i_{|i}. \quad . \quad . \quad . \quad . \quad . \quad (17')$$

From the general rule for covariant differentiation, or more specifically from (5′), we have

$$X^i_{|i} = \frac{\partial X^i}{\partial x_i} + \sum_{1}^{n}{}_j \{ji, i\} X^j.$$

Now sum with respect to i. Substituting from (17′) on the left, and from the identity

$$\sum_{1}^{n}{}_i \{ji, i\} = \frac{1}{\sqrt{a}} \frac{\partial \sqrt{a}}{\partial x_j}$$

(cf. formula (26), Chapter V, p. 112) on the right, and writing l as the index of summation on the right instead of i and j, we get

$$\Theta = \sum_1^n{}_l \left(\frac{\partial X^l}{\partial x_l} + \frac{1}{\sqrt{a}} \frac{\partial \sqrt{a}}{\partial x_l} X^l \right)$$

or, taking the factor $\frac{1}{\sqrt{a}}$ outside the summation sign,

$$\Theta = \frac{1}{\sqrt{a}} \sum_1^n{}_l \frac{\partial}{\partial x_l} (\sqrt{a}\, X^l). \quad . \quad . \quad . \quad (17'')$$

This expression for the divergence is completely equivalent to the formulæ (17) and (17′); it is more useful for purposes of calculation, while (17) and (17′) on the contrary are more suited to theoretical discussions.

In particular, consider the case where the vector in question is the *gradient* of an invariant u, i.e. where

$$X_i = \frac{\partial u}{\partial x_i} \qquad (i = 1, 2, \ldots n).$$

In this case the divergence is denoted by the symbol $\Delta_2 u$ and is called the *second differential parameter* of the function u; the expression for it can be deduced at once from (17) or from (17″), using in the calculations the fact that

$$u^l = \sum_1^n{}_i a^{il} \frac{\partial u}{\partial x_i}.$$

We thus get

$$\Delta_2 u = \sum_1^n{}_{ik} a^{ik} u_{ik} = \frac{1}{\sqrt{a}} \sum_1^n{}_l \frac{\partial}{\partial x_l} (\sqrt{a}\, u^l), \quad . \quad (18)$$

both these expressions being generalizations of the ordinary expression for Δ_2 in Cartesian co-ordinates.

Next, take a contravariant double tensor X^{ik}. We note first of all that if instead the given tensor were covariant (X_{ik}) or mixed (X_i^k), we could always compound it with the a^{ik}'s and so obtain an associated tensor in which both indices are indices of contravariance; so that the choice of a contravariant tensor does not really constitute a restriction. From this tensor, taking the covariant derivative and applying the process of contraction, we get the contravariant simple system

$$Y^i = \sum_1^n {}_k X^{ik}_{|k}, \quad . \quad . \quad . \quad . \quad . \quad (19)$$

which, by an obvious analogy with the former case, is called the *divergence of the given double tensor*. If the process of contraction were applied to the first instead of to the second index, we should plainly get a contravariant system

$$\sum_1^n {}_k X^{ki}_{|k};$$

in general this is distinct from the divergence Y^i, coinciding with it only in the particular case when the given tensor X^{ik} is symmetrical. Vice versa, if X_{ik} is the system reciprocal to X^{ik} (the indices corresponding in the order written), we see at once from the rules in § 5 that the system

$$Y_i = \sum_1^n {}_{kl} a^{kl} X_{ik|l}$$

is merely the covariant system reciprocal to (19). Returning to (19), it should be added that the expression on the right cannot in general be transformed (as was done for the ordinary divergence (17)) into an expression which is convenient for actual calculations. In the case of an antisymmetrical tensor ($X^{ik} + X^{ki} = 0$), however, the analogy in this respect is perfect. In fact, if we substitute in (19) the values of $X^{ik}_{|k}$ given by (9′), the second term on the right vanishes from the antisymmetry of the X's, while the other two give

$$\sum_1^n {}_k \frac{\partial X^{ik}}{\partial x_k} + \sum_1^n {}_{jk} \{jk, k\} X^{ij}.$$

From this expression, by the same method as that just used to pass from (17′) to (17″), we get the equation

$$Y^i = \frac{1}{\sqrt{a}} \sum_1^n {}_k \frac{\partial(\sqrt{a}\, X^{ik})}{\partial x_k}. \quad . \quad . \quad . \quad (19')$$

8. **Some laws of transformation. ϵ-systems. Vector product. Extension of a field.**

Consider a set of n covariant simple systems $\lambda_{\alpha|i}$ (where α is the ordinal number of the system and i the index of covariance) and the determinant of the set

$$\nabla = \| \lambda_{\alpha|i} \|.$$

Changing the co-ordinates from the x's to another set of variables $\bar{x}$, the systems $\lambda_{\alpha|i}$ are transformed (in accordance with the law of covariance) into another set of systems $\bar{\lambda}_{\alpha|i}$ Construct the determinant of these new quantities

$$\bar{\nabla} = \| \bar{\lambda}_{\alpha|i} \|.$$

We shall show that the relation between $\bar{\nabla}$ and ∇ is

$$\bar{\nabla} = \nabla D, \quad . \quad . \quad . \quad . \quad . \quad . \quad (20)$$

where D denotes the Jacobian determinant of the transformation, i.e.

$$D = \begin{pmatrix} x_1 \, x_2 \, \ldots \, x_n \\ \bar{x}_1 \, \bar{x}_2 \, \ldots \, \bar{x}_n \end{pmatrix},$$

which is of course not zero, it being always supposed that we are using a reversible transformation (§ 2, p. 3). The relation (20) can be verified at once if we construct the product by rows of the two determinants on the right, viz.

$$\begin{vmatrix} \lambda_{1|1} & \lambda_{1|2} & \ldots & \lambda_{1|n} \\ \lambda_{2|1} & \lambda_{2|2} & \ldots & \lambda_{2|n} \\ \cdot & \cdot & \cdot & \cdot \\ \lambda_{n|1} & \lambda_{n|2} & \ldots & \lambda_{n|n} \end{vmatrix} \cdot \begin{vmatrix} \dfrac{\partial x_1}{\partial \bar{x}_1} & \dfrac{\partial x_2}{\partial \bar{x}_1} & \ldots & \dfrac{\partial x_n}{\partial \bar{x}_1} \\ \dfrac{\partial x_1}{\partial \bar{x}_2} & \dfrac{\partial x_2}{\partial \bar{x}_2} & \ldots & \dfrac{\partial x_n}{\partial \bar{x}_2} \\ \cdot & \cdot & \cdot & \cdot \\ \dfrac{\partial x_1}{\partial \bar{x}_n} & \dfrac{\partial x_2}{\partial \bar{x}_n} & \ldots & \dfrac{\partial x_n}{\partial \bar{x}_n} \end{vmatrix}.$$

Recalling (§ 11, p. 80) that

$$\bar{\lambda}_{\alpha|i} = \sum_1^n{}_j \lambda_{\alpha|j} \frac{\partial x_j}{\partial \bar{x}_i}, \quad . \quad . \quad . \quad . \quad (21)$$

we see at once that the elements of the product determinant are precisely the quantities $\bar{\lambda}_{\alpha|i}$.

It is also useful to examine the behaviour of the discriminant

$$a = \| a_{ik} \|$$

of the fundamental form when we change from the variables x to the variables $\bar{x}$. For this purpose, we take the transformation law of the coefficients a_{ik} (§ 12, p. 85),

$$\bar{a}_{ik} = \sum_{1}^{n}{}_{jh} \, a_{jh} \frac{\partial x_j}{\partial \bar{x}_i} \frac{\partial x_h}{\partial \bar{x}_k};$$

putting

$$b_{jk} = \sum_{1}^{n}{}_{h} \, a_{jh} \frac{\partial x_h}{\partial \bar{x}_k}, \quad \ldots \ldots \quad (22)$$

we can write this law in the form

$$\bar{a}_{ik} = \sum_{1}^{n}{}_{j} \, b_{jk} \frac{\partial x_j}{\partial \bar{x}_i}.$$

This law, which is completely analogous to (21), enables us to conclude at once, from the example of the preceding case, that the relation between $\bar{a}$ and the determinant b of the quantities b_{jk} is analogous to (20), i.e. that

$$\bar{a} = bD. \quad \ldots \ldots \quad (23)$$

Further, as (22) is of the same type as (21), the determinant b will be connected with a by the relation

$$b = aD,$$

which, combined with (23), gives us the required relation between a and $\bar{a}$, namely,

$$\bar{a} = aD^2. \quad \ldots \ldots \quad (24)$$

It follows from (20) and (24) that the ratio $\frac{\nabla}{\sqrt{a}}$ is an absolute invariant, i.e. that

$$\frac{\bar{\nabla}}{\sqrt{\bar{a}}} = \frac{\nabla}{\sqrt{a}}.$$

Strictly speaking, this equality holds except for sign; but if we agree to change the sign of the radical when a transformation is made for which D is negative, it holds in sign as well as in numerical value.

The remark just made leads us to define a particularly useful tensor whose elements can be expressed in a very simple form.

In fact, we note that the quantity $\frac{\nabla}{\sqrt{a}}$, which we have just seen to be invariant, is merely a multilinear form in the n sets of variables $\lambda_{a|i}$; this is seen at once by expanding the determinant ∇ in the usual way, as the sum of products of its elements n at a time, where no product contains two elements from the same row or column, and with the usual rule as to sign. We may write the result in the form

$$\frac{\nabla}{\sqrt{a}} = \frac{1}{\sqrt{a}} \mathrm{S}_{i_1 \ldots i_n} \pm \lambda_{1|i_1} \lambda_{2|i_2} \ldots \lambda_{n|i_n}, \quad . \quad (25)$$

where the symbol S denotes the sum of all the possible products, subject to the conventions stated as to their structure and sign. Since this form is invariant, its coefficients constitute a contravariant system. If we put $\epsilon^{i_1 i_2 \ldots i_n}$ for the coefficient of the product $\lambda_{1|i_1} \lambda_{2|i_2} \ldots \lambda_{n|i_n}$, we see at once that we have:

$\epsilon^{i_1 i_2 \ldots i_n} = 0$ if at least two of the indices $i_1 \, i_2 \ldots i_n$ are equal;

$\epsilon^{i_1 i_2 \ldots i_n} = \frac{1}{\sqrt{a}}$ if these indices are all different and constitute a permutation of *even* order with respect to the fundamental permutation $1, 2, \ldots n$;

$\epsilon^{i_1 i_2 \ldots i_n} = -\frac{1}{\sqrt{a}}$ if the indices are all different and constitute a permutation of *odd* order.

Hence it follows that the system of order n whose elements are 0, $\frac{1}{\sqrt{a}}$, $-\frac{1}{\sqrt{a}}$, respectively according to the rules just stated, is *contravariant*; we shall call it the *contravariant ϵ-system.*

We can give an analogous definition of the *covariant ϵ-system* by considering the determinant (reciprocal to ∇)

$$\nabla = \| \lambda_a^i \|$$

constructed from the reciprocal elements of the systems $\lambda_{\alpha|i}$ in the determinant ∇; these elements, as we know from Chapter IV, p. 74, constitute a set of n contravariant simple systems. By a well-known theorem, which can at once be verified, we have

$$\nabla \Delta = 1;$$

hence the quantity $\sqrt{a}\Delta$ $\left(\text{the reciprocal of } \frac{\nabla}{\sqrt{a}}\right)$ will be invariant. Expanding the determinant Δ, this can be written in the form

$$\sqrt{a}\, \mathrm{S}_{i_1 \ldots i_n} \pm \lambda_1^{i_1} \lambda_2^{i_2} \ldots \lambda_n^{i_n},$$

where the symbol S as before denotes the sum for all permutations of the indices i.

It follows that the system whose elements $\epsilon_{i_1 i_2 \ldots i_n}$ are zero if the indices $i_1\, i_2 \ldots i_n$ are not all different, and are equal to $\sqrt{a}$ or $-\sqrt{a}$ if the indices are all different according as the permutation $i_1\, i_2 \ldots i_n$ is of even or odd order, is *covariant*.

The use of the same letter ϵ for both is justified by the fact that this covariant system is the reciprocal of the former system. This statement can easily be verified by the reader.[1]

By means of the ϵ-systems, when $n - 1$ vectors $\mathbf{v}_\alpha$ ($\alpha = 1, 2, \ldots n - 1$) are given, we can deduce from them (by invariant processes) an nth vector $\mathbf{w}$, which is called their *vector product*, as in three-dimensional Euclidean space it is identical with the ordinary vector product. If v_α^i and $v_{\alpha|i}$ ($i = 1, 2, \ldots n$) denote the contravariant and covariant components of the $n - 1$ given vectors, the formulæ

$$w_i = \sum_1^n {}_{i_1 i_2 \ldots i_{n-1}} \epsilon_{i\, i_1 i_2 \ldots i_{n-1}} v_1^{i_1} v_2^{i_2} \ldots v_{n-1}^{i_{n-1}},$$

$$w^i = \sum_1^n {}_{i_1 i_2 \ldots i_{n-1}} \epsilon^{i\, i_1 i_2 \ldots i_{n-1}} v_{1|i_1} v_{2|i_2} \ldots v_{n-1|i_{n-1}}$$

define two reciprocal systems, as can easily be verified; hence either separately defines the same vector, which we call $\mathbf{w}$. When $n = 3$ and the space is Euclidean the components of $\mathbf{w}$ do in fact reduce to those of the ordinary vector product.

[1] For this and other properties of the ϵ-systems, cf. an interesting note by LIPKA: "Sui sistemi E nel calcolo differenziale assoluto", in *Rend. della R. Acc. dei Lincei*, Vol. XXXI (first half-year, 1922), pp. 242–245.

In any case it follows from the preceding definition of the components w_i (or w^i) that $\mathbf{w} = 0$ if the vectors $\mathbf{v}_a$ are not all linearly independent, i.e. if the characteristic of the matrix composed of their components $< n - 1$; when they are independent, $\mathbf{w} \neq 0$ and is perpendicular to every $\mathbf{v}_a$. The latter property follows from the consideration of a generic vector product $\mathbf{w} \times \mathbf{v}_a$. Taking, say, the first group of formulæ, we have

$$\mathbf{w} \times \mathbf{v}_a = \sum_1^n{}_i w_i v_a^i = \sum_1^n{}_{i\, i_1 i_2 \ldots i_{n-1}} \epsilon_{i\, i_1 i_2 \ldots i_{n-1}} v_1^{i_1} v_2^{i_2} \ldots v_{n-1}^{i_{n-1}} v_a^i,$$

which is zero from the definition of the ϵ-system, or, in other words, because the sum is the expansion of a determinant with two rows the same.

Lastly, we wish to introduce into the metric of a V_n the notion of the *extension* of a field, i.e. to define, for a given field of V_n, a quantity V analogous to the area of a portion of a surface or to the volume of a three-dimensional field. Evidently we have *a priori* a free choice as to the definition of dV, provided that when $n = 2$ it reduces to the expression already given for the element of area (formula (17), p. 99), and that when $n = 3$, in Cartesian co-ordinates, we have $dV = dx\,dy\,dz$; further, from the geometrical meaning of the term, the extension V of a field must be an invariant.[1] All these conditions are satisfied if we assume

$$dV = \sqrt{a}\, dx_1 \ldots dx_n, \qquad \ldots \qquad (26)$$

where $\sqrt{a}$ denotes the arithmetical value of the radical, and therefore

$$V = \int_C \sqrt{a}\, dx_1 \ldots dx_n.$$

We know in fact that on a change of co-ordinates the product $dx_1\, dx_2 \ldots dx_n$ must be replaced by $|\, D \,|\, d\bar{x}_1\, d\bar{x}_2 \ldots d\bar{x}_n$. From (24), extracting the square root, and taking the absolute values of both sides of the equation, we get

$$|\sqrt{a}\,| \,.\, |\, D \,|\, d\bar{x}_1\, d\bar{x}_2 \ldots d\bar{x}_n = \sqrt{\bar{a}}\, d\bar{x}_1\, d\bar{x}_2 \ldots d\bar{x}_n.$$

[1] A detailed study of the concept of extension and of its analytical expression has recently been made by O. Hölder. Cf. "Das Volumen in einer Riemann'schen Mannigfaltigkeit", in *Math. Zeitschrift*, Vol. 20 (1924), pp. 7–20.

But the left-hand side is $\sqrt{a}\, dx_1\, dx_2 \ldots dx_n$, which is therefore invariant.

9. Rotor of a simple tensor in three dimensions.

We can now give a definition of the *rotor* (or *rotation*, or *curl*) of a vector **X** given as a function of position, which shall hold good both when the space considered is not Euclidean, and also when it is Euclidean but the co-ordinates are not Cartesian. For any value of n, the generalization consists in defining as the rotor the covariant double system

$$p_{il} = X_{i|l} - X_{l|i},$$

which is obviously antisymmetrical, since $p_{il} + p_{li} = 0$ identically. As we saw in § 2, the p's can also be written as the differences of the ordinary derivatives $\frac{\partial X_i}{\partial x_l} - \frac{\partial X_l}{\partial x_i}$; if then we consider the X's as coefficients of a Pfaffian

$$\psi = \sum_1^n {}_i X_i\, dx_i,$$

it will be seen that the p's are merely the coefficients of the bilinear covariant of this Pfaffian (cf. Chapter II, p. 20).

To get the full analogy to the ordinary rotor, however, we should consider a space of only three dimensions. For $n = 3$, there are three different elements $p_{il} = -p_{li}$, corresponding to the pairs of different suffixes 23, 31, 12, pairs of equal suffixes giving zero values of the p's. Each of the pairs 23, 31, 12, can be associated with the absent suffix (1, 2, or 3 respectively), or, in a general formula, the index h can be associated with the pair of the type $h + 1$, $h + 2$, with the convention that suffixes which differ by 3 are to be considered equivalent; for instance, if $h = 2$, $h + 2$ represents the suffix 1. It is therefore easy to understand how when $n = 3$ the rotor can be represented by a simple instead of a double system. If, however, we were to put

$$p_h = p_{h+1,\, h+2},$$

the simple system so defined would be neither covariant nor contravariant. Instead, it will be convenient to apply the term rotor to a vector **R** whose contravariant components R^h are defined as follows (with the help of the ϵ-systems introduced in the preceding section):

$$R^h = \sum_{il}^{3} \epsilon^{hil} X_{l|i} \qquad (h = 1, 2, 3).$$

The contravariance of R^h follows immediately from the principle of contraction. In order to see the analogy between this expression and the ordinary rotor, note that in the double sum i and l can take only the values $h+1$, $h+2$ (since the ϵ corresponding to the value h would be zero); since i and l must also be unequal, there are two possible cases:

$$i = h+1, \quad l = h+2, \quad \text{when } \epsilon^{hil} = \frac{1}{\sqrt{a}},$$

$$i = h+2, \quad l = h+1, \quad \text{when } \epsilon^{hil} = -\frac{1}{\sqrt{a}}.$$

Hence this sum will have only two terms, and R^h can be written in the following form:

$$R^h = \frac{1}{\sqrt{a}}(X_{h+2\,|\,h+1} - X_{h+1\,|\,h+2})$$

or
$$R^h = \frac{1}{\sqrt{a}}\left(\frac{\partial X_{h+2}}{\partial x_{h+1}} - \frac{\partial X_{h+1}}{\partial x_{h+2}}\right);$$

the latter being convenient for actual calculations. In Cartesian co-ordinates $a = 1$, and we get the ordinary expression for the components of a rotor (it being supposed that x_1, x_2, x_3 correspond in order with x, y, z).

10. Sections of a manifold. Geodesic manifolds.

We know that in ordinary space S_3 if we are given two directions $\boldsymbol{\lambda}$, $\boldsymbol{\mu}$ starting from the same point P and defined by their cosines λ^i, μ^i $(i = 1, 2, 3)$, every other direction $\boldsymbol{\xi}$ through P whose cosines ξ^i are linear combinations of those of $\boldsymbol{\lambda}$ and $\boldsymbol{\mu}$, i.e. $\xi^i = \rho\lambda^i + \sigma\mu^i$, lies in the plane determined by $\boldsymbol{\lambda}$ and $\boldsymbol{\mu}$.

The coefficients ρ and σ are of course not independent, as the ξ's must satisfy a quadratic identity; we have in fact

$$\rho^2 + \sigma^2 + 2\rho\sigma \cos\widehat{\boldsymbol{\lambda}\boldsymbol{\mu}} = 1.$$

The directions $\boldsymbol{\xi}$ so defined are therefore simply infinite in number, and their aggregate is called a *section.*

All this can easily be extended to a generic V_n, in which m directions $\boldsymbol{\lambda}_\alpha$ $(\alpha = 1, 2, \ldots m)$ are given.

Take m multipliers ρ_α, for the moment arbitrary, and consider the directions $\boldsymbol{\xi}$ whose parameters are

$$\xi^i = \sum_1^m{}_\alpha \rho_\alpha \lambda^i_\alpha, \quad . \quad . \quad . \quad . \quad . \quad (27)$$

and consequently whose moments are

$$\xi_i = \sum_1^m{}_\alpha \rho_\alpha \lambda_{\alpha|i}. \quad . \quad . \quad . \quad . \quad (27')$$

In order that these expressions may effectively represent parameters and moments respectively, it is necessary and sufficient that they should satisfy the relation

$$\sum_1^n{}_i \xi^i \xi_i = 1,$$

that is to say,
$$\sum_1^m{}_{\alpha\beta} \rho_\alpha \rho_\beta \sum_1^n{}_i \lambda^i_\alpha \lambda_{\beta|i} = 1,$$

or, denoting the angle between the direction $\boldsymbol{\lambda}_\alpha$ and $\boldsymbol{\lambda}_\beta$ by $\widehat{\alpha\beta}$,

$$\sum_1^m{}_{\alpha\beta} \rho_\alpha \rho_\beta \cos\widehat{\alpha\beta} = 1. \quad . \quad . \quad . \quad . \quad (28)$$

Now suppose that the ρ's are connected by this relation but are otherwise arbitrary. We then see that (27) (or (27′)) defines an aggregate of ∞^{m-1} directions (this being the number of arbitrary parameters), including in particular the m given directions; this aggregate is called a *section.*

A section G being defined in this way by means of m of its directions $\boldsymbol{\lambda}_\alpha$, take in it any m directions $\boldsymbol{\lambda}'_\alpha$ whatever $(\alpha = 1, 2, \ldots m)$. It is almost obvious that the section G' determined by these directions is again G itself.

This can of course be verified algebraically. In fact, if a direction ξ belongs to G', this is equivalent to saying that its parameters are linear combinations of the parameters λ'^i_α, and therefore also of the parameters λ^i_α; i.e. the direction ξ also belongs to G; and vice versa.

We saw in Chapter V, p. 130, that a geodesic is uniquely determined if its starting-point and direction are given. Now let us fix a point P in a V_n, and draw from it two directions $\boldsymbol{\lambda}$, $\boldsymbol{\mu}$; these will determine a section of ∞^1 directions drawn from P. Consider the ∞^1 geodesics drawn from P in all these directions: they constitute a surface (∞^2 points) which is called a *geodesic surface with pole* P.

A geodesic surface is therefore determined by a point and two directions.

A similar definition can be given of an m-dimensional geodesic manifold. Take a point in V_n, and m directions drawn from it, which will define a section of ∞^{m-1} directions, and construct the geodesic corresponding to each of these directions. Since each geodesic contains ∞^1 points, the aggregate of all of them will contain ∞^{m-1+1} points; i.e. it will constitute a manifold V_m, which we shall call a *geodesic manifold.*

Particularly important cases are the *geodesic surfaces* ($m = 2$), and the *geodesic hypersurfaces* ($m = n - 1$) determined by $n - 1$ directions drawn from a point; we shall use these in the following section.

11. Locally geodesic (or locally Cartesian) co-ordinates.

In general, a system of co-ordinates in which ds^2 is represented by a form with constant coefficients is called Cartesian. It is not always possible to choose co-ordinates of this kind in a given V_n; it is however always possible to find a system of co-ordinates which behave like Cartesians *in the immediate vicinity of a point* P *assigned beforehand*, or, more precisely, which are such that the derivatives of the coefficients of ds^2 (which would vanish identically if the co-ordinates were Cartesian) all vanish at the point P. Such co-ordinates are called *locally geodesic*, or *locally Cartesian*, co-ordinates.

Their interest from the point of view of parallelism, or more generally of elementary equipollent displacement, appears plainly from equations (52) and (52′) of Chapter V, pp. 138, 139, which define the increments of the contravariant and covariant components respectively. It follows from these equations that when the system of reference is geodesic at P, these increments, in passing to any very near point, are zero, precisely as are those of the ordinary Cartesian components in Euclidean spaces.

Now take a V_n, and in it any system of co-ordinates x; we propose to introduce—if this is possible—a new set of variables

$$\bar{x}_i = f_i(x_1, x_2, \ldots x_n) \qquad (i = 1, 2, \ldots n) \quad . \quad (29)$$

such that the $\bar{x}$'s are geodesic co-ordinates at P, or in other words, putting $\bar{a}_{ik}$ for the coefficients of ds^2 in the new variables, such that

$$\left(\frac{\partial \bar{a}_{ik}}{\partial \bar{x}_l}\right)_P = 0 \qquad (i, k, l = 1, 2, \ldots n), \quad . \quad (30)$$

where the use of the suffix P denotes that after differentiation the $\bar{x}$'s are to be replaced by the co-ordinates $\bar{x}_P$ of P. Remembering the definition of Christoffel's symbols (Chapter V, pp. 109, 110), we see that (30) is equivalent to the condition that these symbols themselves are all zero at P, i.e. that

$$\{\overline{jl, i}\}_P = 0 \qquad (j, l, i = 1, 2, \ldots n). \quad . \quad (30')$$

The following analysis shows the possibility of finding a set of functions f_i to define a transformation of this kind.

The condition (30) consists of $n \cdot \frac{1}{2}n(n+1)$ equations containing the first and second derivatives of the f's (since $\bar{a}_{ik}$, by the law of covariance, can be expressed in terms of the a_{ik}'s and the first derivatives of the f's). Now the number of first derivatives is n^2, and that of second derivatives is $n \cdot \frac{1}{2}n(n+1)$, so that the number of both together is greater than the number of equations. Since, as we shall see, the equations are not algebraically inconsistent, it follows that we can solve the equations (30) for the values at P of the first and second derivatives of the f's, or rather for some of them, the others remaining arbitrary; further, the behaviour of the functions at points other than P is a matter of indifference. Thus the choice of the f's can be made with a wide degree of arbitrariness.

To avoid, however, the direct discussion of the equations (30), we shall start from the ideas contained in § 26 of Chapter V, p. 138. We saw there that the expressions

$$\tau^i = d\xi^i + \sum_{jl}^{n} {}_{1} \{jl, i\} \, \xi^j \, dx_l$$

constitute a contravariant simple tensor, the vector $\boldsymbol{\xi}$ and therefore its contravariant components ξ^i, and also the differentials dx_l, being all arbitrary. This holds in particular for the hypothesis $\xi^i = dx_i$, i.e. when we suppose

$$\tau^i = d^2 x_i + \sum_1^n{}_{jl} \{jl, i\} dx_j dx_l. \quad . \quad . \quad . \quad (31)$$

If on changing the variables we have at a special point P

$$\frac{\partial \bar{x}_i}{\partial x_k} = \delta_k^i, \quad . \quad . \quad . \quad . \quad . \quad (32)$$

then at that point, from the law of contravariance

$$\bar{\tau}^i = \sum_1^n{}_k \tau^k \frac{\partial \bar{x}_i}{\partial x_k},$$

it follows that

$$\bar{\tau}^i = \tau^i. \quad . \quad . \quad . \quad . \quad . \quad . \quad . \quad (33)$$

If we suppose (as we are always free to do, by making a preliminary change of variables from x_i to $x_i +$ a constant) that the x_i's vanish at P, the equations (32) are satisfied provided the formulæ of transformation (29) are of the type

$$\bar{x}_i = x_i + \phi_i(x_1, x_2, \ldots x_n), \quad . \quad . \quad (29')$$

where ϕ_i denotes a function of the x's which is regular at P, and whose expansion in a series of powers of the x's begins with terms of at least the second degree, e.g. a polynomial of the second degree in the x's. In fact, if these conditions are fulfilled, all the first derivatives of the ϕ's vanish at P. The second derivatives $\dfrac{\partial^2 \phi_i}{\partial x_j \partial x_l}$ are identical with the second derivatives $\dfrac{\partial^2 \bar{x}_i}{\partial x_j \partial x_l}$, and give the terms of the second degree (by Maclaurin's theorem) on the right-hand side of the equations (29′). By a suitable choice of the numerical values of these second derivatives at P, we can make all the Christoffel's symbols for the variables $\bar{x}$ vanish, so satisfying the equations (30′), as we shall now show.

In fact, writing out both sides of equation (33) in full by means of (31), and considering the x's, in virtue of (29′), as independent variables (with their second differentials zero) and the $\bar{x}$'s as functions of them, we can write (33) in the form

$$d^2 \bar{x}_i + \sum_{1}^{n}{}_{hk} \overline{\{hk, i\}} d\bar{x}_h d\bar{x}_k = \sum_{1}^{n}{}_{jl} \{jl, i\} dx_j dx_l.$$

Equating the coefficients of $dx_j dx_l$ on both sides and remembering (32) we get

$$\left(\frac{\partial^2 \bar{x}_i}{\partial x_j \partial x_l}\right)_P + \overline{\{jl, i\}}_P = \{jl, i\}_P; \quad . \quad . \quad (34)$$

from which it appears that we need only take

$$\frac{\partial^2 \bar{x}_i}{\partial x_j \partial x_l} = \{jl, i\}_P \qquad (j, l, i = 1, 2, \ldots n)$$

at P in order to have

$$\overline{\{jl, i\}}_P = 0$$

for every possible set of values of j, l, i. Q.E.D.[1]

[1] Prof. Fermi has recently established an important extension of this result by showing that, given any curve whatever, it is also possible to choose co-ordinates which are locally geodesic at every point of the curve. Cf. his notes "Sopra i fenomeni che avvengono in prossimità di una linea oraria" in *Rend. della R. Acc. dei Lincei*, Vol. XXXI (first half-year, 1922), pp. 21–23, 51–52. Fermi's result can be quickly justified as follows, by calculating the number of available unknowns and of conditions to be satisfied.

Take the equations of the curve L in the form

$$x_i = \chi_i(x_n) \qquad (i = 1, 2, \ldots n - 1),$$

as we may always do by considering a suitably limited segment. Note first that if the values z_L of a generic function $z(x_1, x_2, \ldots x_n)$ and of its partial derivatives with respect to $x_1, x_2, \ldots x_{n-1}$ are known at all points of the curve, then the values of $\frac{\partial z}{\partial x_n}$ also are determined at all points of the curve. This is obvious if we take the identity $z(x_1, x_2, \ldots x_n) = z_L(x_n)$ which holds at all points of L, and differentiate it, so getting

$$\frac{\partial z}{\partial x_n} = \frac{dz_L}{dx_n} - \sum_{1}^{n-1}{}_i \frac{\partial z}{\partial x_i} \frac{dx_i}{dx_n}.$$

Now suppose that we make a change of variables of the general type (29), and that we wish to determine, if possible, the n functions $f_i(x_1, x_2, \ldots x_n)$ so as to make every $\frac{\partial \bar{a}_{ik}}{\partial \bar{x}_l} = 0$ along L. As has already been noted in dealing with a single point P, we thus get $n \cdot \frac{1}{2}n(n+1)$ conditions involving the first and second derivatives of the f's. Now the number of first derivatives $\frac{\partial f_i}{\partial x_k}$ is n^2, and that of the second derivatives $\frac{\partial^2 f_i}{\partial x_h \partial x_k}$ is $n \cdot \frac{1}{2}n(n+1)$; but from the preceding remark, the n^2 of the latter, which are of the type $\frac{\partial^2 f_i}{\partial x_n \partial x_k}$ $(i, k = 1, 2, \ldots n)$, can be expressed

It is not inapposite to give a geometrical interpretation of the conditions imposed on the co-ordinates $\bar{x}$ in order that (30′) may be satisfied, or, in other words, in order that they may be geodesic at P. These conditions may be put in the following form:

(*a*) The n co-ordinate hypersurfaces passing through P must behave as geodesic hypersurfaces with respect to points infinitely near P (or, in particular, must be geodesic everywhere).

(*b*) If through a point P', infinitely near to P and on one of the n co-ordinate lines through P—say that along which x_i alone varies—we construct the direction parallel to another of the co-ordinate lines, this parallel must belong to the co-ordinate surface x_i = constant which passes through P'.

(*c*) When the co-ordinate hypersurfaces are fixed in accordance with the foregoing conditions (which, as is geometrically obvious, is always possible), the numbering of these surfaces (i.e. the way in which they are associated with the values of the parameters $\bar{x}_1, \bar{x}_2, \ldots \bar{x}_n$) must be carried out so as to satisfy certain numerical conditions which we shall subsequently specify, and which, as we shall see, can always be satisfied.

That (*a*) and (*b*) are consequences of (30′) follows immediately from the equations of parallelism and of geodesics. Reciprocally, we shall show that a system of co-ordinates which satisfies the conditions (*a*), (*b*), (*c*) is geodesic at P.

We shall begin by expressing the condition (*a*) analytically. Take a direction with parameters $d\bar{x}_k$, drawn from P and lying in the hypersurface $\bar{x}_i$ = constant (so that $d\bar{x}_i = 0$). We have to express the fact that the geodesic in this direction behaves at

at points of L in terms of the others and of the first derivatives. There remain altogether, including both first and second derivatives, $n^2 + \{n \cdot \frac{1}{2}n(n+1) - n^2\} = n \cdot \frac{1}{2}n(n+1)$ unknown functions of x_n to determine by means of the same number of equations $\frac{\partial \bar{a}_{ik}}{\partial \bar{x}_l} = 0$. These last equations, as can at once be seen, contain the second derivatives $\frac{\partial^2 f_i}{\partial x_h \, \partial x_k}(h, k < n)$ in finite terms (in fact linearly), while the unknown values of the first derivatives $\frac{\partial f_i}{\partial x_k}$ appear together with the terms $\frac{d}{dx_n} \frac{\partial f_i}{\partial x_k}$. In any case we have a system of as many equations as there are unknown functions of x_n alone to determine. When the values of these derivatives are known on L, we can determine, with a wide degree of arbitrariness, functions f_i which admit of these values. This can be seen by taking a Taylor expansion of the f's as a function of the $n-1$ arguments $x_1 - x_1^0, x_2 - x_2^0, \ldots x_{n-1} - x_{n-1}^0$, where the quantities x_i^0 $(i = 1, 2, \ldots n-1)$ are the values of the χ_i's on L, i.e. of the functions $\chi_i(x_n)$ which define this curve.

P as if it lay on this hypersurface, i.e. that d^2x_i vanishes along this geodesic. It follows that $d\bar{x}_i = d^2\bar{x}_i = 0$, and therefore, from the equation of geodesics,

$$\sum_1^n {}_{jl} \overline{\{jl, i\}}_P \, d\bar{x}_j \, d\bar{x}_l = 0.$$

Of the terms in this sum, those in which either j, or l, or both, are equal to i vanish, since $d\bar{x}_i = 0$; the other dx's being arbitrary, the necessary condition for the vanishing of the other terms is that

$$\overline{\{jl, i\}}_P = 0 \qquad (j, l \neq i).$$

We thus see the analytical meaning of condition (a).

Next consider (b). We shall take P' on the line i, so that, if dx represents the increments of the co-ordinates from P to P', we shall have $dx_l = 0$ for every value of l other than i. Let $\boldsymbol{\lambda}$ denote the direction of the co-ordinate line j at P, so that $\lambda^k = 0$ for every k other than j, and let $\boldsymbol{\lambda}$ undergo a parallel displacement from P to P'. Applying the usual formula and remembering that dx_i and λ^j are the only components which are not zero, we get

$$d\lambda^i = -\overline{\{ji, i\}}_P \, \lambda^j \, dx_i.$$

In order that the direction $\lambda' = \lambda + d\lambda$ may lie on the hypersurface x_i = constant, we must have $\lambda'^i = 0$, or (since, as we noted, $\lambda^i = 0$ if $i \neq j$) $d\lambda^i = 0$ if $i \neq j$, so that we must have

$$\overline{\{ji, i\}}_P = 0 \qquad (i \neq j).$$

This is the analytical expression of condition (b). We must now use the third condition in order to show that the symbols with three equal indices vanish; we shall thus have exhausted all the types of Christoffel's symbols.

Suppose that the co-ordinates x satisfy the foregoing conditions. Apply a transformation which leaves the co-ordinate surfaces unchanged; this can be done by putting $x_i = f_i(\bar{x}_i)$ (i.e. every x is a function of a single $\bar{x}$), or, which is the same thing,

$$dx_i = X_i(\bar{x}_i) \, d\bar{x}_i,$$

where X_i denotes the derivative of f with respect to its argument.

We shall now calculate the explicit expression of the symbols which we intend shall vanish. We have

$$\overline{[ii, i]} = \sum_1^n {}_j \bar{a}_{ij} \overline{\{ii, j\}},$$

or, remembering that all the symbols are already zero except those with three equal indices,

$$\overline{[ii, i]} = \bar{a}_{ii} \overline{\{ii, i\}}.$$

Substituting on the left-hand side the expression which defines the symbol of the first kind, we get

$$\tfrac{1}{2} \frac{\partial \bar{a}_{ii}}{\partial \bar{x}_i} = \bar{a}_{ii} \overline{\{ii, i\}}.$$

Hence the condition

$$\overline{\{ii, i\}} = 0 \qquad (i = 1, 2, \ldots n)$$

is equivalent to

$$\left(\frac{\partial \bar{a}_{ii}}{\partial \bar{x}_i}\right)_P = 0 \qquad (i = 1, 2, \ldots n).$$

Now from the law of covariance we have

$$\bar{a}_{ik} = \sum_1^n {}_{jh} \frac{\partial x_j}{\partial \bar{x}_i} \frac{\partial x_h}{\partial \bar{x}_k} a_{jh} = a_{ik} X_i X_k,$$

and therefore

$$\frac{\partial \bar{a}_{ii}}{\partial \bar{x}_i} = \frac{\partial a_{ii}}{\partial x_i} X_i^3 + 2\, a_{ii} X_i X_i'.$$

In order that the required condition may be satisfied, the functions X therefore need only satisfy, at P, the n numerical conditions

$$\frac{\partial a_{ii}}{\partial x_i} X_i^3 + 2\, a_{ii} X_i X_i' = 0;$$

otherwise they may be completely arbitrary.

We thus see how to determine a system of co-ordinates $\bar{x}$ which shall be locally geodesic at P.

12. **Severi's theorem.**

The possibility of choosing co-ordinates which are locally Cartesian at a given point enables us to simplify the proof of some geometrical properties which hold in the neighbourhood of a point. As an example we shall prove, without any calculation, a remarkable theorem due to Professor Severi.[1]

In a given V_n consider two infinitely near points, P and P_1, and a direction $\mathbf{u}$ drawn from P. This direction, and the direction PP_1 determine a section of V_n, and therefore a geodesic surface V_2 which passes through P and P_1 and contains $\mathbf{u}$.

We can now give $\mathbf{u}$ a parallel displacement, from P to P_1, in two ways:

(1) by considering $\mathbf{u}$ as a direction in V_n, and therefore using the metric of this variety; this will give a direction $\mathbf{u}_1$, which we shall call the *ambiental parallel*;

(2) by considering $\mathbf{u}$ as a surface direction, belonging to the geodesic surface V_2 just defined, and using the metric of V_2; this will give a direction $\overset{*}{\mathbf{u}}_1$.

Severi's theorem is that $\mathbf{u}_1$ and $\overset{*}{\mathbf{u}}_1$ are identical.

We shall examine first the case in which V_n is Euclidean. In this case the geodesics are straight lines (since, with a system of Cartesian co-ordinates y, Christoffel's symbols are zero and the equations of the geodesics become $d^2y_i = 0$ $(i = 1, 2, \ldots n)$) and the geodesic surfaces are planes; Severi's theorem becomes an immediate consequence of the ordinary theory of parallelism in Euclidean spaces.

Next, if V_n is not Euclidean, we note that in the definitions of the ambiental parallel $\mathbf{u}_1$, the geodesic surface V_2, and the parallel $\overset{*}{\mathbf{u}}_1$ relative to V_2, the only metrical elements used are Christoffel's symbols for the V_n; since all these can be made to vanish by a suitable choice of co-ordinates, the two methods of displacement are applied exactly as if V_n were Euclidean, and therefore lead to the same result.

[1] "Sulla curvatura delle superficie e varietà" in *Rend. del Circolo Mat. di Palermo*, Vol. XLII, 1917, pp. 227–259. Cf. especially § 11.

CHAPTER VII

Riemann's Symbols and Properties relating to Curvature; Ricci's and Einstein's Symbols; Geodesic Deviation

1. Cyclic displacement and the relations between parallelism and curvature.

Schouten,[1] by his vector methods, and independently of him Pérès,[2] by ordinary calculus methods, have demonstrated the great importance, for determining the geometrical properties of a V_n, of the displacement of a direction round a closed circuit; in particular the importance of infinitesimal circuits in investigating local properties at a generic point P.

Consider a generic direction (a unit vector) $\mathbf{u}$ drawn from P, and give it a parallel displacement round a closed curve T of infinitesimal length so that it comes back again to P; after the displacement we shall have a direction $\mathbf{u}_1$, also drawn from P, but not in general coinciding with $\mathbf{u}$. The change in the contravariant components u^r due to the displacement round the circuit will in general depend on the *area* of the circuit, on its *configuration* (i.e. on the orientation in V_n of the element of surface on which the circuit is drawn), and on the *metrical properties* of the V_n at P. The influence of the last-named properties is exerted through the first and second derivatives of the a_{ik}'s; these derivatives occur in certain characteristic groups which are called *Riemann's symbols*, and which are composed of Christoffel's symbols and their first derivatives. In the particular case of a surface, these expressions reduce to one, which is that known in geometry as the (Gaussian) *curvature* of the surface; for any V_n the consideration of Riemann's symbols provides a convenient way of extending the notion of curvature.

In this chapter we shall first consider displacement round a particular form of infinitesimal circuit, namely, an elementary parallelogram. We shall then discuss some of the properties of Riemann's symbols, which occur in the investigation of the

[1] "Die direkte Analysis zur neueren Relativitätstheorie", in *Verh. der Kon. Ak. van Wet. te Amsterdam*, Deel 12, No. 6, 1919. Cf. also the same writer's *Der Ricci-Kalkül* (Berlin, Springer, 1924), II, §§ 12–16.

[2] "Le parallélisme de M. Levi-Civita et la courbure riemannienne", in *Rend. della R. Acc. dei Lincei*, (5), Vol. XXVII (first half-year, 1919), pp. 425–428.

displacement, and shall use these properties to obtain the formula for changing the order of two successive covariant differentiations, by determining the difference between the derivatives. Lastly, we shall return to the question of displacement round any circuit whatever, and shall deduce from it the notion of curvature, first for a surface, then for any V_n whatever.

2. Cyclic displacement round an elementary parallelogram.

Let two elementary vectors, δP, $\delta' P$, be drawn from a generic point P of a V_n. We shall interpret the first as an elementary displacement PP_1, and give the vector $\delta' P$ a parallel displacement along it; let Q be the position of the extremity of $\delta' P$ after this displacement. If we apply the same process to δP, and give it a parallel displacement along the path PP_2, we reach the same point Q (as we have already shown in Chapter V, p. 116), even if we retain terms of the type $\delta\delta' P$, $\delta'\delta P$, while neglecting terms of the second order in δP and $\delta' P$. We can therefore, in any V_n, consider an *elementary parallelogram* PP_1QP_2.

We shall adopt the obvious convention of representing by δq the change in any quantity q (scalar or vector) in passing from P to P_1 and by $\delta' q$ the analogous change in passing from P to P_2. For a vector, we shall calculate these changes by the formulæ of parallelism.

Now let $D_1 q$ be the change in q on passing from P to Q along the path PP_1Q, and $D_2 q$ the analogous change on passing along the other pair of sides PP_2Q which with the first pair make up the circuit.

It will be seen at once that (neglecting second order terms as explained above) the total change Δq on going round the entire circuit in the sense PP_1QP_2P is $D_1 q - D_2 q$. We shall first examine $D_1 q$.

The change denoted by δ corresponds to the displacement along PP_1; hence, if the value of our quantity was q at P, its value at P_1 will be

$$q + \delta q = q_1.$$

The displacement along P_1Q changes q_1 into $q_1 + \delta' q_1$, so that at Q we shall have the quantity

$$\begin{aligned} & q + \delta q + \delta'(q + \delta q) \\ = {} & q + \delta q + \delta' q + \delta'\delta q \end{aligned}$$

so that

$$D_1 q = \delta q + \delta' q + \delta'\delta q.$$

As $D_2 q$, by its definition, differs from $D_1 q$ only by interchanging P_1 and P_2, and therefore δ and δ', we get

$$D_2 q = \delta' q + \delta q + \delta\delta' q.$$

It follows that the change caused by the displacement round the circuit is

$$\Delta q = (\delta'\delta - \delta\delta')q. \quad . \quad . \quad . \quad . \qquad (1)$$

We must now find an explicit form for this expression, supposing that the quantity q is a vector $\mathbf{u}$, and calculating the increments δ and δ' by the formulæ of parallelism. By these formulæ, the changes δu^r of the contravariant components will be given by the Pfaffian (p. 138, equation (52))

$$\delta u^r = -\sum_{1}^{n}{}_{ih}\{ih, r\}\, u^i\, \delta x_h, \quad . \quad . \quad . \qquad (2)$$

while the changes $\delta' u^r$ will be given by the same Pfaffian relative to the increments $\delta' x_h$. From (1) we see that we have to calculate the bilinear covariant (cf. p. 20, § 4) of this Pfaffian.

Differentiating (2) with the symbol δ' we get

$$\delta'\delta u^r = -\sum_{1}^{n}{}_{ih}\, \delta'\{ih, r\}\, u^i\, \delta x_h - \sum_{1}^{n}{}_{ih}\{ih, r\}\, \delta' u^i\, \delta x_h$$
$$- \sum_{1}^{n}{}_{ih}\{ih, r\}\, u^i\, \delta'\delta x_h.$$

To expand the first sum, we note that the expressions $\{ih, r\}$ are functions of the x's, and therefore

$$\delta'\{ih, r\} = \sum_{1}^{n}{}_{k} \frac{\partial}{\partial x_k}\{ih, r\}\, \delta' x_k.$$

The second, on substituting for $\delta' u^i$ the expression analogous to (2), becomes

$$\sum_{1}^{n}{}_{ihkl}\{ih, r\}\{kl, i\}\, u^l\, \delta' x_k\, \delta x_h,$$

or, interchanging i and l in order to get the factor u^i here too,

$$\sum_{1}^{n}{}_{ihkl}\{lh, r\}\{ki, l\}\, u^i\, \delta' x_k\, \delta x_h.$$

We have, therefore, taking out the factors $u^i\,\delta x_h\,\delta' x_k$ in the first two sums,

$$\delta'\delta u^r = -\sum_1^n{}_{ihk}\left[\frac{\partial}{\partial x_k}\{ih,\, r\} - \sum_1^n{}_l\{lh,\, r\}\{ki,\, l\}\right]u^i\,\delta x_h\,\delta' x_k$$
$$-\sum_1^n{}_{ih}\{ih,\, r\}\,u^i\,\delta'\delta x_h.$$

The expression for $\delta\delta' u^r$ can be obtained from this by interchanging δ and δ'; in the first sum we shall also interchange h and k, giving

$$\delta\delta' u^r = -\sum_1^n{}_{ihk}\left[\frac{\partial}{\partial x_h}\{ik,\, r\} - \sum_1^n{}_l\{lk,\, r\}\{hi,\, l\}\right]u^i\,\delta' x_k\,\delta x_h$$
$$-\sum_1^n{}_{ih}\{ih,\, r\}\,u^i\,\delta\delta' x_h.$$

In taking the difference $\delta'\delta u^r - \delta\delta' u^r$ the third sum cancels out, because $\delta\delta' x_h = \delta'\delta x_h$ (cf. pp. 18, 19, § 4), and there remain the terms involving the indices i, h, k, in which $u^i\,\delta x_h\,\delta' x_k$ can be taken out as a common factor. If we introduce *Riemann's symbols of the second kind*,

$$\left.\begin{aligned}\{ir,\, hk\} &= \frac{\partial}{\partial x_k}\{ih,\, r\} - \frac{\partial}{\partial x_h}\{ik,\, r\}\\ &-\sum_1^n{}_l[\{lh,\, r\}\{ik,\, l\} - \{lk,\, r\}\{ih,\, l\}]\end{aligned}\right\}(i,\, r,\, h,\, k = 1, 2, \ldots n), \quad (3)$$

we shall therefore have

$$\Delta u^r = (\delta'\delta - \delta\delta')u^r = -\sum_1^n{}_{ihk}\{ir,\, hk\}\,u^i\,\delta x_h\,\delta' x_k. \quad (4)$$

This formula shows that the required increment $\Delta\mathbf{u}$ depends on the vector $\mathbf{u}$, on the two vectors δP, $\delta' P$ which define the parallelogram, and lastly, on the metric of the manifold, through the quantities $\{ir,\, hk\}$. From (4) it follows as a particular case that for Euclidean spaces Riemann's symbols as just defined are all zero, whatever may be the co-ordinates x chosen for reference. In fact, for such a manifold, we have $\Delta u^r = 0$ $(r = 1, 2, \ldots n)$, since any vector resumes its original value after parallel displacement round any closed circuit whatever. Hence the right-hand side of (4) vanishes for every r, and for any value of the vector **u** and of the displacements δP, $\delta' P$, i.e. for any values of the arguments u^i, δx_h, $\delta' x_k$. The coefficients $\{ir, hk\}$ must therefore vanish separately.

It will be useful to point out at once the following two properties of the operator Δ:

(*a*) When applied to a product, it behaves like a symbol of ordinary differentiation, i.e.

$$\Delta(\psi\phi) = \psi\Delta\phi + \phi\Delta\psi,$$

which can be verified directly, by calculating first $\delta(\psi\phi)$, then $\delta'\delta(\psi\phi)$, &c.;

(*b*) when applied to a function of position, the result is zero, as is obvious from the meaning of the symbol.

If instead of the increments Δu^r we wish to find those of the covariant components, we can use the relation

$$u_j = \sum_1^n{}_r\, a_{jr}\, u^r,$$

and therefore, from the properties of the operator Δ,

$$\begin{aligned} \Delta u_j &= \sum_1^n{}_r\, a_{jr}\, \Delta u^r \\ &= -\sum_1^n{}_{rihk}\, a_{jr}\, \{ir, hk\}\, u^i\, \delta x_h\, \delta' x_k. \end{aligned}$$

If we introduce *Riemann's symbols of the first kind*,

$$(ij, hk) = \sum_1^n{}_r\, a_{jr}\{ir, hk\}, \quad . \quad . \quad . \quad . \quad (5)$$

we can sum with respect to r, and can then write

$$\Delta u_j = -\sum_1^n{}_{ihk}\, (ij, hk)\, u^i\, \delta x_h\, \delta' x_k, \quad . \quad . \quad (4')$$

which is analogous to (4).

Solving (5) we get Riemann's symbols of the second kind in terms of those of the first kind by the formula (the inverse of (5))

$$\{ir, hk\} = \sum_1^n{}_j\, a^{jr}\, (ij, hk.) \quad . \quad . \quad . \quad (5')$$

3. Fundamental properties of Riemann's symbols of the second kind.

As can be seen from their definition (3), Riemann's symbols of the second kind are functions of position, depending on the coefficients a_{ik}, their first derivatives (contained in Christoffel's symbols) and second derivatives (contained in the derivatives of Christoffel's symbols). They have the following fundamental properties:

(*a*) They are antisymmetrical in the last two indices, i.e.

$$\{ir, hk\} = -\{ir, kh\}, \quad . \quad . \quad . \quad (6)$$

whence in particular

$$\{ir, hh\} = 0.$$

This property follows immediately from (3).

(*b*) They constitute a mixed tensor,[1] contravariant with respect to the second index and covariant with respect to the other three, so that the symbol $\{ir, hk\}$ could also be denoted (as is sometimes done) by a^r_{ihk}. To prove this, consider the invariant

$$F = \sum_1^n{}_r\, p_r\, u^r,$$

where the p's are given (but completely arbitrary) functions of position, so that $\Delta p_r = 0$. If we give F a displacement round an infinitesimal circuit, we find (remembering the behaviour of the operator Δ)

$$\begin{aligned} \Delta F &= \sum_1^n{}_r\, (u^r\, \Delta p_r + p_r\, \Delta u^r) \\ &= \sum_1^n{}_r\, p_r\, \Delta u^r. \end{aligned}$$

As F is invariant, this quantity must also be so; replacing Δu^r in it by its expression (4), we get the quadrilinear form

$$\Delta F = -\sum_1^n{}_{ihkr}\, \{ir, hk\}\, p_r\, u^i\, \delta x_h\, \delta' x_k, \quad . \quad . \quad (7)$$

which expresses the required property of the Riemann's symbols, since the simple systems p_r, u^i, δx_h, $\delta' x_k$ are all arbitrary.

[1] Very generally, especially in works on the Theory of Relativity, called the *Riemann-Christoffel tensor.*

We can use the tensor character of Riemann's symbols to obtain a second proof of the fact that Riemann's symbols are all zero for a Euclidean V_n (the first proof is an immediate consequence of (4), as was shown in the preceding section). In fact, the definition (3) shows at once that they vanish in Cartesian co-ordinates, and in consequence they vanish in any other system of co-ordinates.

(*c*) They have an important cyclic property with respect to the three indices of covariance, namely,

$$\{ir, hk\} + \{hr, ki\} + \{kr, ih\} = 0. \quad . \quad . \quad (8)$$

To prove this, we again take F and formula (7), but we suppose that in them the p's are derivatives of an invariant function f of position (whose numerical values are otherwise arbitrary), and we also take as vector $\mathbf{u}$ an infinitesimal displacement with components $u^r = dx_r$, which is also arbitrary. With this choice, F becomes

$$F = \sum_1^n{}_r \frac{\partial f}{\partial x_r} dx_r = df,$$

and (7) becomes

$$(\delta'\delta - \delta\delta')\, df = -\sum_1^n{}_{ihkr}\{ir, hk\}\, p_r\, dx_i\, \delta x_h\, \delta' x_k. \quad . \quad (9)$$

Interchanging cyclically the three infinitesimal vectors denoted by the operators d, δ, δ', we get the other two formulæ:

$$(d\delta' - \delta' d)\, \delta f = -\sum_1^n{}_{ihkr}\{ir, hk\}\, p_r\, \delta x_i\, \delta' x_h\, dx_k, \quad (9')$$

$$(\delta d - d\delta)\, \delta' f = -\sum_1^n{}_{ihkr}\{ir, hk\}\, p_r\, \delta' x_i\, dx_h\, \delta x_k. \quad (9'')$$

Now on the right-hand side of these last two formulæ we can arrange to have the product $dx_i\, \delta x_h\, \delta' x_k$ in the general term, merely by a suitable interchange of the indices of summation. We can then add (9), (9′), and (9″); the left-hand side gives 0, since the terms cancel out in pairs (e.g. $\delta'\delta df - \delta' d\delta f = \delta'(\delta df - d\delta f) = 0$, since f is a function of position); and we get

$$0 = \sum_1^n{}_{ihkr}\,[\{ir, hk\} + \{hr, ki\} + \{kr, ih\}]\, p_r\, dx_i\, \delta x_h\, \delta' x_k.$$

As p_r, dx_i, δx_h, $\delta' x_k$ are arbitrary, (8) follows at once.

4. Fundamental properties and number of Riemann's symbols of the first kind.

Riemann's symbols of the first kind, as defined by (5) (i.e. the quantities obtained by compounding the quadruple system of the symbols of the second kind with the system of coefficients a_{ik}) have the following properties:

(*a*) They are *covariant with respect to all four indices*, so that they may be denoted, as is often done, by $a_{ij,\, hk}$; this follows from the definition, in consequence of the law of contraction.

The remark that a Euclidean V_n has all its Riemann's symbols zero is true here too, whatever may be the system of reference.

(*b*) They are *antisymmetrical* with respect to each pair of indices, so that we have identically

$$(ij,\, hk) = -(ij,\, kh), \quad . \quad . \quad . \quad . \quad (10)$$

$$(ij,\, hk) = -(ji,\, hk). \quad . \quad . \quad . \quad . \quad (11)$$

The identity (10) follows at once from (5), and from the analogous property of the symbols of the second kind. To prove (11) we shall follow a method analogous to that used in § 3 (*b*), taking as invariant the scalar product of two arbitrary vectors **u, v,**

$$F = \sum_{1}^{n}{}_{j}\, v_j\, u^j.$$

Applying the operator Δ and remembering that in a parallel displacement the scalar product does not change (so that $\Delta F = 0$) we get

$$0 = \sum_{1}^{n}{}_{j}\, v_j\, \Delta u^j + \sum_{1}^{n}{}_{j}\, u^j\, \Delta v_j. \quad . \quad . \quad . \quad (12)$$

The expression for Δu^j is given by (4), by writing j in it instead of r; that for Δv_j is given by (4′). Substituting, we get

$$0 = \sum_{1}^{n}{}_{jihk}\, v_j \{ij,\, hk\}\, u^i\, \delta x_h\, \delta' x_k + \sum_{1}^{n}{}_{jihk}\, u^j\, (ij,\, hk)\, v^i\, \delta x_h\, \delta' x_k. \quad (12')$$

In the first sum we express v_j in terms of the contravariant components v_r, and then, remembering (5), we sum with respect to j; we get successively

$$\sum_{1}^{n}{}_{jihkr}\, a_{jr}\, v^r \{ij,\, hk\}\, u^i\, \delta x_h\, \delta' x_k = \sum_{1}^{n}{}_{ihkr}\, v^r\, (ir,\, hk)\, u^i\, \delta x_h\, \delta' x_k;$$

lastly, changing the indices i and r into j and i respectively (to get the term in the same form as the second part of (12′)), we get

$$\sum_{jihk}^{n} {}_1 v^i (ji, hk) u^j \delta x_h \delta' x_k.$$

We can now return to (12′), and taking out the common factor $v^i u^j \delta x_h \delta' x_k$ we get

$$0 = \sum_{1\,jihk}^{n} [(ji, hk) + (ij, hk)] v^i u^j \delta x_h \delta' x_k; \qquad (13)$$

from this, since u^j, v^i, δx_h, $\delta' x_k$ are arbitrary, we get formula (11).

(*c*) Riemann's symbols of the first kind have also a *cyclic property* analogous to that of the symbols of the second kind and immediately deducible from the latter, namely,

$$(ij, hk) + (hj, ki) + (kj, ih) = 0, \quad . \quad . \quad (14)$$

where the second index remains fixed and the other three are permuted cyclically. This formula follows directly from (8), on multiplying by a_{jr} and summing with respect to r.

As each of the terms in this sum is antisymmetrical, we can at once obtain from (14) a similar identity

$$(ij, hk) + (ih, kj) + (ik, jh) = 0, \quad . \quad . \quad (14')$$

in which the first index remains fixed and the other three are permuted cyclically.

(*d*) Lastly, for the symbols of the first kind, there is a property of *permutability*, which is a consequence of the foregoing properties, and according to which we can interchange the two pairs of indices without altering the value of the symbol; namely,

$$(ij, hk) = (hk, ij). \quad . \quad . \quad . \quad . \quad (15)$$

To prove this, take (14′) and the three other identities obtained from it by cyclic permutation of the four indices in the order i, j, h, k, or

$$\underline{(ij, hk)} + (ih, kj) + (ik, jh) = 0,$$

$$(jh, ki) + (jk, ih) + \underline{(ji, hk)} = 0,$$

$$\underline{(hk, ij)} + (hi, jk) + (hj, ki) = 0,$$

$$(ki, jh) + (kj, hi) + \underline{(kh, ij)} = 0.$$

Adding the first and fourth and subtracting the second and third of these identities, and using the property of antisymmetry, we see that the terms cancel out in pairs, except the four underlined, which give

$$2(ij,\, hk) - 2(hk,\, ij) = 0,$$

whence the required property follows.

We shall now calculate the *number of independent Riemann's symbols of the first kind.* A quadruple system (cf. § 2, p. 65) has in general n^4 elements, if there are n independent variables. The number of distinct Riemann's symbols of each kind is smaller, however, as these symbols are connected by the identities we have just proved. We shall determine this number for the symbols of the first kind, dividing them into three classes, and counting separately those in each class, as follows:

(1) Symbols with only two different indices: these are of the type $(ij,\, ij)$, since the other possible arrangements are either reducible to this, or give zero values. Each pair of unequal indices i, j therefore gives a single symbol of this class, which thus contains

$$\frac{n(n-1)}{2};$$

(2) Symbols with three different indices: these are of the type $(ij,\, ih)$, since here too the other possible arrangements are reducible to this or give zero values. Every triplet of unequal indices will give three symbols of this type (since the repeated index may be any one of the three); since there are $\frac{n(n-1)(n-2)}{1\cdot 2\cdot 3}$ triplets, the number of distinct symbols of the class we are considering amounts to

$$\frac{n(n-1)(n-2)}{2};$$

(3) Symbols with four different indices: a set of four different indices i, j, h, k will give the three symbols

$$(ij,\, hk), \quad (ih,\, kj), \quad (ik,\, jh),$$

while every other possible arrangement gives a symbol reducible to one of these. But these three are not independent, on account of the cyclic relation (14′). It follows that each of the

$\frac{n(n-1)(n-2)(n-3)}{1\cdot 2\cdot 3\cdot 4}$ sets of four indices gives two distinct symbols, so that the total number of these is

$$\frac{n(n-1)(n-2)(n-3)}{12}.$$

Adding these three partial totals, and simplifying, we get the total number N of independent Riemann's symbols of the first kind:

$$N = \frac{n^2(n^2-1)}{12}.$$

Thus for an ordinary surface ($n = 2$) we have $N = 1$; for a three-dimensional space, $N = 6$; for a four-dimensional space, $N = 20$.

5. Bianchi's identities.[1]

Bianchi's identities are cyclic relations between the covariant derivatives of Riemann's symbols of both the first and second kinds. They are obtained as follows:

Take formula (3), which defines the symbol of the second kind $\{ir, hk\}$, and differentiate it with respect to x_l. We note, however, that on differentiation the last part, which consists of terms of the second order in Christoffel's symbols, gives terms made up of the product of one such symbol by the derivative of another: the essential point for us is that, with reference to a specified point P, by choosing co-ordinates which are geodesic at that point, we can make all these terms vanish. We cannot, however, treat the first part in the same way, as the geodesic co-ordinates make Christoffel's symbols vanish but not their derivatives. We shall therefore have the formula, valid at the point P for co-ordinates geodesic at P,

$$\frac{\partial}{\partial x_l}\{ir, hk\} = \frac{\partial^2}{\partial x_k \partial x_l}\{ih, r\} - \frac{\partial^2}{\partial x_h \partial x_l}\{ik, r\}.$$

[1] These identities were stated without proof by PADOVA, on the strength of a verbal communication of RICCI (cf. "Sulle deformazioni infinitesime", in *Rend. della R. Acc. dei Lincei*, (4), Vol. V (first half-year, 1889), p. 176). They were then forgotten even by Ricci himself. BIANCHI rediscovered them and published a proof obtained by direct calculation in 1902 (*Ibid.*, (5), Vol. XI (first half-year, 1902), pp. 3–7).

Write down also the two other formulæ obtained from this by cyclic permutation of the indices h, k, l, leaving i and r fixed;

$$\frac{\partial}{\partial x_h}\{ir, kl\} = \frac{\partial^2}{\partial x_l \partial x_h}\{ik, r\} - \frac{\partial^2}{\partial x_k \partial x_h}\{il, r\},$$

$$\frac{\partial}{\partial x_k}\{ir, lh\} = \frac{\partial^2}{\partial x_h \partial x_k}\{il, r\} - \frac{\partial^2}{\partial x_l \partial x_k}\{ih, r\}.$$

Adding the terms on the left and right of these three equations we get

$$\frac{\partial}{\partial x_l}\{ir, hk\} + \frac{\partial}{\partial x_h}\{ir, kl\} + \frac{\partial}{\partial x_k}\{ir, lh\} = 0, \quad (16)$$

which holds at the point P, in the particular system of co-ordinates chosen. Now consider the following mixed tensor of rank five

$$A^r_{ihkl} = \{ir, hk\}_l + \{ir, kl\}_h + \{ir, lh\}_k,$$

in which the suffixes outside the brackets denote covariant differentiation. This system, referred to the point P and to co-ordinates geodesic at P, is identical with the left-hand side of (16), since in these conditions the covariant and ordinary derivatives are identical; all its elements are therefore equal to zero, and it will therefore be identically zero whatever may be the system of reference (cf. Chapter IV, p. 84). We have thus proved the identity

$$\{ir, hk\}_l + \{ir, kl\}_h + \{ir, lh\}_k = 0, \quad . \quad (17)$$

which is a first form of the result established by Bianchi.

For the symbols of the first kind the analogous relation is easily proved from the definition of these symbols given by (5). In fact, taking the covariant derivative of this formula, and using Ricci's lemma, we find that

$$(ij, hk)_l = \sum_{1}^{n}{}_r\, a_{jr}\{ir, hk\}_l.$$

From this, permuting cyclically the indices h, k, l and summing, we get, by (17),

$$(ij, hk)_l + (ij, kl)_h + (ij, lh)_k = 0, \quad . \quad . \quad (17')$$

which is the second form of Bianchi's identity.

6. Commutation rule for the second covariant derivatives.

An important application of Riemann's symbols occurs in the formula which gives the relation between the two systems

$$A^{(j)}_{(i)hk} \quad \text{and} \quad A^{(j)}_{(i)kh},$$

obtained by double covariant differentiation from a generic tensor $A^{(j)}_{(i)}$, where (i) stands for the aggregate of indices of covariance $i_1 \ldots i_m$ and (j) for the aggregate of indices of contravariance $j_1 \ldots j_\mu$. To simplify the formulæ we shall consider a mixed double system A^j_i, with the remark that the procedure is similar if there are more indices.

We start as usual from the bilinear form

$$F = \sum_{1}^{n}{}_{ij} A^j_i \, \xi^i \, u_j,$$

the invariance of which determines the law of transformation of the A's; the ξ^i's are the contravariant components of an arbitrary vector $\boldsymbol{\xi}$, and the u_j's are the covariant components of another arbitrary vector $\mathbf{u}$. The procedure will consist in calculating, in two different ways, the quantity ΔF corresponding to a cyclic displacement round an elementary parallelogram (cf. § 2), and in equating the two expressions so obtained.

A first way of calculating ΔF is as follows. We associate the increments δ, δ' with two sides of the parallelogram, as in § 2, and use (1). Note first that from the definition of the covariant derivative (cf. Chapter VI, p. 146) δF is given by

$$\delta F = \sum_{1}^{n}{}_{ijh} A^j_{i|h} \, \xi^i \, u_j \, \delta x_h.$$

Similarly, applying the symbol δ' to this expression, we shall get

$$\delta'\delta F = \sum_{1}^{n}{}_{ijhk} A^j_{i|hk} \, \xi^i \, u_j \, \delta x_h \, \delta' x_k.$$

From this, interchanging δ and δ', we get

$$\delta\delta' F = \sum_{1}^{n}{}_{ijhk} A^j_{i|hk} \, \xi^i \, u_j \, \delta' x_h \, \delta x_k;$$

subtracting these two equations, after interchanging the indices h and k in the second, we get

$$\Delta F = \sum_{1}^{n}{}_{ijhk} (A^j_{i|hk} - A^j_{i|kh}) \xi^i u_j \delta x_h \delta' x_k. \quad . \quad (18)$$

The other method of calculating this quantity consists in applying the operator Δ directly to the expression for F. Remembering the fundamental properties of Δ (§ 2) we shall get

$$\Delta F = \sum_{1}^{n}{}_{ij} A^j_i (u_j \Delta \xi^i + \xi^i \Delta u_j),$$

or, substituting for $\Delta \xi^i$ and Δu_j the expressions given by (4) and (4′),

$$\Delta F = -\sum_{1}^{n}{}_{ijlhk} A^j_i \{li, hk\} u_j \xi^l \delta x_h \delta' x_k$$

$$- \sum_{1}^{n}{}_{ijphk} A^j_i (pj, hk) u^p \xi^i \delta x_h \delta' x_k.$$

In order to get the second sum in the same form as the first, we shall express it in terms of symbols of the second kind and of the covariant components of $\mathbf{u}$; to do this we first use the property of antisymmetry with respect to the first pair of indices, and express the u^p's in terms of the u_l's, so that the sum becomes

$$+ \sum_{1}^{n}{}_{ijphkl} A^j_i (jp, hk) a^{pl} \xi^i u_l \delta x_h \delta' x_k.$$

Summing with respect to p, and using (5′), we get

$$\sum_{1}^{n}{}_{ijhkl} A^j_i \{jl, hk\} \xi^i u_l \delta x_h \delta' x_k.$$

Now, in order to reconstruct the second expression for ΔF in a suitable form, we shall first interchange some indices, so as to be able to take out the factor $\xi^i u_j \delta x_h \delta' x_k$ which occurs in (18). We must interchange l and i in the first sum, and l and j in the (modified) second; we shall then get, collecting both sums under a single summation sign,

$$\Delta F = -\sum_{1}^{n}{}_{ijhkl} \left[A^j_l \{il, hk\} - A^l_i \{lj, hk\} \right] \xi^i u_j \delta x_h \delta' x_k.$$

Comparing this expression with (18), and remembering that $\boldsymbol{\xi}$, **u**, the δx's, and the $\delta' x$'s are arbitrary, we get the *commutation formula*

$$A^j_{i|hk} - A^j_{i|kh} = -\sum_1^n{}_l \Big[A^j_l \{il, hk\} - A^l_i \{lj, hk\} \Big]. \quad (19)$$

If the system from which we start has m indices of covariance and μ of contravariance, we must consider m vectors $\boldsymbol{\xi}$, determined by their contravariant components, and μ vectors **u**, determined by their covariant components; by an analogous process we shall find

$$\left. \begin{aligned} A^{(j)}_{(i)|hk} - A^{(j)}_{(i)|kh} = -\sum_1^n{}_l \Big[\sum_1^m{}_r A^{(j)}_{i_1 \ldots i_{r-1} l i_{r+1} \ldots i_m} \{i_r l, hk\} \\ - \sum_1^\mu{}_\rho A^{j_1 \ldots j_{\rho-1} l j_{\rho+1} \ldots j_\mu}_{(i)} \{l j_\rho, hk\} \Big]. \end{aligned} \right\} \quad (20)$$

7. Cyclic displacement round any infinitesimal circuit.

We now return to the order of ideas interrupted at § 2. Given a direction **u** at a point P, we shall give **u** a parallel displacement (in V_n) round a closed curve T, infinitesimal, but of any shape whatever, and of course passing through P; we propose to calculate the change Du^r in a generic parameter of **u** caused by the cyclic displacement. The formula we shall find will be merely a generalization of (4), and must reduce to (4) if as a particular case we take for T an infinitesimal parallelogram.

For an elementary displacement dx_h we know that the change in u^r is

$$du^r = -\sum_1^n{}_{ih} \{ih, r\} u^i dx_h,$$

so that it has the form of a Pfaffian

$$\psi_r = \sum_1^n{}_h X_{rh} dx_h, \quad . \quad . \quad . \quad . \quad (21)$$

in which the X's are functions of position (since they contain Christoffel's symbols) and of the u^i's which are defined *along the curve* T by the equations of parallelism $du^r = \psi_r$. We have to calculate the integral

$$Du^r = \int_T \psi_r = \int_T \sum_1^n {}_h X_{rh}\, dx_h, \quad . \quad . \quad (22)$$

where we use the operator D to indicate the increment resulting from displacement right round the circuit. We shall now consider any surface σ containing the curve T, and we shall call Γ the region of this surface which is within T, and is such that T constitutes its complete boundary. We propose to transform the integral round the circuit T, which occurs in (22), into an integral over the surface Γ. To do this, we shall first introduce a system of co-ordinates q_1 and q_2 on the surface in question, defining them by the parametric equations

$$x_i = x_i(q_1, q_2) \qquad (i = 1, 2, \ldots n).$$

The dx's will consequently be linear functions of dq_1 and dq_2; substituting their expressions in the Pfaffian (21), this will take the form

$$du^r = \psi_r = Q_1\, dq_1 + Q_2\, dq_2, \quad . \quad . \quad (21')$$

where the quantities Q_1 and Q_2, like the X's, are defined along the curve T. The integral to be calculated will thus be

$$Du^r = \int_T (Q_1\, dq_1 + Q_2\, dq_2). \quad . \quad . \quad (22')$$

We shall suppose the curvilinear co-ordinates q_1, q_2 so chosen that the sense of integration round T is the same as that determined on Γ (at a generic point) by the rotation (through the angle less than 180°) from the positive direction of the line q_1 (i.e. in the sense of q_1 increasing) to the positive direction of the line q_2.

The transformation of the line integral (22′) into a surface integral taken over Γ could be effected at once if the Q's were defined as functions of position in the interior of Γ as well as on its boundary. But instead of this they contain the u^r's, which are given at P, and at points on T have values resulting from the parallel displacement along T itself, but are not defined for a point M within Γ, their values at M depending on the path followed in the parallel displacement of **u** from P to M. We shall, however, show that if Γ is infinitesimal the influence of the path of displacement on the values of the u^r's at M is negligible, and therefore we may consider the u^r's, and in consequence the X's or the Q's, as functions of position over the whole area Γ, which will enable us to make the required transformation of (22').[1]

We shall make some preliminary remarks on orders of magnitude, using for this purpose the general existence theorem for integrals of systems of ordinary differential equations. Such a system is constituted by the equations

$$du^r = \psi_r \qquad (r = 1, 2, \ldots n),$$

which define the functions u^r, along a generic line T, starting from a given set of initial values at P (cf. Chapter II, p. 23).

Now the existence theorem assures us that in general (i.e. when certain not very restrictive conditions of continuity and differentiability are satisfied) the initial values define the integrals uniquely, and these integrals and their derivatives are continuous functions for values of the independent variable within an interval which is not shorter than some assignable quantity.

In our case, granting, what will naturally be the case, that the coefficients a_{ik} of ds^2 and the reciprocals a^{ik} are finite and continuous, as well as their first and second derivatives, in a certain region round P, and supposing also that the length of the vector **u** at P is limited, i.e. is not greater than some specified, but arbitrary, constant U, it can easily be deduced from the above-mentioned existence theorem that—considering the arc of the curve of displacement as the independent variable—the u^r's are defined (as continuous, differentiable, &c., functions), with P as starting-point, along any curve whatever, for a segment of the curve of length not greater than a certain quantity Λ

[1] We shall in fact here limit the discussion to an indication of the general lines of the argument, without pausing over the details needed to justify the various steps of the proof with complete rigour. There is an exhaustive proof in the article by H. Tietze: "Ueber Parallelverschiebung in Riemann'schen Raümen", in *Math. Zeitschrift*, Vol. 16, 1923, pp. 308–317.

which depends exclusively on the metric of the manifold and on U. Hence it follows in particular that the differences between the u^r's and their initial values are of the same order[1] of magnitude as the length L of the arc along which the integration of the system $du^r = \psi_r$ is effected.

Further, the area Γ which we are considering on the surface σ is infinitesimal, in the sense that we propose to make it diminish indefinitely. It is therefore perfectly legitimate to suppose that it is already so small that every point in it can be reached from P by a line of length not greater than Λ and that the length of the whole contour T is also less than Λ.

We thus have the result that if **u** undergoes a parallel displacement from P to a point M within the area Γ, or on its boundary T, the u^r's at M differ from their values at P by quantities of the same order as L, if $L (\leqslant \Lambda)$ is the maximum length of the lines considered. As a first approximation, in which quantities of the same order as L are neglected, we can therefore take the components u^r as constant and $= u^{r(0)}$ over the whole area Γ, including its boundary.

We can find a closer approximation if we calculate the u^r's at M by integrating the Pfaffian (21) along a curve PQ, substituting, however, for the coefficients X_{rh} their values at P. This process involves an error of order L in the values of these coefficients, and therefore an error of order L^2 in the values found for the u^r's at M; the choice of the curve PM is indifferent, as in this case the Pfaffian becomes an exact differential. As the coefficients are constants, the integration can be at once effected, and will give linear functions of the x's for the u^r's. We shall thus have obtained the u^r's as functions of position at all points (including the boundary) of the area Γ, neglecting quantities of order L^2.

We can get a third approximation by substituting these approximate linear expressions of the u^r's in the coefficients X_{rh}. The X's will thus become functions of position, as was required, defined throughout Γ, including its boundary T, and coinciding on T (if we neglect L^2) with their accurate values as already defined. These values of the X's can be used to

[1] This means that the differences in question are not greater than the product of L by a certain *finite* coefficient, which does not depend on L or on the curve of integration, but only on P, U, and the metric of the manifold.

calculate the integral (22), which will give the value of Du^r, the error being now of order L^3. It was necessary to carry the approximation thus far, since the other two approximations make ψ an exact differential, and would give the value zero for Du^r, the obvious meaning of this being that Du^r is a quantity of order L^2.

We shall therefore return to (22), giving the X's the meaning just explained, and transform it into the form (22′) by using the parametric equations of the surface σ; Q_1 and Q_2 will now represent functions of position defined at all points of Γ. In consequence[1] we can transform the line integral into a surface integral, getting

$$Du^r = \int_\Gamma \left(\frac{\partial Q_2}{\partial q_1} - \frac{\partial Q_1}{\partial q_2}\right) dq_1\, dq_2,$$

or also

$$Du^r = \int_\Gamma \left[\frac{\partial}{\partial q_1}(Q_2\, dq_2)\, dq_1 - \frac{\partial}{\partial q_2}(Q_1\, dq_1)\, dq_2\right]. \qquad (23)$$

We must now find the value of the integrand on the right of (23), which can be done by means of the following considerations, without writing out the expressions for Q_1 and Q_2 at full length.

Let the operator δ denote the increment of a quantity corresponding to a displacement along the line $q_2 =$ constant, when

[1] By the ordinary formulæ,

$$\int_\Gamma \frac{\partial f}{\partial q_1} dq_1\, dq_2 = \int_T f dq_2, \quad \int_\Gamma \frac{\partial f}{\partial q_2} dq_1\, dq_2 = -\int_T f dq_1, \quad . \quad (G)$$

where f is a function of q_1 and q_2 which is continuous, together with its first derivatives. Usually in these formulæ q_1 and q_2 are interpreted as Cartesian co-ordinates in a plane, and the sense in which the curve T is described is defined by the condition that the pair of directions s, n (s the tangent to the boundary in the sense in which it is described, n the normal to T drawn inwards) is congruent in the plane with the pair q_1, q_2. The formulæ obviously hold, however, independently of this interpretation in plane geometry, and can therefore be applied even if q_1, q_2 are any curvilinear co-ordinates whatever, the sense of describing the curve being determined by an analogous criterion to that just explained, provided that we introduce, at a generic point of the boundary curve, the directions tangential to q_1 (i.e. $q_2 =$ constant) and q_2, in the sense in which the respective parameters increase. Now we have already supposed (p. 187) that the auxiliary curvilinear co-ordinates q_1, q_2 behave as regards sense in precisely this way. Hence the equations (G) hold both in magnitude and in sign.

q_1 increases by dq_1; similarly, δ' corresponds to an increase dq_2 in q_2 alone. We shall therefore have

$$\left.\begin{aligned} \delta x_i &= \frac{\partial x_i}{\partial q_1} dq_1, \\ \delta' x_i &= \frac{\partial x_i}{\partial q_2} dq_2, \end{aligned}\right\} \quad . \quad . \quad . \quad . \quad . \quad (24)$$

and analogous expressions for any function of position. Now note that in (21′) the first term represents precisely the increment of u^r due to a displacement of the first type ($dq_2 = 0$), so that

$$\delta u^r = Q_1\, dq_1,$$

and, the second term having an analogous meaning,

$$\delta' u^r = Q_2\, dq_2.$$

We can therefore write (23) in the form

$$\begin{aligned} Du^r &= \int_\Gamma \left[\frac{\partial}{\partial q_1}(\delta' u^r)\, dq_1 - \frac{\partial}{\partial q_2}(\delta u^r)\, dq_2\right] \\ &= \int_\Gamma [\delta\delta' u^r - \delta'\delta u^r]; \end{aligned}$$

and from this, remembering equation (4), we get

$$Du^r = \int_\Gamma \sum_{1}^{n}{}_{ihk} \{ir, hk\}\, u^i\, \delta x_h\, \delta' x_k.$$

Now let us put ξ, η for the parameters (in V_n) of the lines $q_2 =$ constant, $q_1 =$ constant, i.e.

$$\xi^i = \frac{\delta x_i}{\delta s}, \quad \eta^i = \frac{\delta' x_i}{\delta' s},$$

where δs, $\delta' s$ are the lengths of the displacements, along the lines $q_2 =$ constant, $q_1 =$ constant, whose components are δx_i, $\delta' x_i$. Then we can also write

$$Du^r = \int_\Gamma f_r\, \delta s\, \delta' s, \quad . \quad . \quad . \quad . \quad . \quad (25)$$

where for shortness we have put

$$f_r = \sum_{ihk}^{n} {}_1 \{ir, hk\} \, u^i \, \xi^h \, \eta^k .$$

We can now remark that if ϑ is the angle between the coordinate lines q_1, q_2, the element of area (cf. Chapter V, p. 99) is

$$d\Gamma = \delta s \, \delta' s \sin\vartheta,$$

and therefore (25) becomes

$$Du^r = \int_\Gamma \frac{f_r}{\sin\vartheta} \, d\Gamma .$$

By the mean value theorem, if $\left[\frac{f_r}{\sin\vartheta}\right]_{M_0}$ denotes the value of the function of position $\frac{f_r}{\sin\vartheta}$ at a suitable point M_0 (not known *a priori*) within Γ, and $D\Gamma$ the area of the region Γ, we can also write

$$Du^r = \left[\frac{f_r}{\sin\vartheta}\right]_{M_0} D\Gamma .$$

Now the value at M_0 of the function of position $\frac{f_r}{\sin\vartheta}$ differs from its value at P by quantities of order L (since the distance M_0P is of this order); since the area $D\Gamma$ is of order L^2, the error caused by substituting the value at P for that at M is of order L^3, which we have agreed to neglect. We shall therefore have

$$Du^r = \frac{f_r}{\sin\vartheta} D\Gamma ,$$

or

$$Du^r = \frac{D\Gamma}{\sin\vartheta} \sum_{ihk}^{n} {}_1 \{ir, hk\} \, u^i \, \xi^h \, \eta^k . \quad . \quad . \quad (26)$$

In this formula the quantities ξ^h and η^k represent the parameters of the lines q_1, q_2 at P, and ϑ is the angle between these lines; the values of the u^i's and of the Riemann's symbols refer to the point P. It will be seen that the influence of the circuit of displacement appears in this formula in three geometrical elements, which serve substantially to determine the circuit itself, namely: two directions $\boldsymbol{\xi}$, $\boldsymbol{\eta}$ (*a priori* any whatever) which determine the section on which the circuit is supposed drawn, together with the angle between them; and the area $D\Gamma$ of the circuit itself (measured according to the metric of V_n).

8. **Pérès's formula.**

From (26) we immediately get the fundamental formula which serves as a link between parallelism and curvature.

Take any (fourth) direction $\mathbf{v}$ drawn from P. Let α be the angle between $\mathbf{v}$ and $\mathbf{u}$, and $\alpha + D\alpha$ the angle between $\mathbf{v}$ and $\mathbf{u} + D\mathbf{u}$; we propose to calculate $D\alpha$. To do this we take the scalar product

$$\mathbf{u} \times \mathbf{v} = \cos\alpha$$

(assuming that $\mathbf{u}$, like $\mathbf{v}$, is a unit vector), and differentiate with the symbol D, remembering that $D\mathbf{v} = 0$ since $\mathbf{v}$ is a fixed vector. We get

$$\mathbf{v} \times D\mathbf{u} = -\sin\alpha \, D\alpha,$$

or, substituting for the scalar product on the left its expression $\sum_1^n{}_r v_r \, Du^r$, and using (26),

$$\sin\alpha \, D\alpha = -\frac{D\Gamma}{\sin\vartheta} \sum_1^n{}_{irhk} \{ir, hk\} \, u^i v_r \xi^h \eta^k.$$

If in this formula we express the v_r's in terms of the v^j's, and sum with respect to r (remembering (5)), we get

$$\sin\alpha \, D\alpha = -\frac{D\Gamma}{\sin\vartheta} \sum_1^n{}_{ijhk} (ij, hk) \, u^i v^j \xi^h \eta^k.$$

Now if the directions $\mathbf{u}$, $\mathbf{v}$ coincide or are opposite, this formula reduces to an identity of no interest, since on the left we have $\sin\alpha = 0$, and the right-hand side vanishes from the antisymmetry of the Riemann's symbols. Excluding this case, we can divide the whole equation by $\sin\alpha$, and we get *Pérès's formula*

$$D\alpha = \frac{-D\Gamma}{\sin\alpha \sin\vartheta} \sum_1^n{}_{ijhk} (ij, hk) \, u^i v^j \xi^h \eta^k. \quad . \quad (27)$$

9. **Application to surfaces. Gaussian curvature of a V_2.**

In considering the particular case of a V_2, i.e. an ordinary surface, the directions $\mathbf{u}$, $\mathbf{v}$ must of course be contained in the section (the only one there is) defined by $\boldsymbol{\xi}$, $\boldsymbol{\eta}$. Since the only purpose of these last two vectors is to specify the section on which the circuit is drawn, we can make them coincide with $\mathbf{u}$ and $\mathbf{v}$ without loss of generality; (27) will then become

$$Da = \frac{-D\Gamma}{\sin^2\alpha} \sum_{ijhk}^{n}{}_{1} (ij, hk)\, u^i v^j u^h v^k. \quad . \quad . \quad . \qquad (27')$$

Since for $n = 2$ Riemann's symbols which do not vanish are represented by the single arrangement of indices (12, 12), this formula can be further reduced to

$$Da = \frac{-D\Gamma}{\sin^2\alpha} (12, 12)\, (u^1 v^2 - u^2 v^1)^2;$$

or finally, remembering the expression for $\sin\alpha$ in terms of the parameters of the two directions $\mathbf{u}$, $\mathbf{v}$ (cf. Chap. V, p. 94, formula (9′)), we get

$$Da = -\frac{D\Gamma}{a} (12, 12).$$

It is usual to put

$$\frac{(12, 12)}{a} = K, \quad . \quad . \quad . \quad . \quad . \quad . \qquad (28)$$

so that the foregoing formula becomes

$$\frac{Da}{D\Gamma} = -K. \quad . \quad . \quad . \quad . \quad . \quad . \qquad (29)$$

From this it will be seen that the function of position K defined by (28) is an invariant; it depends on the coefficients a_{ik} and their first and second derivatives, and is identical with the quantity which in the theory of surfaces is known as the *total*, or *Gaussian, curvature* (the product of the curvatures of the principal sections).[1] The equation (29) can be put in a more instructive form if we introduce the *angle of parallelism* ϵ, i.e. the angle between $\mathbf{u}$ and $\mathbf{u} + D\mathbf{u}$ (or between $\mathbf{u}$ and its parallel obtained by displacement round the circuit), measured in the sense in which the circuit is described. We can also say that ϵ is the angle through which $\mathbf{u}$ has been rotated as a result of the cyclic dis-

[1] See also below, Chapter IX, p. 261.

placement. It is obvious that ϵ has the same absolute value as $D\alpha$, but we shall see that as regards sign the precise relation is

$$\epsilon = -D\alpha.$$

To show this we need only remember the convention adopted above (§ 7) that the circuit is to be described in the positive sense with respect to the co-ordinate lines q_1, q_2, or from the versor $\boldsymbol{\xi}$ to the versor $\boldsymbol{\eta}$. As $\boldsymbol{\xi}$ and $\boldsymbol{\eta}$ now coincide with $\mathbf{u}$ and $\mathbf{v}$ the sense in which the circuit is described is from $\mathbf{u}$ to $\mathbf{v}$ (through the convex angle). Accordingly (see fig. 3) if $D\alpha > 0$, $\mathbf{u} + D\mathbf{u}$ is outside the convex angle $\widehat{\mathbf{uv}}$, and is therefore reached from $\mathbf{u}$ by moving in the negative sense ($\epsilon < 0$), and vice versa. We can therefore write (29) in the form

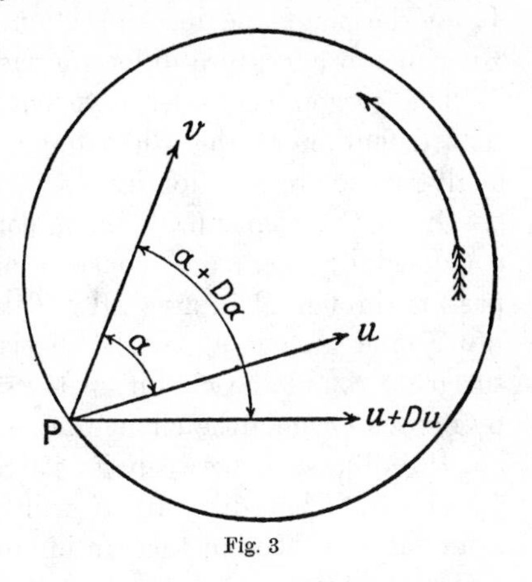

Fig. 3

$$\frac{\epsilon}{D\Gamma} = K, \quad . \quad . \quad . \quad . \quad . \quad . \qquad (29')$$

which gives the following important interpretation of the curvature K: K *is the ratio of the angle of parallelism* (taken with its proper sign in relation to the sense of describing the circuit) *to the area of the circuit.*

In the case of a Euclidean V_2 Riemann's symbol is zero (cf. § 4) and therefore $K = 0$. This can also be deduced from the geometrically obvious fact (already used in § 2) that the parallel displacement is integrable (i.e. that the result does not depend on the curve of displacement).

10. Riemannian curvature of a V_n.

If instead of a surface we consider any V_n whatever, the notion of curvature becomes less simple. If P is a fixed point of the V_n, then with every section through P determined by two arbitrary directions $\boldsymbol{\xi}$, $\boldsymbol{\eta}$ drawn from P we can associate an invariant K, which is called the *Riemannian curvature* of the V_n at P with respect to the section considered. Following Riemann, we construct the geodesic surface determined by the point P and the two directions $\boldsymbol{\xi}$, $\boldsymbol{\eta}$, and then take the Gaussian curvature K of this geodesic surface as the curvature of the V_n at the point and in the section in question. In general the Riemannian curvature differs in the different sections.

The foregoing considerations enable us to give another important definition of the Riemannian curvature, and to find the analytical expression for it.

Given the elements P, $\boldsymbol{\xi}$, $\boldsymbol{\eta}$, construct the geodesic surface g defined by them and consider an infinitesimal circuit on g, passing through P, of area $D\Gamma$. Take one of the given directions, say $\boldsymbol{\xi}$, and give it a parallel displacement with respect to the surface g round the circuit, in the sense $\boldsymbol{\xi} \rightarrow \boldsymbol{\eta}$. Now calculate by Pérès's formula the change $D\alpha$ in the angle between $\boldsymbol{\xi}$ and $\boldsymbol{\eta}$, i.e. the difference between its values before and after the displacement. The curvature K will then be given by (29). Now from Severi's theorem that an infinitesimal parallel displacement with respect to the surface g (using the metric of g) can be replaced by the analogous infinitesimal parallel displacement in V_n, it follows at once that this method of calculating K does not really involve the use of the geodesic surface g. Hence the Riemannian curvature K can be defined as *the ratio* (*with sign changed*) *of* $\mathrm{D}\alpha$ *to* $\mathrm{D}\Gamma$, *where* $\mathrm{D}\alpha$ *is the change in the angle between the given directions* $\boldsymbol{\xi}$, $\boldsymbol{\eta}$ *caused by the parallel displacement in* V_n *of one of these directions round an infinitesimal circuit of area* $\mathrm{D}\Gamma$, *belonging to the section* $\boldsymbol{\xi}$, $\boldsymbol{\eta}$, and described in the sense $\boldsymbol{\xi} \rightarrow \boldsymbol{\eta}$. We therefore have

$$K = -\frac{D\alpha}{D\Gamma}, \quad . \quad . \quad . \quad . \quad . \quad (30)$$

as in the V_2.

The explicit expression for K corresponding to this can be obtained from (27′), and we get

$$K = \frac{1}{\sin^2\alpha} \sum_{1}^{n}{}_{ijhk} \, (ij, hk) \, u^i v^j u^h v^k. \quad . \quad . \quad (31)$$

The symmetry of the right-hand side in $\mathbf{u}$ and $\mathbf{v}$ provides formal confirmation of the fact that it is immaterial whether we displace **u** or **v**, as we get in either case the same value for Da and the same value for K.

A third definition of the Riemannian curvature can be obtained from the proof of the following lemma.

Take any three points P, P', P'' on a surface V_2, and join them in pairs by arcs of geodesics, forming a so-called *geodesic triangle*. If ϕ, ϕ', ϕ'' are the angles of this triangle, the quantity

$$\epsilon = \phi + \phi' + \phi'' - \pi \quad . \quad . \quad . \quad . \quad . \quad (32)$$

is called the *geodesic excess*.

Now take any point on the triangle, and give a parallel displacement round the triangle to the direction of the side at that point, or of one of the sides if the point is a vertex; e.g. starting from P, the direction **u** at P of the side PP' (taken in the sense $P \rightarrow P'$).

We want to show that the angle between the initial and final positions of **u** (measured from the initial position in the sense which at each vertex is dependent on the sense $PP'P''$ in which the circuit is described), i.e. *the angle of parallelism* (relative in this case to a circuit of a special kind, but without the restriction of being infinitesimal), *is the same as the geodesic excess* ϵ.

For the proof we shall follow **u** in a cyclic displacement round the circuit $PP'P''$, noting in the first place that from P to P' **u** remains tangential to the side PP', on account of the autoparallelism of geodesics (cf. Chapter V, p. 104). On arriving at P', **u** is thus inclined at an angle $\pi - \phi'$ (outside the triangle at P') to the side $P'P''$ (in the sense indicated by these letters); more precisely, **u** is behind the tangent to the side $P'P''$ by an angle of $\pi - \phi'$, the sense of rotation at P', as we have already said, being determined by the sense of description $PP'P''$. In the displacement from P' to P'' this angle remains unchanged; at P'' there will be a further loss of $\pi - \phi''$ (with respect to the new side $P''P$); and finally at P yet another loss of $\pi - \phi$ (with respect to PP'). Taking all these together, we see that in its final position the parallel to **u** has been rotated away from its initial position through an angle of $3\pi - (\phi + \phi' + \phi'')$ in the negative direction, or $\epsilon - 2\pi$ in the positive direction. Now in the pencil of directions at a point an angle is determined geometrically by a quantity of the form $\theta \pm 2n\pi$, where n is any integer; hence we have proved that *one value of the angle of parallelism is the geodesic excess* ϵ. Now in a Euclidean space the geodesic excess is zero, and in any manifold whatever, for an infinitesimal triangle, the excess is infinitesimal; we therefore see that for reasons of continuity the value adopted for the angle of parallelism is in fact the most suitable, being that which tends to zero with the triangle.

The lemma just proved is rigorously true in a V_2, whatever may be the magnitude of the geodesic triangle considered. If in particular we apply it to an infinitesimal triangle, we can substitute ϵ for $-Da$ in (29), so getting

$$K = \frac{\epsilon}{D\Gamma},$$

a formula which defines the Riemannian curvature as *the ratio of the geodesic excess to the area of an infinitesimal geodesic triangle lying in the section considered, and having one vertex at the given point* P. It will be seen that this is an obvious extension to n-dimensional manifolds with any metric of an elementary theorem in spherical geometry (the area of a spherical triangle = the spherical excess × the square of the radius); the latter theorem, however, unlike the former, holds also for a finite triangle.

We must confine ourselves to a mere reference to the important researches of Professors Schouten[1] and Bompiani[2] on the simultaneous cyclic displacement of several directions, and even of the whole sheaf of directions drawn from a single point. Their work throws light on the theory of Riemannian curvature under various new aspects.

11. Case of a V_3. The tensors a_{ik} of Ricci and G_{ik} of Einstein.

For a manifold of three dimensions the symbols of the first species (ij, hk) which do not vanish reduce substantially, in virtue of (10), (11), and (14), to the scheme

$$(i+1\ \ i+2,\ k+1\ \ k+2) \qquad (i, k = 1, 2, 3),$$

[1] See references given in note (1), p. 172.

[2] "Studi sugli spazi curvi", in *Atti del R. Ist. Veneto*, Vol. LXXX, 1921, pp. 355–386, 839–859, 1113–1145.

with the convention (p. 161, § 9) of regarding as equivalent two indices which differ by a multiple of 3.

Using this notation, we now introduce after Ricci the double symmetric system

$$a^{ik} = \frac{(i+1 \;\; i+2, \;\; k+1 \;\; k+2)}{a} \qquad (i, k = 1, 2, 3), \quad (33)$$

which constitutes, as we shall now show, a *contravariant tensor.* In fact, if we make use of the contravariant system ϵ (Chapter VI, p. 158) we see that (33) is equivalent to

$$a^{ik} = \tfrac{1}{4}\sum_{pqrs}^{3}{}_{1}\, \epsilon^{ipq}\,\epsilon^{krs}\,(pq, rs) \qquad (i, k = 1, 2, 3): \quad . \; (33')$$

which proves the assertion. The verification of this last formula is immediate, when we remember that of the various determinations *a priori* possible for the pair p, q, there are only two,

$$p = i+1, \;\; q = i+2$$

and

$$p = i+2, \;\; q = i+1,$$

corresponding to which ϵ^{ipq} has a value different from zero, viz. $\frac{1}{\sqrt{a}}$ in the former case, $-\frac{1}{\sqrt{a}}$ in the latter. Similarly, for the other pair r, s, the only determinations to which there corresponds a non-vanishing ϵ^{krs} are

$$r = k+1, \;\; s = k+2$$

and

$$r = k+2, \;\; s = k+1.$$

The sum $\sum_{1}^{3}{}_{pqrs}$ reduces therefore to four terms all expressible as

$$\frac{(i+1 \;\; i+2, \;\; k+1 \;\; k+2)}{a},$$

thus giving the result stated in (33).

The name *Ricci's symbols* is sometimes given to the a^{ik}'s just defined, or, more specifically, to *the elements*

$$a_{ik} = \sum_{1}^{3}{}_{jh}\, a_{ij}\, a_{hk}\, a^{jh} \qquad (i, k = 1, 2, 3) \quad . \quad (34)$$

of the reciprocal covariant tensor.

It is worth noting that equations (33), if we take account of the properties of Riemann's symbols in respect of symmetry and antisymmetry, can be solved so as to give the values of these symbols in the form

$$(ij, hk) = \sum_{\nu\rho}^{3} {}_{1} a^{\nu\rho} \epsilon_{\nu ij} \epsilon_{\rho hk}. \quad . \quad . \quad . \quad (33'')$$

In fact, expanding the second member and attending to the definition of the covariant system ϵ, we find the value zero corresponding to every set of four i, j, h, k for which the first member vanishes, and the value aa^{ik} corresponding to a set of four of the type $i+1, i+2, k+1, k+2$; (33″) therefore follows from (33).

By contracting the Ricci tensor a by means of the fundamental tensor (the coefficients of ds^2, or their reciprocals) we obtain the *linear invariant* of the tensor a

$$\mathcal{M} = \sum_{ik}^{3} {}_{1} a_{ik} a^{ik} = \sum_{ik}^{3} {}_{1} a^{ik} a_{ik}. \quad . \quad . \quad . \quad (35)$$

We may point out another formal relation of which use will be made in Chapter XII. For any V_n whatever we can derive from the Riemannian tensor, by contraction with respect to two indices, the covariant double tensor

$$G_{ik} = \sum_{jh}^{n} {}_{1} a^{jh} (ij, hk), \quad . \quad . \quad . \quad . \quad (36)$$

which, in virtue of equations (5), can also be written in the form

$$G_{ik} = \sum_{h}^{n} {}_{1} \{ih, hk\}. \quad . \quad . \quad . \quad . \quad (36')$$

This tensor was noticed by Ricci, who applied it to the study of the local distribution of curvatures in a V_n; it was afterwards taken up by Einstein, who gave it a fundamental place in the theory of relativity (in which $n = 4$): it is commonly known as the *Einstein tensor*.

For $n = 3$ the a_{ik}'s are related in a simple way to the G_{ik}'s. To bring out the connexion simply and neatly, it is convenient to make use of two properties of the ternary systems ϵ: one expressed by the identity (which can be verified immediately)

$$\sum_{j}^{3} {}_{1} \epsilon^{pqj} \epsilon_{rsj} = \delta_r^p \delta_s^q - \delta_s^p \delta_r^q \qquad (p, q, r, s = 1, 2, 3), \quad . \quad (37)$$

the δ's having the usual meaning (0 for different indices, 1 for equal indices); and the other translating, as it were, the definition of a^{jh} as a reciprocal element

$$a^{jh} = \frac{1}{2}\sum_{1}^{3}{}_{pqrs}\,\epsilon^{pqj}\,\epsilon^{rsh}\,a_{pr}\,a_{qs}.$$

Substitute in equations (36), in place of the a^{jh}'s, the second member here; and, in place of the (ij, hk)'s, the expression for them in (33″), with $\epsilon_{\rho hk}$ changed into $-\epsilon_{\rho kh}$. Taking account of (37) we find

$$G_{ik} = \frac{1}{2}\sum_{1}^{3}{}_{pqrs\nu\rho}\,a_{pr}\,a_{qs}\,a^{\nu\rho}\,(\delta^p_\nu\,\delta^q_i - \delta^p_i\delta^q_\nu)\,(\delta^r_k\,\delta^s_\rho - \delta^r_\rho\,\delta^s_k).$$

Of the four terms obtained by expanding the product of the bracketed expressions, the two which are positive are merely interchanged by interchange of p with q and of r with s, which does not alter the product $a_{pr}\,a_{qs}$; and similarly with the negative terms. We can therefore confine our attention to one term only of each kind, and suppress the factor $\frac{1}{2}$. If we take, e.g.

$$\delta^p_\nu\,\delta^q_i\,(\delta^r_k\,\delta^s_\rho - \delta^r_\rho\,\delta^s_k),$$

and bear in mind the meaning of the symbols δ, we find that the sum reduces to one with respect to ν and ρ only, giving

$$G_{ik} = \sum_{1}^{3}{}_{\nu\rho}\,(a_{\nu k}\,a_{i\rho}\,a^{\nu\rho} - a_{\nu\rho}\,a_{ik}\,a^{\nu\rho}),$$

or finally, having regard to (34), (35),

$$G_{ik} = \alpha_{ik} - \mathcal{M}\,a_{ik} \qquad (i, k = 1, 2, 3). \quad . \quad (38)$$

12. **Curvature of a manifold of three dimensions around a point. Principal directions and invariants.**

Consider in a V_3 a generic section or *facet*, f, defined by two directions (versors) **u**, **v**, whose parameters are u^i, v^i $(i = 1, 2, 3)$, issuing from the same point P. Let **w** be their vector product (Chapter VI, p. 159), the moments of which are

$$w_\nu = \sum_{1}^{3}{}_{ij}\,\epsilon_{\nu ij}\,u^i\,v^j \qquad (\nu = 1, 2, 3). \quad . \quad . \quad (39)$$

Corresponding to the section f in our manifold V_3 we have the

Riemannian curvature K given by (31), with $n = 3$. In the sum $\sum_1^3{}_{ijhk}$ of (31) it is convenient to introduce, in place of Riemann's symbols (ij, hk) the expressions for them given in (33″). Taking account of (39), we find immediately

$$K = \frac{1}{\sin^2\alpha} \sum_1^3{}_{\nu\rho}\, a^{\nu\rho}\, w_\nu\, w_\rho. \quad . \quad . \quad . \quad (40)$$

Deferring for a moment the illustration of this formula, we recall the fact that in general the length w of the vector $\mathbf{u} \wedge \mathbf{v}$ (the vector product of $\mathbf{u}$ and $\mathbf{v}$) is given by the product of the lengths of $\mathbf{u}$, $\mathbf{v}$ by the sine of the angle between them. For the Euclidean V_3 this implies, as already mentioned (Chapter VI, p. 159), the identity of the moments (39) referred to Cartesian co-ordinates with the ordinary components (orthogonal projections) of the vector product $\mathbf{u} \wedge \mathbf{v}$. For a general V_3, it is sufficient to remember that we can choose co-ordinates which are *locally Cartesian* at any assigned point P, so that the a_{ik}'s have the values δ_i^k (and so that also—though this is not important for the present purpose—their first derivatives vanish). Now both in the measures of the lengths and the angular distances of vectors proceeding from the point P, and in the definition (39) of the w_ν's, there enter only the components of the vectors and the values of the a_{ik}'s at P. Locally then everything is the same as for Euclidean space.[1]

Turn now to our case, in which $\mathbf{u}$ and $\mathbf{v}$ are unit vectors. It follows that $w = \sin\alpha$, whence $\frac{w_i}{w} = \frac{w_i}{\sin\alpha}$ constitute the

[1] We can also of course calculate the length of $\mathbf{w}$. Thus we note that, besides (39), we have the equivalent formulæ defining the contravariant components:

$$w^\nu = \sum_1^3{}_{hk}\, \epsilon^{\nu hk}\, u_h\, v_k.$$

Hence, in view of (37) (the index of summation in which we may suppose transferred to the first place),

$$\begin{aligned} w^2 &= \sum_1^3{}_\nu\, w^\nu w_\nu = \sum_1^3{}_{ijhk}\, u^i v^j u_h v_k \left(\delta_i^h \delta_j^k - \delta_j^h \delta_i^k\right) \\ &= \sum_1^3{}_i\, u^i u_i \sum_1^3{}_j\, v^j v_j - \sum_1^3{}_i\, u^i v_i \sum_1^3{}_j\, v^j u_j \\ &= u^2 v^2 - (uv\cos\alpha)^2 = u^2 v^2 \sin^2\alpha. \end{aligned}$$

moments λ_i of the direction $\boldsymbol{\lambda}$ normal to the section f. The *sense* assigned to the normal by (39) is characterized by the system ϵ or, geometrically, by the sense of the trihedron formed at P by the positive directions of the co-ordinate lines. In fact, from these equations (39), if we suppose, for example, that the lines 1, 2 are taken in the directions $\mathbf{u}$, $\mathbf{v}$, there results

$$w_1 = w_2 = 0, \qquad w_3 = \frac{\sqrt{a}}{\sqrt{a_{11}\, a_{22}}} > 0,$$

so that (Chapter V, p. 127) $\mathbf{w}$ makes an acute angle with the positive direction of the line 3, whose parameters are

$$s_3^1 = s_3^2 = 0, \qquad s_3^3 = \frac{1}{\sqrt{a_{33}}} > 0.$$

Thus $\boldsymbol{\lambda}$ is perpendicular to f at P, and directed so that the trihedron $\mathbf{u}$, $\mathbf{v}$, $\boldsymbol{\lambda}$ has the same sense as the trihedron formed by the positive directions of the co-ordinate lines at the same point P.

Equation (40) now takes the form, given by Ricci,

$$K = \sum_1^3{}_{\nu\rho}\, a^{\nu\rho}\, \lambda_\nu\, \lambda_\rho = \sum_1^3{}_{\nu\rho}\, a_{\nu\rho}\, \lambda^\nu\, \lambda^\rho, \quad . \quad . \quad (40')$$

which defines the curvature of a variable section f *through* P *as a homogeneous quadric in the parameters* λ^ν (*or in the moments* λ_ν) *of the normal to* f, the sense of the normal being indifferent, since the expressions in (40′) do not change when we change the signs of the λ's.

The dependence of K on the direction $\boldsymbol{\lambda}$ of the section with centre at P is of purely local nature. We can therefore, in accordance with an observation made above, make use of the elementary criteria of analytical geometry just as if it were a question of ordinary space—we have only to take co-ordinates which are *Cartesian and orthogonal at* P. The λ^ν's then become direction cosines, and we have for K (except for a different signification of the coefficients $a_{\nu\rho}$) the same expression as the one which characterizes the distribution of moments of inertia (of an assigned material system) with respect to the ∞^2 axes coinciding with the lines of the versors $\boldsymbol{\lambda}$ proceeding from P. As we know, the law of variation of K becomes expressible geometrically if we introduce the ellipsoid of inertia E, of centre P: to any direction of $\boldsymbol{\lambda}$ there belongs a value of K, viz. $\frac{1}{PQ^2}$, where Q is the intersection of the line of $\boldsymbol{\lambda}$ with the ellipsoid E; the three axes of E correspond to stationary values of K (with respect to neighbouring directions), the axes of maximum and minimum length in particular corresponding respectively to minimum and maximum values of K.

That being so, the same law of variation will hold for K the curvature; but in the general V_3 interpretation is necessary. This implies merely that the ellipsoid E (like, for example, Dupin's indicatrix for the case of an ordinary surface) takes us in thought outside the V_3, as an auxiliary representative element to be associated with the Euclidean three-dimensional space which is tangential to the V_3 at P when, as is always allowable (Chapter V, p. 121), we imagine the V_3 immersed in a Euclidean S_N ($N \geqslant 6$).

The outstanding result is that *there exist at every point* P *at least three mutually orthogonal directions* $\boldsymbol{\lambda}_1$, $\boldsymbol{\lambda}_2$, $\boldsymbol{\lambda}_3$ *to which* (or rather to the normal sections perpendicular to which) *belong curvatures which are stationary with respect to those of adjacent sections.* These directions are called *principal directions of curvature*, and the corresponding values ω_1, ω_2, ω_3 of K *principal curvatures.* In general, that is when the three ω's are distinct, the principal directions are uniquely determinate (ellipsoid with three unequal axes); when two principal curvatures are equal but differ from the third (ellipsoid of revolution), e.g. $\omega_1 = \omega_2 \neq \omega_3$, only the principal direction $\boldsymbol{\lambda}_3$ is uniquely determinate, while every pair of directions $\boldsymbol{\lambda}_1$, $\boldsymbol{\lambda}_2$ orthogonal to $\boldsymbol{\lambda}_3$ and to each other can be considered principal. If the three principal curvatures coincide (sphere) the curvature K is the same for all sections, and every set of three mutually orthogonal directions is a principal set.

All this of course can be established by purely algebraic methods: we have only to avail ourselves of the theory of quadratic forms and their transformations. Let

$$\phi = \sum_{1}^{n}{}_{ij} a_{ij} \lambda^i \lambda^j, \quad \psi = \sum_{1}^{n}{}_{ij} \alpha_{ij} \lambda^i \lambda^j \quad . \quad . \quad (41)$$

be two quadrics, *one of which at least is definite*, say ϕ; the independent variables being λ^i $(i = 1, 2, \ldots n)$. Then the following facts are well-known[1]:

(1) If we consider the ratio

$$\omega = \frac{\psi}{\phi}$$

as a function of the λ's, and look for the values of these variables which make $\delta\omega = 0$, we are led to the system of n linear homogeneous equations

$$\sum_{1}^{n}{}_j (a_{ij} - \omega a_{ij}) \lambda^j = 0 \qquad (i = 1, 2, \ldots n); \quad . \quad (42)$$

these are satisfied by values not all null of the λ's if and only if the determinant of the coefficients vanishes, so that the ω's are roots of the equation of degree n:

$$\| a_{ij} - \omega a_{ij} \| = 0, \quad . \quad . \quad . \quad . \quad (43)$$

called the *characteristic equation.*

(2) If λ^i, λ'^i are two solutions of equations (42) corresponding to two distinct roots ω, ω' of the characteristic equation, there exists between them the relation (of orthogonality)

$$\sum_{1}^{n}{}_{ij}\, a_{ij}\ \lambda^i \lambda'^j = 0.$$

(3) The characteristic equation (43) has its n roots $\omega_1, \omega_2, \ldots \omega_n$ *all real* (distinct or coincident).

(4) To each simple root ω_h of (43) corresponds, in the manifold V_n the ds^2 of which has the a_{ij}'s as its coefficients, one and only one direction $\boldsymbol{\lambda}_h$ whose parameters λ_h^i satisfy (42) for $\omega = \omega_h$. With each root of multiplicity μ (> 1) we can in an infinite number of ways associate μ mutually orthogonal directions in V_n, whose parameters are independent solutions of equations (42) when we give ω the value of the said multiple root.

From all this there results that it is possible in every case to set up at least one set of n mutually orthogonal directions $\boldsymbol{\lambda}_h$ $(h = 1, 2, \ldots n)$, (uniquely determined in the general case,

[1] For proofs, see for example: BIANCHI, *Lezioni di geometria analitica*, Appendix; (Pisa: Spoerri, 1915): or BROMWICH, *Quadratic Forms and their Classification* . . . (Cambridge tracts on Mathematics . . . No. 3).

in which the roots of (43) are all simple), the parameters λ_h^j of which satisfy the system (42), with ω equal to the corresponding ω_h.

By means of these $\boldsymbol{\lambda}_h$'s we can obtain the so-called canonical expressions. We begin by introducing the moments

$$\lambda_{h|i} = \sum_{1}^{n}{}_j\, a_{ij}\,\lambda_h^j \qquad (h, i = 1, 2, \ldots n), \quad . \quad (44)$$

and observe that, by associating with the n identities

$$\sum_{1}^{n}{}_j\, \lambda_h^j\,\lambda_{h|j} = 1 \qquad (h = 1, 2, \ldots n)$$

the $\frac{1}{2}n(n-1)$ conditions of orthogonality of two different directions $\boldsymbol{\lambda}_h$, $\boldsymbol{\lambda}_k$ of the n-ple set

$$\sum_{1}^{n}{}_j\, \lambda_h^j\,\lambda_{k|j} = 0 \qquad (h \neq k),$$

we obtain altogether the n^2 relations

$$\sum_{1}^{n}{}_j\, \lambda_h^j\,\lambda_{k|j} = \delta_h^k \qquad (h, k = 1, 2, \ldots n),$$

which express the noteworthy fact that *the* n^2 *parameters* λ_h^j *of an* n*-ple orthogonal set are the elements reciprocal (in an algebraic sense) to the* n^2 *moments in the determinant* $\| \lambda_{h|j} \|$ *which they form; and vice versa that the moments are the reciprocals of the parameters in the corresponding determinant* $\| \lambda_h^j \|$ (cf. Chapter IV, p. 74).

Further, besides the relations just written which refer to the columns, their analogues with respect to the rows also hold good; these may be written

$$\sum_{1}^{n}{}_h\, \lambda_h^j\,\lambda_{h|i} = \delta_i^j \qquad (i, j = 1, 2, \ldots n). \quad . \quad (45)$$

Taking account of this, if we multiply (44) by $\lambda_{h|k}$ and sum with respect to h, there results immediately

$$a_{ik} = \sum_{1}^{n}{}_h\, \lambda_{h|i}\,\lambda_{h|k} \qquad (i, k = 1, 2, \ldots n), \quad . \quad (46)$$

which are *expressions for the fundamental tensor* a_{ik} *in terms of the moments of any orthogonal* n*-ple set.*

Consider in particular the n-ple set (or one of the n-ple sets) $\boldsymbol{\lambda}_h$, the parameters of which, λ_h^j, satisfy (42) where, for each index h, $\omega = \omega_h$. These equations, in virtue of (44), may be written

$$\sum_1^n{}_j a_{ij} \lambda_h^j = \omega_h \lambda_{h|i} \qquad (h, i = 1, 2, \ldots n).$$

Multiplying by $\lambda_{h|k}$, and summing with respect to the index h, while attending to (45), we obtain the canonical expression for the a_{ik}'s to be associated with (46), viz.

$$a_{ik} = \sum_1^n{}_h \omega_h \lambda_{h|i} \lambda_{h|k} \qquad (i, k = 1, 2, \ldots n). \quad . \quad (47)$$

After this the simultaneous reduction of the two quadrics to orthogonal form is easily effected by substituting for the original variables λ^i the n linear combinations

$$z_h = \sum_1^n{}_i \lambda_{h|i} \lambda^i \qquad (h = 1, 2, \ldots n). \quad . \quad (48)$$

In fact, when we substitute in (41) the values (46), (47) of the coefficients, taking account of (48), there results

$$\phi = \sum_1^n{}_h z_h^2, \quad \psi = \sum_1^n{}_h \omega_h z_h^2. \quad . \quad . \quad . \quad (49)$$

The condition that the λ^i's should be parameters, that is to say that the expression (41) for ϕ should have the value 1, becomes in the new variables z, $\sum_1^n{}_h z_h^2 = 1$. The mode of variation of ψ, when the λ's are parameters of direction, is identical with that of the ratio $\dfrac{\psi}{\phi}$, when the variables are independent; or, if we wish, of the quadric $\psi = \sum_1^n{}_h \omega_h z_h^2$, when the z's are connected the relation $\sum_1^n{}_h z_h^2 = 1$.

Moreover, the stationary values of ψ in these cases are precisely the roots of the characteristic equation (43).

The form

$$\psi = \sum_1^n{}_{ij} a_{ij} \lambda^i \lambda^j$$

is an obvious generalization, for any value of n, of the expression (40′) for K, which defines the local distribution of curvature in a V_3.

The above considerations manifestly apply also to the behaviour of the curvature K when the direction of the section is supposed to vary, a topic which has already been discussed in a more elementary way.

13. Geodesics infinitely near a given geodesic.

We shall conclude this chapter with a discussion of the extension to n dimensions of a classical formula due to Jacobi, which defines in a very simple way the aggregate of those geodesics g of a surface which are infinitely near a given geodesic B, called the geodesic *base*. Jacobi gives the linear equation

$$\frac{d^2y}{d\sigma^2} + Ky \doteq 0, \qquad \text{(J)}$$

where y denotes the distance (normal) of any point M of g from the base, σ the arc of the base measured from an arbitrary origin O up to the projection P of M upon B, and $K(\sigma)$ the Gaussian curvature of the surface at P.

(J) is simply, in Poincaré's phrase, the *equation of variations* of geodesics starting from B. There can be deduced from it, as we know,[1] some very remarkable consequences with respect to the behaviour of geodesics in the immediate neighbourhood of the base, the nature of the surface intervening only through its total curvature K. This is obviously an *intrinsic* question, depending entirely on the metric of the surface (as defined by its ds^2), and not at all on the different configurations which the surface can present in space.

It naturally suggests itself that we should try to extend the study of this subject of geodesic deviation to a Riemannian manifold V_n of any number of dimensions. We have long had, of course, the equations of Lagrange defining the geodesics of a V_n, in a form the convenience of which, whether from the point of view of theory or of notation, is all that could be desired.

[1] See, for example, Darboux, *Théorie des surfaces*, Vol. III (Paris, Gauthier-Villars, 1894; new impression 1923), Chap. V; or Blaschke, *Vorlesungen über Differentialgeometrie*, Vol. I (2nd edition, Berlin, Springer, 1924), §§ 83–88.

These equations we may therefore use for the purpose of forming the equations of variations. Then with the help of Bianchi's idea (Chapter V, p. 139) of the *derivative* of a vector attached to the points of a line of V_n, we reduce these equations to a condensed form, geometrically suggestive and of course invariant. The actual construction of the equations (linear of the second order, n in all, with the same number of unknowns) requires no further data than the base B and the metric of the manifold (especially Riemann's symbols) along that curve.

We find that this system of equations admits a linear first integral, which in its turn leads to a linear relation in finite terms among the unknowns. We are thus left with a system of $n - 1$ equations or, in the special case of an ordinary surface, with a single equation, as in Jacobi's classical result.

To bring the final system of equations to as simple a form as possible we have to make a suitable choice of variables. Now we have already seen (§ 11, p. 164) that it is possible, beginning with any co-ordinates x, to define new co-ordinates y in terms of which ds^2 becomes locally Cartesian, in the sense that the derivatives of the coefficients a_{ik} all vanish at an assigned point O. We have also seen (Chapter VI, p. 167, footnote) that it is possible, any curve B being given, to choose co-ordinates y for which ds^2 has this locally Cartesian character *at every point of* B. When B is a geodesic, the system of co-ordinates y which (§ 17) will finally be used has the following properties, which we merely state here without proof.[1] Let M be any point in the immediate neighbourhood of B, P the orthogonal projection of M upon B. Then y_n is the arc of the base B measured from an arbitrary origin O up to P; the y_α's ($\alpha = 1, 2, \ldots n - 1$) may be regarded as components of the elementary vector PM in $n - 1$ directions mutually perpendicular and all perpendicular to B, chosen arbitrarily at O and carried by parallelism along B.

14. **Geodesic deviation in an n-dimensional manifold.**

Consider, along with the geodesic B, any other geodesic g (more precisely, an arc of g) belonging to the immediate neighbourhood

[1] For a complete development of this point, compare the paper "Sur l'écart géodésique" in *Mathematische Annalen* (Vol. 97).

bourhood of B.[1] Corresponding to every point M of g, take the point P of B having the same y_n as M and the rest of its y's $= 0$. It is important for what follows to fix precisely the relation between an elementary arc ds of g and the corresponding arc $d\sigma = dy_n$ of B. Throughout the neighbourhood of B we have, for the coefficients b_{ik} of ds^2 in terms of the co-ordinates y, the Euclidean values

$$b_{ik} = 0 \quad (i \neq k), \quad b_{ii} = 1,$$

neglecting quantities of the second order.

It follows that for any curve whatever

$$ds = dy_n \sqrt{1 + \sum_1^{n-1}{}_\alpha \left(\frac{dy_\alpha}{dy_n}\right)^2},$$

the quantity under the radical differing from its exact value only by terms of the second order. For g, both the y_α and the $\frac{dy_\alpha}{dy_n}$ may be regarded as infinitely small. It follows that

$$\frac{ds}{dy_n} = \frac{ds}{d\sigma} = 1$$

to a second approximation—a generalization of the elementary fact that a segment infinitely near (with respect to direction) a given right line differs from its projection on the line by an infinitesimal of the second order.

15. **Invariant form of the equations defining geodesic deviation.**

We have now to form the equations defining in general co-ordinates the behaviour of any geodesic g infinitely near B.

Put

$$x_i = \phi_i(\sigma) + \xi^i, \quad . \quad . \quad . \quad . \quad . \quad (50)$$

where the ξ^i's and their derivatives with respect to σ are infinitely small.

The ξ^i's represent the increments of the co-ordinates x_i of a point P of B which passes to a corresponding point M of g: they can be regarded as the contravariant components of the

[1] In the *strict* sense, that is to say, with the understanding that not only are changes of position of corresponding points on B and g to be small, but also changes of direction of corresponding tangents.

elementary vector $PM = \boldsymbol{\xi}$. Adopting a more general point of view, we shall now regard this vector as not necessarily perpendicular to B, its orientation depending on the law of correspondence between the points P and M of our two geodesics. We can therefore no longer assume that the elementary arc ds of g is (as in § 14) equal to the corresponding arc $d\sigma$ of B; but, the displacement in question being always infinitely small, we can foresee all the same that, if we put

$$\frac{ds}{d\sigma} = 1 + \lambda, \quad . \quad . \quad . \quad . \quad . \quad . \qquad (51)$$

the *elongation* (or coefficient of dilatation) λ remains infinitely small with the ξ^i's and their derivatives. This will be proved formally in a moment. Meantime, differentiate the formulæ (50) with respect to σ. We have

$$\frac{ds}{d\sigma}\dot{x}_i = \dot{\phi}_i + \frac{d\xi^i}{d\sigma}, \quad . \quad . \quad . \quad . \quad . \qquad (50')$$

where $\dot{x}_i$ is written for $\dfrac{dx_i}{ds}$, and $\dot{\phi}_i$ for $\dfrac{d\phi_i}{d\sigma}$.

A second differentiation gives

$$\left(\frac{ds}{d\sigma}\right)^2 \ddot{x}_i + \frac{d^2s}{d\sigma^2}\dot{x}_i = \ddot{\phi}_i + \frac{d^2\xi^i}{d\sigma^2}.$$

From (51),

$$\frac{d^2s}{d\sigma^2} = \frac{d\lambda}{d\sigma},$$

and (50′) can be written

$$\dot{x}_i - \dot{\phi}_i = -\lambda\dot{x}_i + \frac{d\xi^i}{d\sigma};$$

all this holds without any assumption concerning the smallness of λ.

Considering now λ and $\dfrac{d\lambda}{d\sigma}$ as infinitesimal, and neglecting all terms of the second order, we may replace $\dfrac{d^2s}{d\sigma^2}\dot{x}_i = \dfrac{d\lambda}{d\sigma}\dot{x}_i$ by $\dfrac{d\lambda}{d\sigma}\dot{\phi}_i$, and consequently take the equation for $\ddot{x}_i$ to be

$$\left(\frac{ds}{d\sigma}\right)^2 \ddot{x}_i = \ddot{\phi}_i + \frac{d^2\xi^i}{d\sigma^2} - \frac{d\lambda}{d\sigma}\dot{\phi}_i. \quad . \quad . \qquad (50'')$$

Now (Chapter V, p. 134) for any geodesic g we have

$$\ddot{x}_i = - \sum_1^n{}_{jh} \{jh, i\} \dot{x}_j \dot{x}_h \qquad (i = 1, 2, \ldots n).$$

Multiply these equations by $\left(\frac{ds}{d\sigma}\right)^2$, and make the following substitutions: (50) in the $\{jh, i\}$, (50′) in the $\frac{ds}{d\sigma} \dot{x}_j$, (50″) in the $\left(\frac{ds}{d\sigma}\right)^2 \ddot{x}_i$.

Remembering that

$$\ddot{\phi}_i = - \sum_1^n{}_{jh} \{jh, i\} \dot{\phi}_j \dot{\phi}_h,$$

since B is a geodesic, and of course retaining terms of the first order only, we find

$$\frac{d^2 \xi^i}{d\sigma^2} - \frac{d\lambda}{d\sigma} \dot{\phi}_i = - \sum_1^n{}_{jhk} \frac{\partial \{jh, i\}}{\partial x_k} \xi^k \dot{\phi}_j \dot{\phi}_h - 2 \sum_1^n{}_{jh} \{jh, i\} \dot{\phi}_j \frac{d\xi^h}{d\sigma}.$$

Next, for the sake of showing explicitly the contravariance of the $\dot{\phi}_i$ (parameters of the direction tangential to B), put

$$\dot{\phi}_i(\sigma) = b^i \qquad (i = 1, 2, \ldots n). \quad . \quad . \quad (52)$$

The equations just obtained (if we also change the indices i and j into r and i) become

$$\frac{d^2 \xi^r}{d\sigma^2} - \frac{d\lambda}{d\sigma} b^r + 2 \sum_1^n{}_{jh} \{jh, r\} b^j \frac{d\xi^h}{d\sigma}$$
$$= - \sum_1^n{}_{ihk} \frac{\partial \{ih, r\}}{\partial x_k} b^i b^h \xi^k. \quad (53)$$

We proceed to transform these equations for the sake of showing their invariant structure with reference to any change of coordinates. For this purpose we introduce Bianchi's conception of the *derivative* vector of a vector $\boldsymbol{\xi}$ given as a function of the points of a line (Chapter V, p. 139).

If B is the line, the contravariant components of the vector $D\boldsymbol{\xi}$, the derivative of $\boldsymbol{\xi}$, are given by the equations

$$(D\boldsymbol{\xi})^r = \frac{\partial \xi^r}{\partial \sigma} + \sum_1^n{}_{ih} \{ih, r\} b^i \xi^h \qquad (r = 1, 2, \ldots n). \quad (54)$$

For co-ordinates which are Cartesian either rigorously, or else locally along the line B, we have simply

$$(D\boldsymbol{\xi})^r = \frac{d\xi^r}{d\sigma}.$$

With the help of the vector $\boldsymbol{\xi}$, it is easy to give an explicit expression for the elongation λ, without making any hypothesis as to its order of magnitude, the equations (50) and (50′), which we shall use, being exact.

On g we have identically

$$\sum_{1}^{n}{}_{ir}\, a_{ir}\, \dot{x}_i\, \dot{x}_r = 1.$$

If we multiply on both sides by $\left(\frac{ds}{d\sigma}\right)^2$, and replace (in the a_{ir}'s) the x_i's by their values (50) and the $\frac{ds}{d\sigma}\dot{x}_i$'s by (50′), we find at once (neglecting terms of the second order with respect to ξ)

$$\left(\frac{ds}{d\sigma}\right)^2 = 1 + 2\sum_{1}^{n}{}_{ir}\, a_{ir}\, \dot{\phi}_i \frac{d\xi^r}{d\sigma} + \sum_{1}^{n}{}_{irh} \frac{\partial a_{ir}}{\partial x_h} \dot{\phi}_i\, \dot{\phi}_r\, \xi^h.$$

We have, moreover, the well-known identities (Chapter V, p. 111)

$$\frac{\partial a_{ik}}{\partial x_l} = \sum_{1}^{n}{}_{h} [a_{hi}\{kl, h\} + a_{hk}\{il, h\}].$$

If in the last sum of the expression for $\left(\frac{ds}{d\sigma}\right)^2$ we replace the indices r, h by k, l and take account of the identity just written down, replacing also, as in (52),

$$\sum_{1}^{n}{}_{i}\, a_{ir}\, \dot{\phi}_i, \quad \sum_{1}^{n}{}_{i}\, a_{hi}\, \dot{\phi}_i, \quad \sum_{1}^{n}{}_{k}\, a_{hk}\, \dot{\phi}_k,$$

by b_r, b_h, b_h,

we find, on account of (54),

$$\left(\frac{ds}{d\sigma}\right)^2 = 1 + 2\sum_{1}^{n}{}_{r}\, (D\boldsymbol{\xi})^r\, b_r.$$

The vector $D\xi$ is infinitely small at the same time as the ξ^i's and their derivatives. We have, therefore, by extracting the root, neglecting terms of the second order, and attending to the definition (51) of the elongation λ,

$$\lambda = \sum_1^n {}_r \, (D\xi)^r \, b_r, \quad . \quad . \quad . \quad . \quad (51')$$

which shows its infinitesimal character.

Naturally $D\xi$ admits in its turn a derivative vector $D^2\xi$. Its contravariant components are defined, from (54), by

$$(D^2\xi)^r = \frac{d(D\xi)^r}{d\sigma} + \sum_1^n {}_{kl} \{kl, r\} \, b^k \, (D\xi)^l.$$

Introducing on the right-hand side the expressions (54) for $(D\xi)^r$ and $(D\xi)^l$ (after making some literal changes in the indices), we obtain

$$(D^2\xi)^r = \frac{d^2\xi^r}{d\sigma^2} + \frac{d}{d\sigma}\left(\sum_1^n {}_{ih} \{ih, r\} \, b^i \, \xi^h\right)$$

$$+ \sum_1^n {}_{kl} \{kl, r\} \, b^k \frac{d\xi^l}{d\sigma} + \sum_1^n {}_{ihkl} \{kl, r\} \{ih, l\} \, b^k \, b^i \, \xi^h, \quad (55)$$

the second members, like the first, constituting a contravariant system. Bringing out explicitly the differentiation with respect to σ, and making some changes of indices, we can write

$$(D^2\xi)^r = \frac{d^2\xi^r}{d\sigma^2} + 2\sum_1^n {}_{jh} \{jh, r\} \, b^j \frac{d\xi^h}{d\sigma} + \Xi^{(r)}, \quad (55')$$

where we put for brevity

$$\Xi^{(r)} = \sum_1^n {}_{ihk} \frac{\partial\{ik, r\}}{\partial x_h} \, b^i \, b^h \, \xi^k + \sum_1^n {}_{lk} \{lk, r\} \, \xi^k \frac{db^l}{d\sigma}$$

$$+ \sum_1^n {}_{ihkl} \{lh, r\} \{ik, l\} \, b^i \, b^h \, \xi^k.$$

We may note in passing that, in the auxiliary quantities $\Xi^{(r)}$, the index r is purely ordinal: we have placed it above, but in brackets, so as to avoid the suggestion that the $\Xi^{(r)}$'s form a contravariant system, which they do not.

For our purpose it is sufficient to replace in $\Xi^{(r)}$ the derivatives $\frac{db^l}{d\sigma} = \ddot{\phi}^l$ by their values

$$-\sum_1^n{}_{ih} \{ih, l\}\, b^i b^h,$$

as given by the equations for geodesics (Chapter V, p. 134), so that we may write

$$\Xi^{(r)} = \sum_1^n{}_{ihk} \frac{\partial\{ik, r\}}{\partial x_h} b^i b^h \xi^k$$
$$+ \sum_1^n{}_{ilhk} [\{lh, r\}\{ik, l\} - \{lk, r\}\{ih, l\}]\, b^i b^h \xi^k. \quad (56)$$

If then we add $\Xi^{(r)}$ to the two members of equations (53), attending to (55′) and to the definition of Riemann's symbols (§ 2), the equations take the *invariant form*

$$(D^2\boldsymbol{\xi})^r - \frac{d\lambda}{d\sigma} b^r = -\sum_1^n{}_{ihk} \{ir, hk\}\, b^i b^h \xi^k \qquad (r = 1, 2, \ldots n). \quad (57)$$

16. **Geodesic deviation. Specification of the differential system. First integral. Linear relation in finite terms.**

The system (57), taken along with the value of λ given in (51′), contains $n + 1$ equations, with the same number of unknowns; but it is easy to foresee, from the method of obtaining it, that it cannot by itself determine completely all the unknowns; there ought to remain an element of indeterminateness arising from the arbitrary nature of the law of correspondence between the points P of B and M of g. More definitely, we can prove that the definition (51′) of λ, or rather the relation derived from it by differentiation

$$\frac{d}{d\sigma}\left\{\lambda - \sum_1^n{}_r (D\boldsymbol{\xi})^r b_r\right\} = 0 \quad . \quad . \quad . \quad (51'')$$

is a consequence of equations (57) themselves. To establish this, note first that, for any vector $\mathbf{v}$ whose contravariant components are v^r, we have

$$\frac{d}{d\sigma}\sum_1^n{}_r v^r b_r = \sum_1^n{}_r \frac{dv^r}{d\sigma} b_r + \sum_1^n{}_r v^r \frac{db_r}{d\sigma}.$$

But the derivatives $\frac{db_r}{d\sigma}$ of the moments of a geodesic B satisfy the equations (Chapter V, p. 134 and p. 139 (52′))

$$\frac{db_r}{d\sigma} = \sum_1^n{}_{ih} \{ir, h\} b^i b_h,$$

which express, we may say, the autoparallelism of the geodesic B in terms of its moments b_r. The preceding identity, after interchange of the indices h and r in the last sum, therefore takes the form

$$\frac{d}{d\sigma} \sum_1^n{}_r v^r b_r = \sum_1^n{}_r (D\mathbf{v})^r b_r. \quad . \quad . \quad . \quad (58)$$

In virtue of this equation, the first member of (51″) now becomes (if we replace the vector $\mathbf{v}$ by $D\boldsymbol{\xi}$)

$$\frac{d\lambda}{d\sigma} - \sum_1^n{}_r (D^2\boldsymbol{\xi})^r b_r.$$

That this expression vanishes we can easily prove from equation (57), making use of the properties of Riemann's symbols, as follows. Multiply both sides of (57) by b_r and sum with respect to r, noting that $\sum_1^n{}_r b_r b^r = 1$. The right-hand member can be written

$$- \sum_1^n{}_{ihkrp} \{ir, hk\} a_{pr} b^p b^i b^h \xi^k.$$

We now sum with respect to r, thus changing the Riemann symbol to the symbol of the first kind, and then make use of the antisymmetry of (ip, hk) in i and p (p. 179), from which it follows that the sum is zero.

Equation (51′) is therefore simply a particular integral (or invariant relation) of the system (57); its role reduces to that of fixing one of the constants of integration. As the system (57) contains $n + 1$ unknowns, to make it determinate it is necessary to associate with it some other condition—a circumstance easy to understand from the geometrical point of view, since we have still to fix the law of punctual correspondence between g and the base.

From a formal point of view the easiest way to complete the system (57) is to cut out the unknown λ by putting

$$\frac{d\lambda}{d\sigma} = 0.$$

The equations (57) are thus reduced to the normal form (Chapter III, p. 36)

$$(D^2\boldsymbol{\xi})^r = -\sum_{1}^{n}{}_{ihk}\{ir, hk\}\, b^i\, b^h\, \xi^k \qquad (r = 1, 2, \ldots n); \quad \text{(I)}$$

and we see that, on account of the identity (51″), *the system* (I) *admits the first integral*

$$\sum_{1}^{n}{}_{r}\,(D\boldsymbol{\xi})^r\, b_r = \lambda = \text{constant}, \quad . \quad . \quad . \quad \text{(II)}$$

expressing the fact that there is a constant linear dilatation when we pass from any arc of B to the corresponding arc of g.

Since, on account of the identity (58), the first member of the integral (II) is simply the derivative of $\sum_{1}^{n}{}_{r}\,\xi^r\, b_r$, it follows that every solution of the differential system (I) gives also

$$\sum_{1}^{n}{}_{r}\,\xi^r\, b_r = \lambda\sigma + C, \quad . \quad . \quad . \quad . \quad \text{(III)}$$

where C is a second constant.

If in particular we take $\lambda = 0$, we see that we can associate with the differential system the relation

$$\sum_{1}^{n}{}_{r}\,\xi^r\, b_r = C.$$

This gives the translation into analysis of the obvious geometrical fact that we can assign the correspondence between the points M of g and P of B in such a way that the (infinitely small) vector PM will have its orthogonal projection upon the tangent $\mathbf{t}$ to the base at the point P equal to a constant C. Such a law of correspondence implies, in virtue of (II), that there is no alteration of length ($\lambda = 0$) as between the arcs of B and the corresponding arcs of g. To particularize still further, if $C = 0$, we arrive at the orthogonal law of correspondence (PM perpendicular to B) considered in § 13.

It is scarcely necessary to add that, in order to substitute other geometrical laws of correspondence, we have only to associate with the system (57) the analytical translation of the law chosen, instead of the law $\frac{d\lambda}{d\sigma} = 0$. For example, if we wish PM to be inclined to B at an angle ψ (constant or a given function of σ) the additional equation will be

$$\sum_1^n {}_r\, \xi^r b_r = \xi \cos\psi,$$

where
$$\xi = \left| \sqrt{\sum_1^n {}_{ik}\, a_{ik}\, \xi^i \xi^k} \right|$$

represents the length of the vector $\boldsymbol{\xi}$.

In a case like this, some slight supplementary discussion of the complete system will be needed—its reduction to the normal form, determination of the number of constants of integration, &c.

17. Reduced form of the differential system (I) in terms of the co-ordinates y.

We now return to the co-ordinates y, and fix definitely on the orthogonal law of correspondence between the base B and any geodesic in its neighbourhood. As we have just seen, such a correspondence is expressed analytically by the differential system (I), with the specifications $\lambda = C = 0$ of the constants of integration connected with (II) and (III).

As remarked in § 14, the co-ordinate y_n of M is identical with that of P. Since the other co-ordinates y_α $(\alpha = 1, 2, \ldots n - 1)$ of P are 0, the variations η^i of the co-ordinates y are respectively

$$\eta^\alpha = y_\alpha \quad (\alpha = 1, 2, \ldots n - 1), \quad \eta^n = 0,$$

thus justifying the name of Cartesian components of the normal displacement or *deviation* $PM = \boldsymbol{\eta}$ which we give to the y_α's.

Moreover, the parameters $b^i = \frac{dy_i}{d\sigma}$ of the base B vanish for $i = 1, 2, \ldots n - 1$; and $b^n = 1$. Christoffel's symbols also vanish along B, as well as their first derivatives with respect to

σ (or, what comes to the same thing, to y_n), and consequently

$$(D\boldsymbol{\eta})^r = \frac{d\eta^r}{d\sigma}.$$

Equations (I) thus become

$$\frac{d^2 y_\alpha}{d\sigma^2} = -\sum_1^{n-1}{}_\beta \{n\alpha, n\beta\} y_\beta \quad (\alpha = 1, 2, \ldots n-1), \quad \text{(I')}$$

$$0 = -\sum_1^{n-1}{}_\beta \{nn, n\beta\} y_\beta,$$

where, in both sums, we have suppressed the term corresponding to the value n of the index, since every Riemann's symbol which has its last indices equal vanishes (p. 177).

The first group (I') (comprising $n-1$ linear equations of the second order) *defines the* n — 1 *Cartesian components of the* (*normal*) *deviation* PM. The last equation reduces to an identity, as may be seen as follows. Riemann's symbols of the second kind are in all cases connected to those of the first kind by the relations

$$(ij, hk) = \sum_1^n{}_r\, a_{jr} \{ir, hk\};$$

we have, moreover, for the symbols of the first kind (§ 4),

$$(ij, hk) = -(ji, hk).$$

In our case the coefficients a_{jr} of ds^2 reduce, *on* B, to 0 (for $r \neq j$) and to 1 (for $r = j$). We have therefore, on the base, equality between symbols of the two kinds whose indices are the same; and, in particular,

$$\{nn, n\beta\} = (nn, n\beta) = 0. \qquad \text{Q.E.D.}$$

Of course, the integral relations (II) and (III) are of no further account, being now identities on account of the vanishing of the b^i's ($i = 1, 2, \ldots n-1$) and of the nth component $\xi^n = \eta^n$ of the displacement PM.

18. **Case of $n = 2$ — formula of Jacobi.**

For $n = 2$, that is to say for an ordinary surface, if B is the geodesic base, $y_2 =$ the arc σ, and y_1 $(= y)$ the normal distance from M to B, the system (I') reduces to the single equation

$$\frac{d^2 y}{d\sigma^2} = -\{21, 21\} y.$$

Now (§ 9) for any co-ordinates whatever, the Gaussian curvature K of a V_2 is expressed by the ratio

$$\frac{(12, 12)}{a} = \frac{(21, 21)}{a},$$

a denoting the discriminant of the ds^2 of V_2.

For our co-ordinates, which are Cartesian along B, $a = 1$, and Riemann's symbols of the second kind are (§ 17) the same as their homologues of the first kind. The equation defining y is therefore none other than the equation of Jacobi (§ 13)

$$\frac{d^2y}{d\sigma^2} + Ky = 0.$$

CHAPTER VIII

Relations between two Different Metrics referred to the same Parameters; Manifolds of Constant Curvature

1. Differences between Christoffel's symbols relative to two different metrics assigned to the same analytical manifold.

We introduced in Chapter IV notions of *tensor*, *covariance*, &c., relative to an analytical manifold V_n, i.e. to the aggregate of n variables $x_1, x_2, \ldots x_n$; we then, in the third part of the following chapter, considered the metrical manifolds obtained by associating with an analytical manifold V_n a specified (but arbitrary) positive and definite differential quadratic form.

There is clearly no reason against assigning in turn to the same analytical manifold two distinct metrical determinations, defined by the two quadratic forms [1]

$$ds^2 = \sum_{1}^{n}{}_{ik}\, a_{ik}\, dx_i\, dx_k, \quad . \quad . \quad . \quad . \quad (1)$$

$$ds'^2 = \sum_{1}^{n}{}_{ik}\, a'_{ik}\, dx_i\, dx_k \quad . \quad . \quad . \quad . \quad (1')$$

[1] As a geometrical interpretation, we can think of two distinct V_n's whose points are in one-to-one correspondence, so that a set of n values assigned to $x_1, x_2, \ldots x_n$ can be represented either by a point P of one, or by the corresponding

From each of these forms we can obtain a set of Christoffel's symbols, which we shall denote by

$$\{ih, r\} \quad \text{and} \quad \{ih, r\}'$$

respectively, and from these we can construct Riemann's symbols

$$\{ir, hk\} \quad \text{and} \quad \{ir, hk\}',$$

and the analogous symbols of the first kind. In this chapter we propose to find the relations between the symbols relative to the two metrics, and then to apply the results to geometrical considerations.

We shall begin by forming the differences

$$\{ih, r\}' - \{ih, r\} = \rho_{ih}^{r}; \quad . \quad . \quad . \quad (2)$$

we shall justify the positions of the indices on the right by showing that the ρ's constitute a tensor covariant with respect to i and h and contravariant with respect to r.

To prove this, consider an arbitrary contravariant system ξ^r whose elements are functions of position, and a system (also arbitrary) of increments dx_h of the independent variables. We know (cf. Chapter V, p. 138) that the expressions of the type

$$\tau^r = d\xi^r + \sum_{1}^{n}{}_{ih} \{ih, r\} \xi^i dx_h$$

constitute a contravariant system. The same result is of course true for the analogous expressions corresponding to ds'^2

$$\tau'^r = d\xi^r + \sum_{1}^{n}{}_{ih} \{ih, r\}' \xi^i dx_h$$

and also for the differences

$$\tau'^r - \tau^r = \sum_{1}^{n}{}_{ih} \rho_{ih}^{r} \xi^i dx_h.$$

The fact that this expression is contravariant means that, denoting by u_r an arbitrary covariant simple system, the expression

$$\sum_{1}^{n}{}_{r} (\tau'^r - \tau^r) u_r = \sum_{1}^{n}{}_{ihr} \rho_{ih}^{r} \xi^i dx_h u_r$$

point P' of the other. E.g. a map and the surface of the earth are two V_2's with different metrics (one is Euclidean, the other not), and to every pair of values, ϕ (for the latitude), λ (for the longitude), correspond one point on the map and one point on the earth.

is an invariant; and if we examine the right-hand side of this equality we see that its invariance requires that ρ_{ih}^{r} should be a tensor of the kind stated.

It will be convenient for subsequent purposes to introduce also the associated covariant system

$$\rho_{ihj} = \sum_{1}^{n}{}_{r}\, a_{jr}\, \rho_{ih}^{r}. \quad . \quad . \quad . \quad . \quad . \quad (2')$$

2. **Differences between the covariant derivatives.**

Given a generic tensor $A_{(i)}^{(h)}$, where as usual (i) and (h) denote the aggregates of m indices $i_1 \ldots i_m$ and μ indices $h_1 \ldots h_\mu$ respectively, we can consider its covariant derivatives with reference to either the first or the second fundamental form, i.e. with respect to either ds^2 or ds'^2. A generic element of the system obtained by differentiation relative to the first form will be denoted as usual by $A_{(i)|k}^{(h)}$, and the analogous expression relative to the second by $\left(A_{(i)}^{(h)}\right)_k'$. We wish to evaluate the difference

$$\left(A_{(i)}^{(h)}\right)_k' - A_{(i)|k}^{(h)}.$$

To find it we can use the explicit expression (p. 146, formula (4)) for the covariant derivatives of a generic mixed system. These are linear in Christoffel's symbols, so that the differences in question will be linear in the ρ's; the expression for them will be

$$\left.\begin{aligned}\left(A_{(i)}^{(h)}\right)_k' - A_{(i)|k}^{(h)} = &- \sum_{1}^{m}{}_{\alpha} \sum_{1}^{n}{}_{j}\, \rho_{i_\alpha k}^{j}\, A_{i_1 \ldots i_{\alpha-1} j\, i_{\alpha+1} \ldots i_m}^{(h)} \\ &+ \sum_{1}^{\mu}{}_{\beta} \sum_{1}^{n}{}_{j}\, \rho_{jk}^{h_\beta}\, A_{(i)}^{h_1 \ldots h_{\beta-1} j\, h_{\beta+1} \ldots h_\mu}.\end{aligned}\right\} \quad (3)$$

These general formulæ can also be obtained, without using any special *memoria technica*, from the original definition of covariant differentiation, with respect to a given fundamental form, of a generic tensor. It is therefore well to remind the reader that, for an arbitrary displacement dx_i, we assigned to the symbol d, when prefixed to a function of position, the usual meaning of the infinitesimal increment (the differential) caused by the displacement (cf. Chapter VI, p. 145); while for a generic vector $\boldsymbol{\xi}$ and its contravariant components ξ^r we assumed

$$d\xi^r = -\sum_{1}^{n}{}_{ih} \{ih, r\} \xi^i dx_h; \quad . \quad . \quad . \quad (4)$$

i.e. we defined the expressions $d\xi^r$ as the increments dependent on parallelism. All of this referred to the metric (1), which was then supposed fixed once and for all. We can of course follow the same procedure taking (1′) as the fundamental form; but to avoid ambiguity it will be well to denote by d' the increments of the ξ^r's due to the same displacement as before, so that we shall have

$$d'\xi^r = -\sum_{1}^{n}{}_{ih} \{ih, r\}' \xi^i dx_h. \quad . \quad . \quad . \quad (4')$$

It will also be useful to introduce the operator

$$d^* = d' - d.$$

Since for functions of position d' and d have the same meaning, we have

$$d^*f = 0$$

for any function of position f; for the contravariant components of a generic vector $\boldsymbol{\xi}$ we have, subtracting (4) from (4′),

$$d^*\xi^r = -\sum_{1}^{n}{}_{ih} \rho^r_{ih} \xi^i dx_h \qquad (r = 1, 2, \ldots n). \quad . \quad (5)$$

Similarly, given the covariant components u_h of some other vector $\mathbf{u}$, we find that

$$d^*u_h = \sum_{1}^{n}{}_{jl} \rho^j_{hl} u_j dx_l. \quad . \quad . \quad . \quad . \quad (5')$$

Now, in order to prove (3), we need only consider the invariant multilinear form F whose coefficients are the elements of the given tensor $A^{(h)}_{(i)}$: we know that the covariant derivatives $A^{(h)}_{(i)|k}$, $\left(A^{(h)}_{(i)}\right)'_k$ are merely the coefficients of dF and $d'F$ respectively. If now we take the identity

$$d'F - dF = d^*F, \quad . \quad . \quad . \quad . \quad . \quad (6)$$

and apply the operator d^* on the right, using the property $d^*f = 0$ for any function of position f, and (5), (5′), then equating coefficients of like terms on each side we get formula (3).

As a very simple application of this process, we shall determine directly the values of the differences $(A_r)'_k - A_{r|k}$ of the derivatives of a covariant simple system A_r. We start from the invariant form

$$F = \sum_1^n{}_r A_r \xi^r \quad . \quad . \quad . \quad . \quad . \quad (7)$$

and consider the usual generic displacement, determined by the increments dx_i of the independent variables. We shall have, with reference to ds^2,

$$dF = \sum_1^n{}_{rk} A_{r|k} \xi^r dx_k; \quad . \quad . \quad . \quad . \quad (8)$$

and with reference to ds'^2,

$$d'F = \sum_1^n{}_{rk} (A_r)'_k \xi^r dx_k. \quad . \quad . \quad . \quad . \quad (8')$$

Further, applying the operator d^* to F, and remembering that d^*A_r is zero, and that $d^*\xi^r$ is given by (5), we get

$$d^*F = - \sum_1^n{}_{rih} A_r \rho^r_{ih} \xi^i dx_h.$$

The identity (6) therefore takes the form

$$\sum_1^n{}_{rk} [(A_r)'_k - A_{r|k}] \xi^r dx_k = - \sum_1^n{}_{rih} A_r \rho^r_{ih} \xi^i dx_h.$$

Replacing on the right h by k, and interchanging i and r, we get the typical term on the right also in a form involving $\xi^r dx_k$; hence, equating coefficients, we have

$$(A_r)'_k - A_{r|k} = - \sum_1^n{}_i \rho^i_{rk} A_i. \quad . \quad . \quad . \quad (9)$$

This is the particular case of (3) which we shall require in the next section.

3. Differences between Riemann's symbols.

We propose in this section to calculate the differences

$$R^r_{ihk} = \{ir, hk\}' - \{ir, hk\},$$

which, being differences of two like tensors, are by definition tensors of the kind indicated. The calculations could be effected directly on the expressions defining Riemann's symbols (Chapter VII, p. 175), but the long formal expansion can be avoided by the following method.

Let A_r be any covariant simple system, and ξ^r any contravariant simple system, and consider the invariant form (7). Applying to it the operator[1] $\Delta = \delta d - d\delta$ with reference to ds^2 (cf. Chapter VII, p. 173), and remembering the fundamental properties of this operator, we shall get

$$\Delta F = \sum_1^n {}_r A_r \Delta \xi^r,$$

or, expanding $\Delta \xi^r$ by formula (4) of Chapter VII,

$$\Delta F = -\sum_1^n {}_{irhk} \{ir, hk\} \xi^i A_r dx_h \delta x_k.$$

Similarly, with reference to ds'^2, we can introduce the operator $\Delta' = \delta' d' - d'\delta'$, and write

$$\Delta' F = -\sum_1^n {}_{irhk} \{ir, hk\}' \xi^i A_r dx_h \delta x_k;$$

and subtracting the former equation from this we get

$$(\Delta' - \Delta)F = -\sum_1^n {}_{irhk} R^r_{ihk} \xi^i A_r dx_h \delta x_k. \quad . \quad (10)$$

We shall now obtain by another method the expression for the same quantity as a quadrilinear form in the quantities $\xi^i A_r dx_h \delta x_k$, and hence, equating corresponding coefficients of the two forms, we shall find the expression for R^r_{ihk}.

Note first that

$$\begin{aligned}\Delta' - \Delta &= (\delta' d' - d'\delta') - (\delta d - d\delta) \\ &= (\delta' d' - \delta d) - (d'\delta' - d\delta).\end{aligned}$$

Since the second expression in brackets is obtained from the first by interchanging d and δ, we need only calculate the expression

[1] To avoid ambiguity we have here replaced the symbols δ and δ', used in Chapter VII to denote two distinct systems of increments, by d and δ respectively.

for $\delta' d' - \delta d$. Further, in making this calculation we can ignore all those terms which remain unchanged on interchanging d and δ, since they will disappear when we take the difference; we shall denote them collectively by $X(d, \delta)$.

Introducing the notation

$$d' - d = d^*, \quad \delta' - \delta = \delta^*,$$

we have

$$\begin{aligned} \delta' d' &= (\delta + \delta^*) d' = \delta d' + \delta^* d' \\ &= \delta(d + d^*) + \delta^* d' \\ &= \delta d + \delta d^* + \delta^* d', \end{aligned}$$

so that

$$\delta' d' - \delta d = \delta d^* + \delta^* d'.$$

We therefore get

$$(\Delta' - \Delta) F = \delta d^* F + \delta^* d' F - (d \delta^* F + d^* \delta' F).$$

To calculate the first term we first apply the operator d^* to the form F, remembering $d^* A_r = 0$ and (5). We get

$$\begin{aligned} d^* F &= \sum_1^n{}_r A_r \, d^* \xi^r \\ &= - \sum_1^n{}_{ihr} A_r \, \rho^r_{ih} \, \xi^i \, dx_h. \end{aligned}$$

Applying the operator δ to this form we get, from the definition of covariant differentiation,

$$\begin{aligned} \delta d^* F &= - \sum_1^n{}_{ihrk} (A_r \, \rho^r_{ih})_k \, \xi^i \, dx_h \, \delta x_k \\ &= - \sum_1^n{}_{ihrk} \rho^r_{ih \mid k} \, \xi^i A_r \, dx_h \, \delta x_k - \sum_1^n{}_{irhk} \rho^r_{ih} \, A_{r \mid k} \, \xi^i \, dx_h \, \delta x_k. \end{aligned}$$

Observing that the second sum can be written in the form

$$\sum_1^n{}_{rk} A_{r \mid k} \, \delta x_k \sum_1^n{}_{ih} \rho^r_{ih} \, \xi^i \, dx_h,$$

we have, applying (5),

$$\delta d^* F = - \sum_1^n{}_{ihrk} \rho^r_{ih \mid k} \, \xi^i A_r \, dx_h \, \delta x_k + \sum_1^n{}_{rk} A_{r \mid k} \, \delta x_k \, d^* \xi^r. \quad (11)$$

To calculate the second term $\delta^* d' F$, we apply the operator δ^* to (8′), and get

$$\delta^* d' F = \sum_1^n {}_{rk} (A_r)'_k \xi^r \delta^* dx_k + \sum_1^n {}_{rk} (A_r)'_k \delta^* \xi^r dx_k.$$

In the second sum, we can substitute for $(A_r)'_k$ the expression given by (9), and we get

$$\delta^* d' F = \sum_1^n {}_{rk} (A_r)'_k \xi^r \delta^* dx_k$$
$$+ \sum_1^n {}_{rk} A_{r|k} \delta^* \xi^r dx_k - \sum_1^n {}_{irk} \rho^i_{rk} A_i \delta^* \xi^r dx_k. \quad (12)$$

We must now add (11) and (12). In doing this, we notice that the first sum in (12) is symmetrical in d and δ, since expanding $\delta^* dx_k$ by (5) it can be written as

$$- \sum_1^n {}_{rkih} (A_r)'_k \xi^r \rho^k_{ih} dx_i \delta x_h;$$

while the second sum in (12) and the second in (11) change one into the other if we interchange d and δ, so that their sum is symmetrical. There remains therefore

$$\delta d^* F + \delta^* d' F$$
$$= X(d, \delta) - \sum_1^n {}_{ihrk} \rho^r_{ih|k} \xi^i A_r dx_h \delta x_k - \sum_1^n {}_{irk} \rho^i_{rk} A_i \delta^* \xi^r dx_k.$$

In the last term we substitute for $\delta^* \xi^r$ its explicit expression, so that it becomes

$$- \sum_1^n {}_{irklh} \rho^i_{rk} \rho^r_{lh} \xi^l A_i \delta x_h dx_k.$$

In order to be able to collect the terms in the two sums with the common factor $\xi^i A_r dx_h \delta x_k$, we apply the substitution formula $\begin{pmatrix} r & l & i & h & k \\ l & i & r & k & h \end{pmatrix}$ to the indices in the last sum, so getting

$$\delta d^* F + \delta^* d' F = X(d, \delta) - \sum_1^n {}_{ihrk} [\rho^r_{ih|k} - \sum_1^n {}_l \rho^r_{lh} \rho^l_{ik}] \xi^i A_r dx_h \delta x_k.$$

The expression obtained by interchanging d and δ on the right (remembering the definition of X) is

$$X(d, \delta) - \sum_1^n {}_{ihrk} [\rho^r_{ih|k} - \sum_1^n {}_l \rho^r_{lh} \rho^l_{ik}] \xi^i A_r \delta x_h dx_k.$$

Interchanging h and k in the second of these, and subtracting, we get finally

$$(\Delta' - \Delta)F$$
$$= -\sum_1^n{}_{ihrk}\,[\rho^r_{ih\,|\,k} - \rho^r_{ik\,|\,h} - \sum_1^n{}_l\,(\rho^r_{lh}\,\rho^l_{ik} - \rho^r_{lk}\,\rho^l_{ih})]\,\xi^i\,A_r\,dx_h\,\delta x_k.$$

Comparing this with (10) we get the required formula:

$$R^r_{ihk} = \{ir,\, hk\}' - \{ir,\, hk\}$$
$$= \rho^r_{ih\,|\,k} - \rho^r_{ik\,|\,h} - \sum_1^n{}_l\,(\rho^r_{lh}\,\rho^l_{ik} - \rho^r_{lk}\,\rho^l_{ih}), \quad . \quad (13)$$

which expresses the differences between Riemann's symbols of the second kind in terms of the differences between Christoffel's symbols. The analogy between this formula and that defining Riemann's symbols (p. 175, formula (3)) should be noted.

If we contract (13) by multiplying by a_{jr} and summing with respect to r, and then use the formula

$$\rho_{ihj\,|\,k} = \sum_1^n{}_r\, a_{jr}\,\rho^r_{ih\,|\,k}$$

obtained by covariant differentiation of (2′) with respect to ds^2, we get the covariant system

$$\sum_1^n{}_r\, a_{jr}\, R^r_{ihk} = \rho_{ihj\,|\,k} - \rho_{ikj\,|\,h} - \sum_1^n{}_l\,(\rho_{lhj}\,\rho^l_{ik} - \rho_{lkj}\,\rho^l_{ih}). \quad (14)$$

It is to be noted that this does not give the differences between Riemann's symbols of the first kind. In fact, substituting on the left the expression for R^r_{ihk}, (14) becomes

$$\sum_1^n{}_r\, a_{jr}\{ir,\, hk\}' - (ij,\, hk)$$
$$= \rho_{ihj\,|\,k} - \rho_{ikj\,|\,h} - \sum_1^n{}_l\,(\rho_{lhj}\,\rho^l_{ik} - \rho_{lkj}\,\rho^l_{ih}), \quad (14')$$

and the first sum is not the same as $(ij,\, hk)'$, which would be

$$\sum_1^n{}_r\, a'_{jr}\{ir,\, hk\}'.$$

4. Case of two metrics in conformal representation.

We shall now apply the formula (14) to the case in which the two fundamental forms (1) and (1′) differ only by a factor. As both forms are positive, this factor must also be positive, so that we can denote it by $e^{2\tau}$; we shall therefore suppose that

$$ds'^2 = e^{2\tau}\, ds^2, \quad . \quad . \quad . \quad . \quad . \quad (15)$$

or

$$ds' = e^{\tau}\, ds.$$

The geometrical interpretation of this condition is quite simple, namely, that the correspondence between the two manifolds is such that infinitesimal segments are proportional, or there is *similarity of infinitesimal parts.* It follows that the angle between two curves (the angle between their tangents at the point of intersection of two infinitesimal elements) is equal to the angle between the corresponding curves; hence the name of *conformal representation.*

In order to calculate (14), we shall obtain in turn, first, Christoffel's symbols of the first kind for the two forms, then those of the second kind, from which we shall get the ρ'^{r}_{ih}'s, and lastly the ρ_{ihj}'s and their derivatives.

We start from the relations equivalent to (15)

$$a'_{ik} = e^{2\tau}\, a_{ik} \qquad (i, k = 1, 2, \ldots n) \quad . \quad (15')$$

and shall calculate the symbol $[ih, l]'$. We get

$$\begin{aligned}[ih, l]' &= \tfrac{1}{2}\left[\frac{\partial a'_{il}}{\partial x_h} + \frac{\partial a'_{lh}}{\partial x_i} - \frac{\partial a'_{ih}}{\partial x_l}\right] \\ &= \tfrac{1}{2}e^{2\tau}\left[\frac{\partial a_{il}}{\partial x_h} + \frac{\partial a_{lh}}{\partial x_i} - \frac{\partial a_{ih}}{\partial x_l}\right] \\ &\qquad + e^{2\tau}\left[a_{il}\frac{\partial \tau}{\partial x_h} + a_{lh}\frac{\partial \tau}{\partial x_i} - a_{ih}\frac{\partial \tau}{\partial x_l}\right] \\ &= e^{2\tau}\left([ih, l] + a_{il}\,\tau_h + a_{lh}\,\tau_i - a_{ih}\,\tau_l\right),\end{aligned}$$

where τ_h stands for $\dfrac{\partial \tau}{\partial x_h}$, &c.

To construct the symbols of the second kind, or

$$\{ih, r\}' = \sum_{1}^{n}{}_l\, a'^{rl}\,[ih, l]',$$

we observe that the coefficients a'^{rl} are, by definition, expressible as the quotient of a determinant A'_{rl} of order $n - 1$ (the complementary minor of a'_{rl} in the determinant $\| a'_{rl} \|$) by a determinant a' (namely, $\| a'_{rl} \|$) of order n. Remembering (15′), we see that in these determinants we can take out a factor $e^{2\tau}$ which is common to every element, so that we can write

$$A'_{rl} = e^{2\tau(n-1)} A_{rl}, \quad a' = e^{2\tau n} a,$$

where A_{rl} and a denote the determinants corresponding to A'_{rl} and a', but relative to the coefficients a_{rl}. We thus have

$$a'^{rl} = e^{-2\tau} a^{rl},$$

and therefore

$$\begin{aligned} \{ih, r\}' &= \sum_1^n {}_l\, a^{rl} ([ih, l] + a_{il} \tau_h + a_{lh} \tau_i - a_{ih} \tau_l) \\ &= \{ih, r\} + \delta_i^r \tau_h + \delta_h^r \tau_i - a_{ih} \tau^r, \end{aligned}$$

where the δ's as usual denote a factor which is 0 or 1 according as the indices are the same or different.

The difference $\{ih, r\}' - \{ih, r\}$ is therefore given by

$$\rho_{ih}^r = \delta_i^r \tau_h + \delta_h^r \tau_i - a_{ih} \tau^r. \quad . \quad . \quad . \quad (16)$$

Multiplying this by a_{jr} and summing with respect to r, we get (using (2′))

$$\rho_{ihj} = a_{ij} \tau_h + a_{hj} \tau_i - a_{ih} \tau_j. \quad . \quad . \quad . \quad (16')$$

By covariant differentiation with respect to ds^2 we get

$$\rho_{ihj|k} = a_{ij} \tau_{hk} + a_{hj} \tau_{ik} - a_{ih} \tau_{jk};$$

subtracting from this formula the analogous one obtained by interchanging h and k (so as to form the first part of (14)), and remembering that $\tau_{hk} = \tau_{kh}$, we find that

$$\rho_{ihj|k} - \rho_{ikj|h} = a_{hj} \tau_{ik} - a_{kj} \tau_{ih} - a_{ih} \tau_{jk} + a_{ik} \tau_{jh}. \quad (17)$$

The second part of (14) can be constructed with the help of (16) and (16′). We shall first calculate

$$\begin{aligned} \sum_1^n {}_l\, \rho_{lhj} \rho_{ik}^l &= \sum_1^n {}_l\, (a_{lj} \tau_h + a_{hj} \tau_l - a_{lh} \tau_j) (\delta_i^l \tau_k + \delta_k^l \tau_i - a_{ik} \tau^l) \\ &= \underline{a_{ij} \tau_h \tau_k} + \underline{a_{kj} \tau_i \tau_h} - a_{ik} \tau_h \tau_j + a_{hj} \tau_k \tau_i + \underline{a_{hj} \tau_i \tau_k} \\ &\quad - a_{ik} a_{hj} \Delta\tau - a_{ih} \tau_j \tau_k - \underline{a_{kh} \tau_j \tau_i} + a_{ik} \tau_j \tau_h, \end{aligned}$$

where $\Delta\tau = \sum_1^n{}_l \tau_l \tau^l$ (the first differential parameter). The third term and the last term cancel out. Denoting by X the aggregate of the four terms underlined, which are unchanged if we interchange the indices h and k, we can write

$$\sum_1^n{}_l \rho_{lhj} \rho^l_{ik} = -a_{ik} a_{hj} \Delta\tau + a_{hj} \tau_k \tau_i - a_{ih} \tau_j \tau_k + X.$$

We shall now subtract from this the formula obtained by interchanging h and k. We shall get

$$\sum_1^n{}_l (\rho_{lhj} \rho^l_{ik} - \rho_{lkj} \rho^l_{ih}) = (a_{ih} a_{jk} - a_{ik} a_{jh}) \Delta\tau$$
$$+ a_{hj} \tau_k \tau_i - a_{kj} \tau_h \tau_i - a_{ih} \tau_j \tau_k + a_{ik} \tau_j \tau_h.$$

Using this and (17), we get the right-hand side of (14′) in the form

$$-a_{ih}(\tau_{jk} - \tau_j \tau_k) + a_{ik}(\tau_{jh} - \tau_j \tau_h) + a_{jh}(\tau_{ik} - \tau_i \tau_k)$$
$$- a_{jk}(\tau_{ih} - \tau_i \tau_h) - (a_{ih} a_{jk} - a_{ik} a_{jh}) \Delta\tau.$$

The left-hand side of (14′), using (15′), can be written as

$$e^{-2\tau} \sum_1^n{}_r a'_{jr} \{ir, hk\}' - (ij, hk)$$

or

$$e^{-2\tau} (ij, hk)' - (ij, hk).$$

Finally, formula (14′), for two metrics in conformal representation, can be written in the form

$$\left. \begin{aligned} e^{-2\tau}(ij, hk)' - (ij, hk) = -a_{ih}(\tau_{jk} - \tau_j \tau_k) + a_{ik}(\tau_{jh} - \tau_j \tau_h) \\ + a_{jh}(\tau_{ik} - \tau_i \tau_k) - a_{jk}(\tau_{ih} - \tau_i \tau_h) - (a_{ih} a_{jk} - a_{ik} a_{jh}) \Delta\tau. \end{aligned} \right\} \quad (18)$$

This formula was found by Finzi, by another method, as early as 1903.[1]

[1] Cf. "Le ipersuperficie a tre dimensioni che si possono rappresentare conformemente sullo spazio euclideo", in *Atti del R. Ist. Veneto*, Vol. LXII, pp. 1049–1062. The later researches of Finzi and Schouten on the manifolds of any number of dimensions which can be conformally represented in a Euclidean space of the same number of dimensions, should also be mentioned. Cf. *Rend. della R. Acc. dei Lincei*, Vol. XXX (first half-year, 1922), pp. 8–12, and Vol. XXXI (first half-year, 1923), pp. 215–218, and Schouten's book cited at Chap. VII, p. 172. Cf. also D. J. Struik: *Grundzüge der mehrdimensionalen Differentialgeometrie* (Berlin, Springer, 1922), Ch. IV, § 13, p. 150, where Schouten's results are given with bibliographical notes.

This formula can be given a simpler form by putting

$$u = e^{-\tau},$$

so that (15) becomes

$$ds'^2 = \frac{1}{u^2} ds^2.$$

We then have

$$u_i = - e^{-\tau} \tau_i,$$
$$u_{ik} = - e^{-\tau} (\tau_{ik} - \tau_i \tau_k),$$

from which we get

$$\tau_i = - u_i e^{\tau}, \quad \tau_{ik} - \tau_i \tau_k = - \frac{u_{ik}}{u},$$

$$\Delta \tau = \sum_{ik}^{n} {}_1 a^{ik} \tau_i \tau_k = e^{2\tau} \sum_{ik}^{n} {}_1 a^{ik} u_i u_k = e^{2\tau} \Delta u = \frac{\Delta u}{u^2}.$$

Thus (18) becomes

$$u^2 (ij, hk)' - (ij, hk)$$
$$= a_{ih} \frac{u_{jk}}{u} - a_{ik} \frac{u_{jh}}{u} - a_{jh} \frac{u_{ik}}{u} + a_{jk} \frac{u_{ih}}{u} - (a_{ih} a_{jk} - a_{ik} a_{jh}) \frac{\Delta u}{u^2}. \quad (18')$$

At the end of this chapter we shall have occasion to point out an interesting geometrical application of this result.

5. Isotropic manifolds.

Leaving aside for a moment this order of ideas, we propose to study those V_n's in which the Riemannian curvature, as defined in Chapter VII, pp. 195–198, does not depend on the section, but only (if it is variable) on the point. This is expressed analytically by the fact that the expression for K given by (31) of the preceding chapter is independent of the u's and the v's. We shall see that these V_n's, which we shall call *isotropic*, i.e. with (locally) constant curvature, are characterized by a particularly simple expression for Riemann's symbols.

We observe first that a fairly simple algebraic combination of the coefficients a_{ik}, which possesses the fundamental properties of Riemann's symbols, is the following:

$$b_{ij,hk} = \gamma (a_{ih} a_{jk} - a_{ik} a_{jh}),$$

where γ is *a priori* any function whatever of position. Everything reduces to proving that when these quantities are substituted for the symbols (ij, hk) in (31) of the preceding chapter, the resulting value of K is independent of $\mathbf{u}$ and $\mathbf{v}$. In fact, making the substitution, we have

$$K = \frac{\gamma}{\sin^2\alpha} \sum_{1}^{n}{}_{ijhk} (a_{ih} a_{jk} - a_{ik} a_{jh}) u^i v^j u^h v^k$$

$$= \frac{\gamma}{\sin^2\alpha} \left[\sum_{1}^{n}{}_{ih} a_{ih} u^i u^h \sum_{1}^{n}{}_{jk} a_{jk} v^j v^k - \sum_{1}^{n}{}_{ik} a_{ik} u^i v^k \sum_{1}^{n}{}_{jh} a_{jh} v^j u^h \right],$$

and since

$$\sum_{1}^{n}{}_{ih} a_{ih} u^i u^h = \sum_{1}^{n}{}_{jk} a_{jk} v^j v^k = 1,$$

$$\sum_{1}^{n}{}_{ik} a_{ik} u^i v^k = \sum_{1}^{n}{}_{jh} a_{jh} v^j u^h = \cos\alpha,$$

it follows that

$$K = \frac{\gamma}{\sin^2\alpha} [1 - \cos^2\alpha] = \gamma.$$

Hence the Riemannian curvature of a V_n whose Riemann's symbols are the expressions $b_{ij,\,hk}$, is γ, and is therefore independent of the section. But we can also show that this is the most general expression of Riemann's symbols which will make $K = \gamma$. In fact, if we put

$$(ij, hk) = b_{ij,\,hk} + B_{ij,\,hk},$$

where $b_{ij,\,hk}$ has the meaning assigned to it above, we shall show that $B_{ij,\,hk} = 0$. To do this, we insert this expression for (ij, hk) in (31); the right-hand side can then be broken up into two parts, the first of which, containing the symbols $b_{ij,\,hk}$, is, as we have seen, equal to γ, and the second, which is

$$\frac{1}{\sin^2\alpha} \sum_{1}^{n}{}_{ijhk} B_{ij,\,hk} u^i v^j u^h v^k,$$

must vanish if we are to have $K = \gamma$.

The sum just written can be simplified if we observe that since Riemann's symbols are antisymmetrical, the two terms

$$B_{ij,\,hk} u^i v^j u^h v^k,$$
$$B_{ij,\,hk} u^i v^j u^k v^h$$

can be collected into a single term; putting

$$u^h v^k - u^k v^h = p_{hk},$$

this term becomes

$$B_{ij,\,hk} u^i v^j p_{hk}.$$

Thus the sum for all the *permutations* hk becomes merely a sum for all the *simple combinations* (h, k) of unlike indices, since it is useless to consider terms with a repeated index $(k = h)$, because $p_{hh} = 0$. We shall denote the sum extended only to simple combinations of the indices by $\mathrm{S}_1^n{}_{hk}$ instead of $\Sigma_1^n{}_{hk}$. The quadruple sum thus becomes

$$\Sigma_1^n{}_{ij}\, u^i v^j\, \mathrm{S}_1^n{}_{hk}\, B_{ij,\,hk}\, p_{hk}.$$

Proceeding in the same way for the indices i, j, we get ultimately

$$\mathrm{S}_1^n{}_{ij}\, \mathrm{S}_1^n{}_{hk}\, B_{ij,\,hk}\, p_{ij}\, p_{hk},$$

i.e. an expression bilinear in the p's. Each summation will extend to $m = \frac{1}{2}n(n - 1)$ pairs; we shall number these (in any order) from 1 to m, and put

$$p_{ij} = z_\beta, \quad p_{hk} = z_\gamma, \quad B_{ij,\,hk} = B_{(\beta),\,(\gamma)}$$
$$(i, j, h, k = 1, 2, \ldots n; \ \beta, \gamma = 1, 2, \ldots m),$$

where β is the ordinal number of the pair ij and γ that of hk, so that the sum can be written

$$\Sigma_1^m{}_{\beta\gamma}\, B_{(\beta),\,(\gamma)}\, z_\beta\, z_\gamma.$$

It will now be clear that this expression cannot vanish for arbitrary values of the z's, unless all the B's are zero; which is precisely what we wished to prove.

We may therefore conclude that for a V_n whose curvature is *locally* constant (i.e. independent of the section) and equal to a given function of position K, Riemann's symbols are necessarily given by the formula

$$(ij, hk) = K(a_{ih}\, a_{jk} - a_{ik}\, a_{jh}). \quad . \quad . \quad (19)$$

Multiplying by a^{jr} and summing with respect to j, we get the expression for the symbols of the second kind:

$$\{ir, hk\} = K(a_{ih}\, \delta^r_k - a_{ik}\, \delta^r_h). \quad . \quad . \quad (19')$$

The function K, however, cannot be arbitrarily assigned;

we shall show in the following section that for $n > 2$ it must be a constant.

6. **Schur's theorem.**

This theorem states that *if the curvature is locally constant, it is also the same at all points.* The case $n = 2$ is not considered, as there is only one section at each point, so that we cannot properly speak of locally constant curvature.

We shall therefore show that the K of formula (19) is constant, or that

$$K_l = 0 \qquad (l = 1, 2, \ldots n),$$

where K_l represents a generic covariant derivative, identical (cf. Chapter VI, p. 147) with the ordinary derivative.

To prove this, we take the covariant derivative of (19), remembering Ricci's lemma. This gives

$$(ij, hk)_l = K_l (a_{ih} a_{jk} - a_{ik} a_{jh}).$$

Taking three distinct values for h, k, l (which is possible, since $n > 2$), the other two relations obtained from this one by cyclic permutation of h, k, l can be written in the form

$$(ij, kl)_h = K_h (a_{ik} a_{jl} - a_{il} a_{jk}),$$
$$(ij, lh)_k = K_k (a_{il} a_{jh} - a_{ih} a_{jl}).$$

Adding the terms on the left and on the right of these three equations, and remembering Bianchi's identity (Chapter VII, p. 183), we get

$$0 = K_l (a_{ih} a_{jk} - a_{ik} a_{jh}) + K_h (a_{ik} a_{jl} - a_{il} a_{jk}) + K_k (a_{il} a_{jh} - a_{ih} a_{jl}). \quad (20)$$

By varying i and j, we thus get $\frac{1}{2}n(n-1)$ relations, of which we shall now make a suitable linear combination. Multiplying (20) by $a^{ih} a^{jk}$ and summing with respect to i, j, we find that the coefficients of K_l, K_h, K_k, are all of the type

$$\sum_{ij}^{n}{}_{1} a_{i\alpha} a_{j\beta} a^{ih} a^{jk},$$

where α, β denote two of the indices h, k, l; or, making the two summations in turn, of the type

$$\sum_{i}{}_{1}^{n} a_{i\alpha} a^{ih} \sum_{j}{}_{1}^{n} a_{j\beta} a^{jk} = \delta^h_\alpha \delta^k_\beta,$$

where the δ's as usual denote either 0 or 1. These quantities are therefore always zero, unless we have simultaneously $\alpha = h$, $\beta = k$ (which happens in the coefficient of the first term), in which case the value is 1. Our linear combination of the equations (20) thus reduces to

$$K_l = 0. \qquad \text{Q.E.D.}$$

7. Canonical form of ds^2 for a manifold of constant curvature.

Given a Euclidean space S_n, we propose to find, if it exists, a manifold V_n' with constant curvature K, which can be conformally represented on S_n, or in other words (cf. § 4) such that its linear element is given by

$$ds'^2 = \frac{ds^2}{u^2},$$

where ds is the element of S_n. We shall see that this is always possible, and the solution of the problem will lead us to assign two important forms for the ds^2 of a manifold with constant curvature.

Keeping the notation of § 4, we shall have for Riemann's symbols for the V_n' the expression (cf. formula (19))

$$(ij, hk)' = K(a'_{ih}\, a'_{jk} - a'_{ik}\, a'_{jh}) = \frac{K}{u^4}(a_{ih}\, a_{jk} - a_{ik}\, a_{jh}),$$

and for S_n

$$(ij, hk) = 0,$$

since for a Euclidean space all Riemann's symbols are zero (cf. pp. 173–178).

We must now substitute these values in the equations (18'); these constitute a system of differential equations the integration of which will give the function u. Making the substitution, (18') becomes

$$\left.\begin{aligned} a_{ih}\frac{u_{jk}}{u} - a_{ik}\frac{u_{jh}}{u} - a_{jh}\frac{u_{ik}}{u} + a_{jk}\frac{u_{ih}}{u} & \\ - (a_{ih}\, a_{jk} - a_{ik}\, a_{jh})\frac{\Delta u + K}{u^2} &= 0 \\ (i, j, h, k = 1, 2, \ldots n) & \end{aligned}\right\}. \tag{21}$$

These $\frac{1}{12}n^2(n^2-1)$ equations can be satisfied by putting

$$u_{ik} = ca_{ik} \quad (i, k = 1, 2, \ldots n), \quad . \quad . \quad (22)$$

where c is a constant; in fact, substituting these values, they take the form

$$\frac{1}{u^2}(a_{ih}\, a_{jk} - a_{ik}\, a_{jh})\,(2cu - K - \Delta u) = 0$$

$$(i, j, h, k = 1, 2, \ldots n);$$

and in order that they may all be satisfied, we need only make the common factor vanish, i.e. put

$$2cu - K - \Delta u = 0. \quad . \quad . \quad . \quad . \quad (22')$$

We have therefore substituted for (21) the system composed of the equations (22) and (22′), which holds whatever may be the co-ordinates x. If then we suppose, as we always may, that the x's are orthogonal Cartesian co-ordinates of S_n, so that

$$a_{ik} = \delta_i^k, \quad \Delta u = \sum_1^n{}_\nu \left(\frac{\partial u}{\partial x_\nu}\right)^2, \quad u_{ik} = \frac{\partial^2 u}{\partial x_i\, \partial x_k},$$

our system will take the simpler form

$$\frac{\partial^2 u}{\partial x_i\, \partial x_k} = c\delta_i^k, \quad . \quad . \quad . \quad . \quad . \quad . \quad . \quad (23)$$

$$2cu - K - \sum_1^n{}_\nu \left(\frac{\partial u}{\partial x_\nu}\right)^2 = 0. \quad . \quad . \quad (23')$$

We shall examine separately the two cases $c = 0$, $c \neq 0$. If $c = 0$, the system becomes

$$\frac{\partial^2 u}{\partial x_i\, \partial x_k} = 0, \quad . \quad . \quad . \quad . \quad . \quad (24)$$

$$K + \sum_1^n{}_\nu \left(\frac{\partial u}{\partial x_\nu}\right)^2 = 0, \quad . \quad . \quad . \quad (24')$$

and from the second of these it follows that $K < 0$. Such a solution is therefore possible only for manifolds of constant negative curvature, since we do not consider the case $K = 0$, which has no special interest, i.e. the case when V_n' is itself Euclidean. Equations (24) then give, by an immediate integration,

$$u = \sum_{1}^{n}{}_{\nu} b_\nu x_\nu + b, \quad . \quad . \quad . \quad . \quad (25)$$

where the b_ν's, and b are constants and therefore, substituting in (24′),

$$K + \sum_{1}^{n}{}_{\nu} b_\nu^2 = 0. \quad . \quad . \quad . \quad . \quad (25')$$

This shows that the b_ν's are not all zero, and that therefore, by applying an orthogonal substitution to the co-ordinates,[1] (25) can be put in the form

$$u = k\,x_n \qquad (k \text{ constant})$$

so that (25′) becomes

$$K + k^2 = 0,$$

or $k = \sqrt{-K}$. We therefore have

$$u = \sqrt{-K}\,x_n,$$

and therefore

$$ds'^2 = \frac{dx_1^2 + dx_2^2 + \ldots + dx_n^2}{-K\,x_n^2}. \quad . \quad . \quad (26)$$

This is the *canonical form of the line element of a manifold of constant negative curvature.* It was found by Beltrami[2] in 1868 by another method.

Another type of solution which holds for any value of K whatever is obtained by supposing $c \neq 0$. (23) gives us the two groups of equations

$$\frac{\partial^2 u}{\partial x_i \partial x_k} = 0, \quad i \neq k, \quad . \quad . \quad . \quad . \quad (27)$$

$$\frac{\partial^2 u}{\partial x_i^2} = c. \quad . \quad . \quad . \quad . \quad . \quad . \quad (27')$$

[1] The hypersurfaces u = constant, i.e. $\sum_{1}^{n}{}_{\nu} b_\nu x_\nu$ = constant, are a set of parallel hyperplanes; we need therefore only choose the axis x_n in the direction perpendicular to them in order that their equations may take the form x_n = constant, and therefore that $u = k\,x_n$.

[2] *Opere matematiche*, Vol. I, p. 419. Milan, Hoepli, 1902.

The first group has for its general integral

$$u = \sum_1^n {}_i X_i, \quad . \quad . \quad . \quad . \quad . \quad (28)$$

where X_i is a function of x_i alone.

The second group gives

$$X_i'' = c,$$

where differentiation is denoted without ambiguity by dashes, since the argument of X_i is x_i only.

From this, integrating once, we get

$$X_i' = c\,(x_i - x_i^0),$$

where the arbitrary constant of integration has been put in the form $-cx_i^0$, using the hypothesis $c \neq 0$; and integrating a second time

$$X_i = \frac{c}{2}(x_i - x_i^0)^2 + b_i$$

where b_i is a constant. Substituting from this in (28) and putting $b = \sum_1^n {}_i b_i$ we get the following expression for u:

$$u = \frac{c}{2}\sum_1^n {}_i (x_i - x_i^0)^2 + b, \quad . \quad . \quad . \quad (29)$$

containing $n + 2$ arbitrary constants.

We still have to consider (23′); substituting in it this value of u, it becomes

$$2cb - K = 0 \quad . \quad . \quad . \quad . \quad (23'')$$

and therefore merely establishes a relation between the two constants c and b.

We have therefore obtained a solution containing $n + 1$ arbitrary constants; we can choose these to satisfy specified conditions at a generic but fixed point O of S_n. E.g. suppose we wish to take the x_i^0's in such a way that all the u_j's are zero at the origin. We have from (29)

$$u_j = c\,(x_j - x_j^0);$$

hence every x_j^0 must vanish, so that (29) becomes (substituting for b from (23″))

$$u = \frac{c}{2}\sum_1^n {}_i x_i^2 + \frac{K}{2c}. \quad . \quad . \quad . \quad . \quad (30)$$

We can then determine c so that at the origin $u = 1$; for this we must have $c = \frac{K}{2}$, and we thus get finally

$$u = 1 + \frac{K}{4} \sum_1^n{}_i x_i^2. \quad . \quad . \quad . \quad . \quad (30')$$

This value of u makes ds'^2 take the form given by Riemann:

$$ds'^2 = \frac{\sum_1^n{}_\nu dx_\nu^2}{u^2} = \frac{dx_1^2 + dx_2^2 + \ldots + dx_n^2}{\left(1 + \frac{K}{4} \sum_1^n{}_i x_i^2\right)^2}. \quad (31)$$

We shall show farther on (§ 2, p. 246) that the ds^2 of any V_n whatever of constant curvature K can be put in the form (31), and also if $K < 0$, in the form (26); this will justify the choice of the term *canonical forms* for these expressions.

Here we shall also prove the almost obvious property that a hypersphere of radius R in Euclidean space of $n + 1$ dimensions constitutes a V_n of constant positive curvature $K = \frac{1}{R^2}$. To do this we shall take $y_0, y_1, y_2, \ldots y_n$ to denote orthogonal Cartesian co-ordinates in S_{n+1}, so that

$$ds^2 = \sum_0^n{}_\nu dy_\nu^2. \quad . \quad . \quad . \quad . \quad . \quad . \quad (32)$$

Without loss of generality we can consider only the hypersphere which has its centre at the origin, and is therefore represented by the equation

$$\sum_0^n{}_\nu y_\nu^2 = R^2. \quad . \quad . \quad . \quad . \quad . \quad . \quad (33)$$

We shall prove the assertion in the most direct way, by expressing the $n + 1$ co-ordinates y of the points of the hypersphere, connected by the relation (33), in terms of n suitable curvilinear co-ordinates x, and showing that ds^2 takes the required canonical form (31) when these parametric expressions of the y's in terms of the x's are substituted in (33).

The parametric representation of the y's referred to is an immediate generalization of that given for an ordinary spherical surface by stereographic projection. In this case ($n = 2$), if we project from a point whose co-ordinates are $y_0 = -R$, $y_1 = y_2 = 0$ upon the tangent plane at the diametrically opposite point, every point (y_0, y_1, y_2) of the sphere projects into a point on the plane whose co-ordinates x_1, x_2 are connected with the y's by the relations[1]

$$y_0 = \frac{1}{\sqrt{K}}\left(\frac{2}{u} - 1\right), \quad y_\nu = \frac{x_\nu}{u} \qquad (\nu = 1, 2), \tag{34}$$

where

$$u = 1 + \frac{K}{4}\rho^2, \quad K = \frac{1}{R^2}, \quad \rho^2 = \sum_1^2{}_\nu x_\nu^2 . \tag{35}$$

For *any* value of n, we shall adopt the same formulæ, with the obvious modification that ν is to vary from 1 to n. This does in fact give a parametric representation of our hypersphere; for squaring and adding the equations (34), and substituting for $\sum_1^n{}_\nu x_\nu^2$ its expression in terms of u as given by (30′), namely, $\frac{4}{K}(u - 1)$, we get back to equation (33). We have then, differentiating,

$$dy_0 = -\frac{1}{\sqrt{K}}\frac{2du}{u^2},$$

$$dy_\nu = \frac{dx_\nu}{u} - \frac{x_\nu\, du}{u^2} \qquad (\nu = 1, 2, \ldots n).$$

Squaring and adding, and substituting $\frac{4}{K}(u - 1)$ and $\frac{4}{K}du$ for $\sum_1^n{}_\nu x_\nu^2$ and $2\sum_1^n{}_\nu x_\nu\, dx_\nu$ respectively on the right-hand side, we get finally

$$ds^2 = \frac{\sum_1^n{}_\nu dx_\nu^2}{u^2},$$

which is the required result.

[1] These relations can easily be shown to be the same as those ordinarily used if we replace ρ, the radius vector of the projection, by the colatitude ϑ of the point on the sphere. As by definition $\sqrt{K}\,y_0 = \cos\vartheta$, it follows that $\frac{1}{u} = \cos^2\frac{1}{2}\vartheta$, $\rho = 2R\tan\frac{1}{2}\vartheta$.

CHAPTER IX

Differential Quadratic Forms of Class Zero and Class One

1. Forms of class zero (or Euclidean forms).

In Chapter V, p. 123, we defined the *class* of a given V_n (or of the quadratic form ds^2 which characterizes it) as the number $N - n$, where N is the minimum number of dimensions of a Euclidean space in which the V_n can be immersed.

We shall consequently say that a quadratic differential form

$$ds^2 = \sum_{1}^{n}{}_{ik}\, a_{ik}\, dx_i\, dx_k \qquad (1)$$

is *of class zero* (or is *Euclidean*) if it is possible to substitute for the n variables x a set of n variables y (since $N = n$), connected with the x's by the relations

$$y_\nu = y_\nu(x_1, x_2, \ldots x_n) \qquad (\nu = 1, 2, \ldots n), \qquad (2)$$

and such that (1) assumes the Cartesian form

$$ds^2 = \sum_{1}^{n}{}_{\nu}\, dy_\nu^2 \qquad (1')$$

Given (1) we wish to find a criterion which will enable us to recognize whether such a transformation is possible. We shall show that it is sufficient to construct Riemann's symbols relative to (1), and to determine whether they vanish identically or not. We have already seen (Chapter VII, p. 178) that this condition is necessary; we wish to prove that if, inversely, all Riemann's symbols relative to (1) are identically zero, then (1) can be transformed into (1'); or in other words that the n functions (2) can be so determined as to satisfy the $\frac{1}{2}n(n+1)$ equations

$$a_{ik} = \sum_{1}^{n}{}_{\nu}\, y_{\nu|i}\, y_{\nu|k} \qquad (i, k = 1, 2, \ldots n), \qquad (3)$$

where

$$y_{\nu|k} = \frac{\partial y_\nu}{\partial x_k} \qquad (4)$$

(cf. Chapter V, p. 122, formula (35)).

By covariant differentiation of (3), we get

$$0 = \sum_{1}^{n}{}_{\nu} (y_{\nu|il}\, y_{\nu|k} + y_{\nu|i}\, y_{\nu|kl}).$$

By cyclic permutation of the indices i, k, l, we get from this the two further equations

$$0 = \sum_{1}^{n}{}_{\nu} (y_{\nu|ki}\, y_{\nu|l} + y_{\nu|k}\, y_{\nu|li}),$$

$$0 = \sum_{1}^{n}{}_{\nu} (y_{\nu|lk}\, y_{\nu|i} + y_{\nu|l}\, y_{\nu|ik}).$$

Now add the last two of these equations and subtract the first. From the commutation rule (§ 6, p. 184), combined with the vanishing of Riemann's symbols, it follows that the second derivatives are permutable, so that we get

$$\sum_{1}^{n}{}_{\nu}\, y_{\nu|ik}\, y_{\nu|l} = 0.$$

Keeping i and k fixed, and making l vary from 1 to n, this formula gives us n linear homogeneous equations in the n unknowns $y_{\nu|ik}$ $(\nu = 1, 2, \ldots n)$. The determinant of the system is certainly not zero, since it is composed of the terms $y_{\nu|l}$, i.e. is the functional determinant of the transformation (2); we therefore conclude that

$$y_{\nu|ik} = 0 \qquad (\nu, i, k = 1, 2, \ldots n). \quad . \quad . \quad (5)$$

These equations, which we have deduced from (3), can be put in the form

$$\frac{\partial y_{\nu|i}}{\partial x_k} = \sum_{1}^{n}{}_{j} \{ik, j\}\, y_{\nu|j} = f_{\nu|ik}(x \,|\, y_{\nu|1}, \ldots y_{\nu|n}), \qquad (5')$$

in which we are concerned only to the extent of observing that the right-hand side is a known function of position and of the terms $y_{\nu|j}$.

It is now easy to see that the problem is reduced to that of a mixed system of total differential equations and equations in finite terms which we have already considered in § 8, p. 29.

In fact, considering as unknowns the n quantities y_ν, and the

n^2 quantities $y_{\nu|k}$, we can collect together the equations (4) and (5′) into a system of total differential equations

$$\left.\begin{aligned} dy_\nu &= \sum_1^n {}_k\, y_{\nu|k}\, dx_k, \\ dy_{\nu|i} &= \sum_1^n {}_k\, f_{\nu|ik}(x\,|\,y_{\nu|1}, \ldots y_{\nu|n})\, dx_k \end{aligned}\right\} (\nu, i = 1, 2, \ldots n), \quad (S)$$

while the group (3) constitutes $\frac{1}{2}n(n+1)$ relations in finite terms between the $n^2 + n = n\,(n+1)$ unknowns.

The conditions for complete integrability, by the usual rule, are as follows:

$$\left.\begin{aligned} &(a) \quad \frac{\partial y_{\nu|k}}{\partial x_h} = \frac{\partial y_{\nu|h}}{\partial x_k} \\ &(b) \quad \frac{\partial f_{\nu|ik}}{\partial x_h} = \frac{\partial f_{\nu|ih}}{\partial x_k} \end{aligned}\right\} (\nu, i, h, k = 1, 2, \ldots n);$$

(c) the equations obtained by differentiating the equations (3) must be identically satisfied in virtue of the equations (S).

Introducing the covariant derivatives and once more applying the commutation rule for the second derivatives, the conditions (a) can be written in the form

$$y_{\nu|kh} - y_{\nu|hk} = \text{linear combinations of Riemann's symbols},$$

and it will then at once be seen that they are satisfied identically, since the left-hand side vanishes in virtue of (5), and the right-hand side also vanishes, since by hypothesis Riemann's symbols are zero.

A similar argument holds for the conditions (b), which are equivalent to

$$y_{\nu|ikh} - y_{\nu|ihk} = \text{linear combinations of Riemann's symbols.}$$

Lastly, taking the covariant derivatives of (3), we find the conditions (c) in the form

$$a_{ik|l} = \sum_1^n {}_\nu\, (y_{\nu|il}\, y_{\nu|k} + y_{\nu|i}\, y_{\nu|kl}),$$

and it can at once be verified that all these are satisfied, in virtue of Ricci's lemma and the equations (5).

The mixed system is therefore *complete*, and it will be possible to find the functions (2), which will contain $\frac{1}{2}n(n+1)$ arbitrary constants, this being the difference between the number of unknowns and the number of equations in finite terms. In geometrical terms, if the manifold is Euclidean there are in it $\infty^{\frac{1}{2}n(n+1)}$ (orthogonal) Cartesian systems. If we can find a particular solution $\eta_1, \eta_2, \ldots \eta_n$, we can get the most general solution by a substitution of the type

$$y_i = c_i + \sum_{1}^{n}{}_j\, a_{ij}\eta_j \qquad (i = 1, 2, \ldots n), \quad . \quad (6)$$

where the a's are the coefficients of an orthogonal substitution, i.e. are connected by the $\frac{1}{2}n(n+1)$ equations

$$\sum_{1}^{n}{}_i\, a_{ij}\, a_{i\nu} = \delta_j^\nu \qquad (j, \nu = 1, 2, \ldots n), . \quad . \quad (7)$$

while the c's are n completely arbitrary constants.

This can be immediately proved from the characteristic properties of orthogonal substitutions. In fact, from equations (6), differentiating, squaring, and adding, we get

$$dy_i = \sum_{1}^{n}{}_j\, a_{ij}\, d\eta_j,$$

$$(dy_i)^2 = \sum_{1}^{n}{}_{j\nu}\, a_{ij}\, a_{i\nu}\, d\eta_j\, d\eta_\nu,$$

$$\sum_{1}^{n}{}_i\, (dy_i)^2 = \sum_{1}^{n}{}_{ij\nu}\, a_{ij}\, a_{i\nu}\, d\eta_j\, d\eta_\nu;$$

summing the last of these with respect to i and using (7), we get

$$\sum_{1}^{n}{}_i\, (dy_i)^2 = \sum_{1}^{n}{}_{j\nu}\, \delta_j^\nu\, d\eta_j\, d\eta_\nu = \sum_{1}^{n}{}_\nu\, d\eta_\nu^2.$$

The hypothesis that the η's are a particular solution of the system is expressed algebraically by the equation $\sum_{1}^{n}{}_\nu\, d\eta_\nu^2 = ds^2$; hence we can write

$$\sum_{1}^{n}{}_i\, (dy_i)^2 = ds^2,$$

which proves that the y's also constitute a solution. An easy calculation shows that the number of independent constants in

(6) is $\frac{1}{2}n(n + 1)$, and hence the solution so obtained is the most general.

It is obvious that the equations (6) are a generalization of the formulæ for changing the co-ordinate axes in ordinary analytical geometry.

2. Conformal representation of a manifold of constant curvature on a Euclidean space. Mutual applicability of all V_n's with the same constant curvature.

In the preceding chapter (p. 236) we solved the following problem: given a Euclidean space S_n, to find a manifold V'_n of given constant curvature which can be conformally represented on S_n. We now propose to prove that conversely, given a manifold V_n of constant curvature, it is always possible to represent it conformally on a Euclidean space S'_n. In other words, if ds^2 is the line element of a V_n of constant curvature K, we wish to prove that a suitable function $U = e^{-\tau}$ can be so chosen that

$$ds'^2 = e^{2\tau}\, ds^2 = \frac{1}{U^2}\, ds^2$$

is Euclidean.

The necessary and sufficient condition for this is that the equations (18′) of Chapter VIII, p. 232, should all be satisfied by putting

$$(ij, hk)' = 0,$$
$$(ij, hk) = K\,(a_{ih}\, a_{jk} - a_{ik}\, a_{jh}),$$

and writing U instead of u. U must therefore satisfy the $\frac{1}{12}n^2(n^2 - 1)$ equations

$$\left(\frac{\Delta U}{U^2} - K\right)(a_{ih}\, a_{jk} - a_{ik}\, a_{jh}) = a_{ih}\frac{U_{jk}}{U} - a_{ik}\frac{U_{jh}}{U} - a_{jh}\frac{U_{ik}}{U} + a_{jk}\frac{U_{ih}}{U}.$$

Putting

$$U_{ik} = a_{ik}\,(\alpha U + \beta), \quad . \quad . \quad . \quad . \quad (8)$$

where α and β are two constants, and following the same method as that used in § 7 of Chapter VIII, p. 236, we see that these equations are satisfied provided ultimately

$$\frac{\Delta U}{U^2} = K + 2\alpha + \frac{2\beta}{U}. \quad . \quad . \quad . \quad . \quad (9)$$

If we consider the equations (8) as defining all the derivatives of the n quantities U_i, then together with the identity

$$dU = \sum_1^n U_i\, dx_i \quad . \quad . \quad . \quad . \quad . \qquad (8')$$

they constitute a total differential system in the $n + 1$ functions U_i, U; the equation in finite terms (9) is to be associated with it. It is easy to verify that we need only take $\alpha = -K$ in order that this mixed system may be completely integrable (cf. Chapter II, p. 29).

In fact, the conditions of integrability of the equations (8) are expressed by the commutation formulæ (§ 6, p. 184)

$$U_{i|kl} - U_{i|lk} = -\sum_1^n{}_r \{ir, kl\}\, U_r, \quad . \quad . \qquad (C)$$

and those of the equations (8′) by

$$U_{ik} = U_{ki}.$$

These latter conditions are at once satisfied, on account of equations (8). The left-hand side of (C), also by (8), reduces to

$$\alpha(a_{ik}\, U_l - a_{il}\, U_k),$$

and the right-hand side, using the expression (19′) of the preceding chapter for Riemann's symbols for manifolds of constant curvature, becomes

$$-K(a_{ik}\, U_l - a_{il}\, U_k).$$

The equations (C) therefore reduce to identities provided, as stated above, we take $\alpha = -K$.

Lastly, there is the equation in finite terms (9); putting $\alpha = -K$, this becomes

$$\sum_1^n{}_j U^j\, U_j = -KU^2 + 2\beta U. \quad . \quad . \quad . \qquad (9')$$

Differentiating this, using formula (16′) of p. 152, and taking out the factor 2, we get the conditions

$$\sum_1^n{}_j U^j\, U_{ji} = -KUU_i + \beta U_i,$$

which are also identically satisfied in virtue of (8).

Remark.—Having thus seen that the system is completely integrable, we know (§ 8, pp. 29–33) that the solution contains n arbitrary constants which we can choose in such a way that at a specified (but perfectly general) point O of the manifold the n functions U_i take values arbitrarily fixed in advance. Further, the constant β is still at our disposal.

We get a first class of solutions if we take $\beta = 0$, which makes (8) into

$$U_{ik} = -K a_{ik} U.$$

The hypothesis $\beta = 0$ is therefore admissible *in the real field* only when $K < 0$; in fact, for $\beta = 0$ the equation (9′) reduces to

$$\Delta U = -KU^2.$$

In the real field the left-hand side is always essentially positive, excluding the case when the function U is a pure constant, or in other words (on account of equations (8), which now reduce to $U_{ik} = -K a_{ik} U$) retaining the conditions $K \neq 0$. Since the right-hand side has the opposite sign to K, it follows that the equality is possible only if $K < 0$.

In order to have a generally valid solution, we must suppose $\beta \neq 0$. We shall then choose β and the n other constants so that

$$U_i = 0, \quad U = 1 \qquad (i = 1, 2, \ldots n)$$

at the point O, so that from (9′) we see that $\beta = \frac{K}{2}$, and U will be completely determined.

With the notation of the present problem (i.e. using dashes to denote quantities relative to the *Euclidean* space) we proved in Chapter VIII, § 7, p. 236, that if a factor u exists such that the manifold for which

$$ds^2 = \frac{ds'^2}{u^2}$$

has constant curvature K, and if the conditions $u = 1$, $u_i = 0$ are satisfied at a specified point O (which may always be supposed taken as origin of Cartesian co-ordinates), then the expression for u is

$$u = 1 + \frac{K}{4} \sum_1^n {}_i\, x_i^2.$$

Further, we have now found that the quantity $\frac{1}{U}$ satisfies all these conditions (in fact $\frac{1}{U} = 1$ at O, $\left(\frac{1}{U}\right)_j = -\frac{U_j}{U^2} = 0$ at O), and therefore we must have

$$U = \frac{1}{u} = \frac{1}{1 + \frac{K}{4}\sum_1^n{}_i \, x_i{}^2}.$$

An extremely important corollary can be deduced from the foregoing results. Given two n-dimensional manifolds with the *same* constant curvature K, both their ds^2, as we have seen, can be reduced by suitable changes of variables to the *same* canonical form

$$\frac{\sum_1^n{}_\nu \, dx_\nu{}^2}{u^2}$$

where

$$u = 1 + \frac{K}{4}\sum_1^n{}_i \, x_i{}^2.$$

It is therefore possible by a change of variables to transform one form into the other; or in other words, if two manifolds of the same number of dimensions satisfy the single condition of having the same constant curvature, then either can be conformally represented on the other.

3. **General remarks on hypersurfaces in Euclidean space. Second fundamental form.**

Let S_{n+1} be a Euclidean space and $y_1, y_2, \ldots y_{n+1}$ a system of Cartesian co-ordinates in it, so that

$$ds^2 = \sum_1^{n+1}{}_\nu \, dy_\nu{}^2.$$

Consider a hypersurface V_n (frequently called merely a "surface" when there is no danger of ambiguity) immersed in S_{n+1} and defined by the parametric equations

$$y_\nu = y_\nu(x_1, x_2, \ldots x_n) \qquad (\nu = 1, 2, \ldots n+1). \quad . \quad (10)$$

As usual, the functional matrix of these equations must have n as its characteristic (cf. p. 87).

As an obvious extension of the ordinary case ($n = 2$) we shall first define the direction of S_{n+1} which is *normal* to V_n at any given point P.

Let α_ν ($\nu = 1, 2, \ldots n + 1$) denote the cosines of the direction we are in search of, relative to the axes y (i.e. the parameters or moments, which are indistinguishable in a Euclidean space). These cosines will be connected by the usual quadratic identity

$$\sum_1^{n+1}{}_\nu\, \alpha_\nu^2 = 1. \qquad \ldots \ldots \quad (11)$$

The geometrical property which we have to express is that the direction whose cosines are α_ν is perpendicular to any tangent to V_n at P, or, which is the same thing, to any elementary displacement dP which is a tangent to V_n and therefore (neglecting infinitesimals of higher order than the first) does not move outside the surface. For every such displacement the equations (10) must still be satisfied, but the increments dx_l of the x's will be otherwise arbitrary. If dy_ν denotes the corresponding increments of the Cartesian co-ordinates y_ν, the α's must satisfy the equation

$$\sum_1^{n+1}{}_\nu\, \alpha_\nu\, dy_\nu = 0 \qquad \ldots \ldots \quad (11')$$

for every system of dy's given by (10), i.e. by

$$dy_\nu = \sum_1^{n}{}_l\, y_{\nu|l}\, dx_l, \qquad \ldots \ldots \quad (12)$$

with the dx's arbitrary.

Substituting in (11′) we have

$$\sum_1^{n}{}_l\, dx_l \sum_1^{n+1}{}_\nu\, \alpha_\nu\, y_{\nu|l} = 0;$$

and since the dx's are arbitrary this means that the α's must satisfy the n equations

$$\sum_1^{n+1}{}_\nu\, \alpha_\nu\, y_{\nu|l} = 0 \qquad (l = 1, 2, \ldots n). \quad . \quad (12')$$

These equations, together with (11), determine the α's except as to sign. The ambiguity of the sign is natural, as we are dealing with a direction and have made no hypothesis as to its sense. In what follows we shall suppose the sense fixed in advance as may be most convenient.

We know that the metric of V_n is defined by the quadratic form

$$\phi = ds^2 = \sum_{1}^{n}{}_{ik} a_{ik}\, dx_i\, dx_k.$$

In addition to this it is useful to consider a second differential quadratic form ψ which differs from the first in that it depends on the configuration of V_n in S_{n+1} (or in other words is *not* an intrinsic element), or rather completely determines this configuration.

To find this function we suppose an infinitesimal segment of constant length ϵ measured off along the positive sense (as defined in advance) of the normal at every point of the given V_n. The extremities of all these segments will lie on a hypersurface V'_n, which is said to be *parallel to* V_n; there is an obvious one-to-one correspondence between points on one and points on the other. We wish to consider two infinitely near points of V_n, and to compare their distance apart ds with the distance ds' of the two corresponding points of V'_n.

If the co-ordinates of a generic point of V_n are y_ν $(\nu = 1, 2, \ldots n+1)$, those of the corresponding point of V'_n will be

$$y'_\nu = y_\nu + \alpha_\nu\, \epsilon.$$

From this, differentiating and remembering that ϵ is a constant, we get

$$dy'_\nu = dy_\nu + \epsilon\, d\alpha_\nu.$$

Squaring and adding, we get ds'^2; denoting it by ϕ', we have

$$\phi' = \sum_{1}^{n+1}{}_\nu (dy_\nu{}^2 + \epsilon^2\, d\alpha_\nu{}^2 + 2\epsilon\, dy_\nu\, d\alpha_\nu).$$

Now $\sum_{1}^{n+1}{}_\nu dy_\nu{}^2 = \phi$, and since ϵ is infinitesimal it follows that $\epsilon^2\, d\alpha_\nu{}^2$ is negligible compared with the other terms; hence

$$\phi' - \phi = -2\epsilon\psi, \quad . \quad . \quad . \quad . \quad (13)$$

where we have put

$$\psi = -\sum_{1}^{n+1}{}_{\nu}\, dy_\nu\, d\alpha_\nu. \quad . \quad . \quad . \quad . \qquad (14)$$

Formula (13) gives the increment of the first fundamental form ϕ in passing from the given V_n to an infinitely near parallel surface; this increment is expressed in terms of the quantity ψ, which, as we shall now see, is a quadratic form in the dx's. To show this, we note that

$$dy_\nu = \sum_{1}^{n}{}_{i}\, y_{\nu|i}\, dx_i,$$

$$d\alpha_\nu = \sum_{1}^{n}{}_{k}\, \alpha_{\nu|k}\, dx_k;$$

hence, substituting in (14), and putting

$$b_{ik} = b_{ki} = -\tfrac{1}{2}\sum_{1}^{n+1}{}_{\nu}\, (y_{\nu|i}\, \alpha_{\nu|k} + y_{\nu|k}\, \alpha_{\nu|i}), \quad . \qquad (15)$$

we get

$$\psi = \sum_{1}^{n}{}_{ik}\, b_{ik}\, dx_i\, dx_k. \quad . \quad . \quad . \quad . \qquad (14')$$

This is what is called the *second fundamental form.* Its coefficients b_{ik}, given by (15), can also be expressed in another way, which will be useful farther on. Differentiating (12′), we get

$$\sum_{1}^{n+1}{}_{\nu}\, (\alpha_{\nu|k}\, y_{\nu|l} + \alpha_\nu\, y_{\nu|lk}) = 0,$$

or, interchanging the indices l and k,

$$\sum_{1}^{n+1}{}_{\nu}\, (\alpha_{\nu|l}\, y_{\nu|k} + \alpha_\nu\, y_{\nu|kl}) = 0.$$

Taking the half sum of these two identities, and remembering the symmetry of the second derivatives, we get

$$-\tfrac{1}{2}\sum_{1}^{n+1}{}_{\nu}\, (\alpha_{\nu|k}\, y_{\nu|l} + \alpha_{\nu|l}\, y_{\nu|k}) = \sum_{1}^{n+1}{}_{\nu}\, \alpha_\nu\, y_{\nu|lk}.$$

Changing l into i, the left-hand side of this equality becomes the same as the right-hand side of (15), and therefore

$$b_{ik} = \sum_{1}^{n+1}{}_{\nu}\, \alpha_\nu\, y_{\nu|ik}. \quad . \quad . \quad . \quad . \qquad (15')$$

4. Forms of class 1 (hypersurfaces in Euclidean space).

We now wish to find a criterion to determine whether a given differential quadratic form

$$ds^2 = \sum_{1}^{n}{}_{ik} a_{ik}\, dx_i\, dx_k$$

is of class 1, i.e. whether we can find $n + 1$ functions (10) which will reduce it to the Cartesian type. We shall follow a method similar to that used in § 1, taking as unknowns the $n + 1$ functions y_ν and their $n(n + 1)$ derivatives $y_{\nu|i}$, making $(n + 1)^2$ unknowns in all. By definition these must reduce the given ds^2 to the Euclidean form $\sum_{1}^{n+1}{}_\nu\, dy_\nu{}^2$; this is expressed by the $\frac{1}{2}n(n + 1)$ conditions

$$a_{ik} = \sum_{1}^{n+1}{}_\nu\, y_{\nu|i}\, y_{\nu|k}. \quad . \quad . \quad . \quad . \quad . \qquad (16)$$

From these by covariant differentiation we get the equations

$$0 = \sum_{1}^{n+1}{}_\nu\, (y_{\nu|il}\, y_{\nu|k} + y_{\nu|i}\, y_{\nu|kl}). \quad . \quad . \quad . \qquad (17)$$

We have also the condition that the principal unknowns y_ν and the auxiliary unknowns $y_{\nu|i}$ are not independent but are connected by the differential relations

$$\frac{\partial y_\nu}{\partial x_i} = y_{\nu|i}. \quad . \quad . \quad . \quad . \quad . \qquad (18)$$

We have to determine the conditions of integrability of the system composed of (16), (17), (18).

First, suppose written down the two equations obtained from (17) by cyclic interchange of i, k, l; from these three equations, by adding two of them and subtracting the third, we find, as in § 1,

$$\sum_{1}^{n+1}{}_\nu\, y_{\nu|ik}\, y_{\nu|l} = 0 \qquad (i, k, l = 1, 2, \ldots n).$$

Keeping i and k fixed, we have n linear homogeneous equations in the $n + 1$ unknowns $y_{\nu|ik}$. The matrix of the coefficients $y_{\nu|l}$, as in the preceding section, has n for its characteristic; hence the equations have $(n + 1) - n = 1$ independent solution; the others differ from it by a multiplier. Now we see from (12′) that we get one solution by taking $y_{\nu|ik} = a_\nu$; hence, introducing a multiplier b_{ik}, we can write the most general solution in the form

$$y_{\nu|ik} = b_{ik}\, a_\nu. \quad . \quad . \quad . \quad . \quad . \quad . \qquad (19)$$

To find the significance of these b's, multiply (19) by a_ν and sum with respect to ν from 1 to $n + 1$, using (11); we get

$$\sum_1^{n+1}{}_\nu\, a_\nu\, y_{\nu|ik} = b_{ik};$$

comparing this with (15′), we find that the b's just introduced (which have the property $b_{ik} = b_{ki}$) are identical with the coefficients of the second fundamental form.

We have now to express the fact that the second covariant derivatives of the quantities $y_{\nu|i}$ satisfy the commutation formula

$$y_{\nu|ihk} - y_{\nu|ikh} = -\sum_1^n{}_l\, \{il, hk\}\, y_{\nu|l}, \quad . \quad . \qquad (20)$$

which takes the place of the ordinary condition of symmetry of the second derivatives. To calculate the left-hand side we must start from (19). By covariant differentiation we get

$$y_{\nu|ikh} = a_\nu\, b_{ikh} + b_{ik}\, a_{\nu|h}, \quad . \quad . \quad . \qquad (21)$$

and we have to calculate $a_{\nu|h}$. To do this, we note that on differentiating (11) we have

$$\sum_1^{n+1}{}_\nu\, a_\nu\, a_{\nu|h} = 0, \quad . \quad . \quad . \quad . \quad . \quad . \qquad (22)$$

and also that the coefficients b_{ih} can also be expressed in the form

$$\sum_1^{n+1}{}_\nu\, y_{\nu|i}\, a_{\nu|h} = -\, b_{ih} \qquad (i, h = 1, 2, \ldots n), \ . \qquad (23)$$

which is at once verified by covariant differentiation of the identity

$$\sum_1^{n+1}{}_\nu\, y_{\nu|i}\, a_\nu = 0$$

combined with the expression (15′) for the b's. If h is fixed, the formula (23) represents n linear equations in the $n + 1$ unknowns $a_{\nu|h}$; combined with (22) they form a system which can determine these unknowns. The determinant of this system is in fact

$$\begin{vmatrix} \alpha_1 & \alpha_2 & \dots & \alpha_{n+1} \\ y_{1|1} & y_{2|1} & \dots & y_{n+1|1} \\ \cdot & \cdot & \cdot & \cdot \\ y_{1|n} & y_{2|n} & \dots & y_{n+1|n} \end{vmatrix};$$

squaring this, and remembering (11), (12′), and (16), we get the determinant $\| a_{ik} \|$, which is certainly not zero.

It is easy to verify that the solution of the system (22), (23) is

$$\alpha_{\nu|h} = -\sum_1^n {}_j\, b_{jh}\, y_\nu^j, \quad . \quad . \quad . \quad . \quad (24)$$

where we have put

$$y_\nu^j = \sum_1^n {}_k\, a^{jk}\, y_{\nu|k}. \quad . \quad . \quad . \quad . \quad (25)$$

Hence (21) becomes

$$y_{\nu|ikh} = \alpha_\nu\, b_{ikh} - \sum_1^n {}_j\, b_{ik}\, b_{jh}\, y_\nu^j.$$

The expression for $y_{\nu|ihk}$ is obtained from this by interchanging the indices h and k. We can therefore write (20) in the form

$$\alpha_\nu\,(b_{ikh} - b_{ihk}) - \sum_1^n {}_j\,(b_{ik}\, b_{jh} - b_{ih}\, b_{jk})\, y_\nu^j = \sum_1^n {}_l \{il, hk\}\, y_{\nu|l}. \quad (26)$$

In order to express the right-hand side too in terms of y_ν^j we apply Cramer's usual rule to (25), which gives

$$y_{\nu|l} = \sum_1^n {}_j\, a_{jl}\, y_\nu^j,$$

and substitute this result in the sum; summing with respect to l we have

$$\sum_1^n {}_l \{il, hk\}\, y_{\nu|l} = \sum_1^n {}_j\, (ij, hk)\, y_\nu^j,$$

and therefore (26) becomes

$$\left.\begin{array}{c} -\alpha_\nu\,(b_{ikh} - b_{ihk}) + \sum_1^n {}_j\, y_\nu^j\, [(b_{ik}\, b_{jh} - b_{ih}\, b_{jk}) + (ij, hk)] = 0 \\ (\nu = 1, 2, \dots n+1;\ i, k, h = 1, 2, \dots n). \end{array}\right\} . \quad (27)$$

These conditions can be expressed in a considerably simpler form. Multiply the equality just written by α_ν and sum with respect to ν from 1 to $n+1$; remembering (11) and observing that from (25) and (12′)

$$\sum_{1}^{n+1}{}_\nu \alpha_\nu y_\nu^j = \sum_{1}^{n}{}_k a^{jk} \sum_{1}^{n+1}{}_\nu y_{\nu|k} \alpha_\nu = 0,$$

we get $$(b_{ihk} - b_{ikh}) = 0,$$

or in other words the coefficients b must satisfy the condition

$$b_{ihk} = b_{ikh}. \quad . \quad . \quad . \quad . \quad . \quad (28)$$

The condition (26) can then be written in the form

$$\sum_{1}^{n}{}_j y_\nu^j p_{ij,\,hk} = 0 \qquad (\nu = 1, 2, \ldots n+1), \qquad (29)$$

where we have put

$$p_{ij,\,hk} = (b_{ik} b_{jh} - b_{ih} b_{jk}) + (ij,\, hk).$$

Keeping i, h, k fixed, the equations (29) constitute $n+1$ linear homogeneous equations in the n unknowns $p_{ij,\,hk}$ $(j = 1, 2, \ldots n)$. The characteristic of the matrix of the coefficients y_ν^j is n; in fact, taking any one of its determinants of order n, e.g.

$$\begin{vmatrix} y_1^1 & y_1^2 & \cdots & y_1^n \\ y_2^1 & y_2^2 & \cdots & y_2^n \\ \cdot & \cdot & \cdots & \cdot \\ y_n^1 & y_n^2 & \cdots & y_n^n \end{vmatrix},$$

it will easily be seen with the help of (25) that it is equal to the product of the two determinants

$$\begin{vmatrix} y_{1|1} & y_{1|2} & \cdots & y_{1|n} \\ y_{2|1} & y_{2|2} & \cdots & y_{2|n} \\ \cdot & \cdot & \cdots & \cdot \\ y_{n|1} & y_{n|2} & \cdots & y_{n|n} \end{vmatrix} \cdot \begin{vmatrix} a^{11} & a^{12} & \cdots & a^{1n} \\ a^{21} & a^{22} & \cdots & a^{2n} \\ \cdot & \cdot & \cdots & \cdot \\ a^{n1} & a^{n2} & \cdots & a^{nn} \end{vmatrix}$$

the second of which is certainly not zero. It follows that the characteristic of the matrix y_ν^j is the same as that of the matrix $y_{\nu|j}$, which is n. From a well-known theorem on linear equations it follows that the system (29) has no solutions except

$$p_{ij,\,hk} = 0 \qquad (i, j, h, k = 1, 2, \ldots n),$$

which is the same as

$$(ij, hk) = b_{ih}\, b_{jk} - b_{ik}\, b_{jh}. \quad . \quad . \quad . \quad (30)$$

A more rigorous discussion would show that the formulæ (28) and (30) express *all* the conditions of integrability of the system. We can therefore conclude that:

The necessary and sufficient conditions that a given differential quadratic form may be of class 1 *are that it shall be possible to determine a (real) symmetrical double system* b_{ik} *such that Riemann's symbols for the given form can be expressed by formula* (30), *and also such that the system* b_{ikh} (*the covariant derivative of* b_{ik} *with respect to the given differential form*) *is symmetrical* (*formula* (28)).

At the end of last chapter (§ 7, p. 236) we found directly, by assigning suitable explicit expressions to the functions $y_\nu(x_1, x_2, \ldots x_n)$, that every ds^2 of constant *positive* curvature is of class 1. The necessary and sufficient conditions just enumerated must of course be satisfied.

To verify this, we need only take the auxiliary quantities b_{ik} in the form $\sqrt{K}\, a_{ik}$, and remember that, as the manifold in question by hypothesis is of constant curvature, Riemann's symbols (ij, hk) take the form $K(a_{ih}\, a_{jk} - a_{ik}\, a_{jk})$. The conditions (30) are therefore automatically satisfied. Further, by Ricci's lemma, the covariant derivatives of the quantities b_{ik}, i.e. of $\sqrt{K}\, a_{ik}$, vanish, so that the conditions (28) are also satisfied.

For $K < 0$, the hypothesis $b_{ik} = \sqrt{K}\, a_{ik}$ is of no use, as it would take us out of the real field, so that we cannot assert that the analogous property holds. We can in fact prove that for $n > 2$ a ds^2 of constant *negative* curvature is not of class 1.[1] For $n = 2$ we know already (§ 21, p. 123) that any ds^2, and therefore in particular a ds^2 of constant curvature, is of class 1

[1] Cf. BIANCHI: *Lezioni di geometria differenziale*, 2nd edition (Pisa, Spoerri, 1902), Vol. I, Ch. XIV, § 205, p. 471.

(at most), or in other words belongs certainly to some surface of ordinary space. There are an infinite number of surfaces of this kind (pseudospherical surfaces), with constant negative K, including surfaces of revolution of three types.[1]

5. Hyperspherical representation and curvature of a hypersurface.

Take any hypersurface V_n, and consider it as immersed in a Euclidean space S_{n+1}; and consider also a hypersphere of unit radius and centre the origin.[2]

We can make each point P of the V_n correspond to a point P' of the hypersphere by drawing from the centre of the latter the parallel to the normal to the V_n at P, and taking the intersection of this parallel with the hypersphere as P'; V_n is then said to be represented on the hypersphere.

The chief interest of this representation is as follows. Let V denote the extension (Chapter VI, p. 160) of a region ϕ of V_n, and V' the extension of the corresponding hyperspherical region ϕ'. Then the ratio $\frac{V'}{V}$ is closely related to the curvature properties of V_n, and is called the *mean curvature of* V_n *in the region* ϕ. If this region reduces to the infinitesimal region round a point P—or in other words if the maximum dimension of ϕ tends to zero—then (if P is not a singular point) the ratio $\frac{V'}{V}$ tends to a positive limit Γ, which is called the *hyperspherical* (if $n = 2$, the spherical) *curvature* of the V_n at P.

To find an expression for this quantity, we shall first establish a system of intrinsic co-ordinates on the hypersphere. The most obvious way of doing this is to assign to each point P' of the hypersphere the co-ordinates $x_1, x_2, \ldots x_n$ of the corresponding point P of V_n. We shall call the line element of the hypersphere $d\sigma$, and shall try to find an expression for it in terms of the dx's.

[1] *Ibid.*, Ch. VII, § 103, p, 225; or 3rd edition (Bologna, Zanichelli, 1922), Vol. I, Ch. VII, § 127, p. 338.

[2] That is to say, as explained in § 7, Chapter VIII, p. 240, a hypersurface Ω_n whose equation in Cartesian co-ordinates is

$$\sum_1^{n+1} {}_\nu\, y_\nu^2 = 1.$$

If we denote the direction cosines of the normal to V_n at a generic point P by α_ν ($\nu = 1, 2, \ldots n+1$), as in preceding sections, then the direction cosines of the parallel through the origin (the centre of the hypersphere) to this normal will also be α_ν. The point P' lies on this line, at unit distance from the origin; its Cartesian co-ordinates are therefore α_ν.

We then have at once

$$d\sigma^2 = \sum_{1}^{n+1}{}_\nu \, d\alpha_\nu^{\,2}$$

and putting
$$\sum_{1}^{n+1}{}_\nu \, \alpha_{\nu|h} \, \alpha_{\nu|k} = e_{hk}, \quad . \quad . \quad . \quad . \quad (31)$$

it follows that
$$d\sigma^2 = \sum_{1}^{n}{}_{hk} \, e_{hk} \, dx_h \, dx_k. \quad . \quad . \quad . \quad . \quad (32)$$

This is the first fundamental form relative to the hypersphere; it is sometimes called the *third* fundamental form of the given V_n. By means of it we can at once calculate the extension V' of a hyperspherical region ϕ':

$$V' = \int_{\phi'} \sqrt{e} \, dx_1 \, dx_2 \ldots dx_n,$$

where e represents the determinant of the quantities e_{hk}. Analogously, for the corresponding field of V_n, we have the extension V of ϕ:

$$V = \int_{\phi} \sqrt{a} \, dx_1 \, dx_2 \ldots dx_n.$$

If the regions considered are infinitesimal, each integral reduces to a single element; taking the ratio of these, we get

$$\Gamma = \operatorname*{Lt}_{\phi=0} \frac{V'}{V} = \sqrt{\frac{e}{a}}, \quad . \quad . \quad . \quad . \quad (33)$$

where in every case the radicals are of course supposed to have their absolute values.

The coefficients e_{hk} can be expressed in terms of the derivatives of the y's by means of (31), which on substituting for $\alpha_{\nu|h}$ the values given by (24) becomes

$$e_{hk} = \sum_{1}^{n}{}_{ij} \, b_{ih} \, b_{jk} \sum_{1}^{n+1}{}_\nu \, y_\nu^i \, y_\nu^j,$$

and by (25) and (16)

$$e_{hk} = \sum_{1}^{n}{}_{ijuv} b_{ih} b_{jk} a^{iu} a^{jv} \sum_{1}^{n+1}{}_{\nu} y_{\nu|u} y_{\nu|v}$$

$$= \sum_{1}^{n}{}_{ijuv} b_{ih} b_{jk} a^{iu} a^{jv} a_{uv}$$

$$= \sum_{1}^{n}{}_{ij} b_{ih} b_{jk} a^{ij}.$$

From this expression of the e_{ik}'s in terms of the b_{ik}'s and the a^{ik}'s it is easy to obtain an expression for the determinant e in terms of the determinants a and b. To find it, put

$$\beta_k^i = \sum_{1}^{n}{}_{j} b_{jk} a^{ij}, \quad . \quad . \quad . \quad . \quad . \quad (34)$$

so that the last of the formulæ just given for e_{hk} may be written as

$$e_{hk} = \sum_{1}^{n}{}_{i} b_{ih} \beta_k^i. \quad . \quad . \quad . \quad . \quad . \quad (34')$$

Comparing (34) and (34′) with the formulæ for the general term in the product of two determinants, we see that from them follow the two equations

$$\beta = b \frac{1}{a}, \quad . \quad . \quad . \quad . \quad . \quad . \quad (35)$$

$$e = b\beta, \quad . \quad . \quad . \quad . \quad . \quad . \quad (35')$$

where $\frac{1}{a} = \| a^{ik} \|$ and we have put $\beta = \| \beta_k^i \|$, as can easily be verified. Multiplying together (35) and (35′) term by term we have

$$e = \frac{b^2}{a}. \quad . \quad . \quad . \quad . \quad . \quad . \quad (36)$$

Hence (33) becomes

$$\Gamma = \left| \frac{b}{a} \right|, \quad . \quad . \quad . \quad . \quad . \quad . \quad (33')$$

a formula expressing the hyperspherical curvature Γ in terms of the discriminants of the two fundamental forms.

It will be seen that the curvature defined here is not an intrinsic property, as it depends on the coefficients b_{ik}.

Let us apply these remarks to an ordinary surface V_2 immersed in a three-dimensional Euclidean space. In this case, as we know, there is one distinct Riemann's symbol (12, 12), and (30) gives

$$(12, 12) = b_{11} b_{22} - b_{12}^2 = b.$$

Hence (33′) can be written in the form

$$\Gamma = \left| \frac{(12, 12)}{a} \right|.$$

Comparing this with formula (28) on p. 194, we see that for $n = 2$, the curvature Γ coincides in absolute value with the Gaussian curvature K.

CHAPTER X

Some Applications of Intrinsic Geometry

1. General remarks on congruences. Geodesic and normal congruences.

Consider a metric manifold V_n, and suppose that at every point of V_n (or of a region of V_n) there is fixed a direction $\boldsymbol{\lambda}$, defined e.g. by its parameters λ^i; i.e. that there is given a contravariant system of regular functions $\lambda^i(x_1, x_2, \ldots x_n)$, connected only by the usual quadratic identity and otherwise arbitrary. On account of this identity one at least of the parameters λ^i is certainly not zero.

If then we consider the following system of $n - 1$ differential equations

$$\frac{dx_1}{\lambda^1} = \frac{dx_2}{\lambda^2} = \ldots = \frac{dx_n}{\lambda^n} \quad . \quad . \quad . \quad (1)$$

(considering e.g. one of the x's as the independent variable and the other $n - 1$ as unknown functions of the first), we see at once that the integrals of this system represent lines of V_n which at every point are in the previously fixed direction $\boldsymbol{\lambda}$; in fact, for an infinitesimal displacement along one of these lines, the dx's are proportional to the parameters of $\boldsymbol{\lambda}$. Through every point of the region considered there passes one (and only one) of these lines; this follows from the fact that the general integral of (1) contains $n - 1$ arbitrary constants, which can be so determined that for an arbitrarily assigned value of the independent variable the other $n - 1$ variables have values which are also arbitrarily assigned. To fix the ideas suppose that in the field considered λ^n is not zero; then (1) can be written

$$\frac{dx_i}{dx_n} = \frac{\lambda^i}{\lambda^n} \qquad (i = 1, 2, \ldots n - 1),$$

considering x_n as the independent variable.

It follows from the existence theorem that the integral equations

$$x_i = x_i(x_n) \qquad (i = 1, 2, \ldots n - 1)$$

of the line can be satisfied by an arbitrary set of values of the n variables, which is equivalent to saying that the line can be made to pass through a point arbitrarily fixed in advance.[1]

Such a system of lines is called a *congruence*. The quantities $\lambda^i = \frac{dx_i}{ds}$, where ds denotes the element of arc of the line passing through the generic point $(x_1, x_2, \ldots x_n)$, are called the *parameters* of the congruence, and the elements λ_i of the reciprocal system are its *moments*.

If all the lines of a congruence are geodesics, the congruence is said to be *geodesic*; e.g. congruences of straight lines in ordinary space. It is easy to determine the analytical condition which expresses this property. We know that the characteristic equations of a geodesic can be put in the following form (cf. Chapter V, formula (53), p. 141):

$$p^i = \frac{d\lambda^i}{ds} + \sum_{1}^{n}{}_{jl} \{jl, i\} \lambda^j \dot{x}_l = 0,$$

where $\lambda^i = \frac{dx_i}{ds} = \dot{x}_i$.

[1] The argument may be made clearer by considering the example of a field of force in ordinary physics. In this case, when a direction λ (that of the force) is physically defined at every point of the space considered, then a system of lines (the lines of force) is determined which have at every point the direction of the force at that point and which, so to speak, fill all space, as through every point there passes one (and only one) line of the system.

Now we have
$$\frac{d\lambda^i}{ds} = \sum_1^n {}_l \frac{\partial \lambda^i}{\partial x_l} \dot{x}_l;$$
substituting in the previous equation and writing everywhere λ^l instead of $\dot{x}_l$, we get
$$p^i = \sum_1^n {}_l \left(\frac{\partial \lambda^i}{\partial x_l} + \sum_1^n {}_j \{jl, i\} \lambda^j \right) \lambda^l = 0,$$
from which, by (5′), p. 147, we get
$$p^i = \sum_1^n {}_l (\lambda^i)_l \lambda^l = 0 \qquad (i = 1, 2, \ldots n). \quad . \quad (2)$$

These are the required conditions. We can express them partly in terms of the *moments* by multiplying by a_{ik} and summing with respect to i, which gives
$$p_k = \sum_1^n {}_{il} a_{ik} (\lambda^i)_l \lambda^l = 0;$$
and as by Ricci's lemma
$$a_{ik} (\lambda^i)_l = (a_{ik} \lambda^i)_l,$$
we get finally
$$p_k = \sum_1^n {}_l \lambda_{kl} \lambda^l = 0 \qquad (k = 1, 2, \ldots n). \quad . \quad (2')$$

Another important special property which a congruence may have is that of being *normal*, i.e. that of being composed of the orthogonal trajectories of a family of surfaces. It should be noted here that, given a family of surfaces, there always exists a congruence of curves which cut all the surfaces of the family at right angles and are called *orthogonal trajectories*; while there does not always exist a family of surfaces which cut at right angles all the curves of a congruence. This can be shown as follows.[1]

First, let there be given a generic family of surfaces whose equation is
$$f(x) = \text{constant}.$$

[1] It may be noted incidentally that in chapter V, p. 127, we have already recognized the existence of the directions normal to the families of co-ordinate surfaces x_i = constant, and determined their moments. These results could have been used here, as any family of surfaces f = constant can always be turned into co-ordinate surfaces by a change of variables. The line of argument followed in the text has the advantage of giving directly the explicit expression for the moments of the normal directions when the equation of the family of surfaces is general.

Consider the surface which passes through a specified point P whose co-ordinates are $x_1, x_2, \ldots x_n$; it is understood that P is regular, i.e. that the first derivatives $\frac{\partial f}{\partial x_i}$ are finite and continuous at P and are not all zero. We wish to show that a direction perpendicular to the surface, i.e. to every displacement δx_i belonging to the surface, is uniquely associated with P.

We first note that for every such displacement δx_i we have

$$f(x + \delta x) = f(x)$$

or

$$\sum_1^n {}_i \frac{\partial f}{\partial x_i} \delta x_i = 0. \quad . \quad . \quad . \quad . \quad . \quad (3)$$

If we denote by λ_i the moments of the hypothetical perpendicular direction, then the condition of perpendicularity to every displacement in the surface is expressed by the relation

$$\sum_1^n {}_i \lambda_i \delta x_i = 0, \quad . \quad . \quad . \quad . \quad . \quad (4)$$

which must hold for all values of the δx_i's which satisfy (3). The coefficients in (3) and (4) of each δx_i must therefore be proportional (cf. § 3, p. 250). In virtue of the quadratic identity

$$\sum_1^n {}_{ik} a^{ik} \lambda_i \lambda_k = 1$$

the moments cannot all be zero, so that we can suppose that one of them, say λ_n, is not zero, and put $\frac{1}{\lambda_n} \frac{\partial f}{\partial x_n} = \rho$. Writing f_i instead of $\frac{\partial f}{\partial x_i}$ for shortness, the explicit relations equivalent to (4) take the form

$$f_i = \rho \lambda_i \qquad (i = 1, 2, \ldots n). \quad . \quad . \quad (5)$$

The f_i's being known, these equations determine the λ_i's, except for a factor, which in turn is determined (except in sign) by the above-mentioned quadratic identity, which gives $\sum_1^n {}_{ik} a_{ik} f_i f_k = \rho^2$. The left-hand side cannot vanish, as by hypothesis one at least of the f_i's is not zero; we are therefore sure that $\rho \neq 0$. Thus given the family of surfaces $f =$ constant, the orthogonal direction at each point is uniquely determined; the positive sense on this direction can be chosen at will (corresponding to the double sign of ρ). The λ_i's being known as functions of position, the reciprocal elements λ^i can be obtained from them, and thence, by (1), we get a congruence of lines which cut orthogonally the surfaces of the given family.

Vice versa, given *a priori* a congruence of lines by means of their moments λ_i (to be considered as given functions of position), then in order that the lines of the congruence may be considered as orthogonal trajectories of a family of surfaces $f =$ constant the necessary condition is that the derivatives of the function $f(x_1, x_2, \ldots x_n)$ (which is *a priori* unknown) should satisfy (5), in which ρ denotes a factor which is not zero, but is *a priori* undetermined. Such an f does not always exist; we have indeed already seen that the necessary and sufficient conditions for its existence are (Chapter II, p. 29, formula (23))

$$\left.\begin{array}{c} X_i\left(\dfrac{\partial X_j}{\partial x_k}-\dfrac{\partial X_k}{\partial x_j}\right)+X_j\left(\dfrac{\partial X_k}{\partial x_i}-\dfrac{\partial X_i}{\partial x_k}\right)+X_k\left(\dfrac{\partial X_i}{\partial x_j}-\dfrac{\partial X_j}{\partial x_i}\right)=0 \\ (i, j, k = 1, 2, \ldots n), \end{array}\right\} \quad (6)$$

where we must now take $X_i = \lambda_i$, $X_j = \lambda_j$, $X_k = \lambda_k$. Only some of these conditions are distinct, e.g. those in which the index k has the fixed value n (the conditions (20) of p. 27), the others being deducible from them.

2. Sets of n congruences. Determination of a vector by n invariants.

We shall now consider n congruences of lines in a generic V_n; thus n directions $\boldsymbol{\lambda}_1, \boldsymbol{\lambda}_2, \ldots \boldsymbol{\lambda}_n$ will be fixed at each point. We shall further suppose that every two of these directions are orthogonal, and we shall then say that we have fixed in V_n a *set of* n *orthogonal congruences.*

The parameters and moments of these congruences will of course have two indices, the first of which represents the ordinal number of the congruence. We shall use the term *the congruence* (h) to denote the congruence whose parameters are $\lambda_h^1, \lambda_h^2, \ldots \lambda_h^n$, and whose moments are therefore the reciprocal elements $\lambda_{h|1}$, $\lambda_{h|2}, \ldots \lambda_{h|n}$ (with respect to the ds^2 of the manifold).

In addition to the usual quadratic identities we shall here have the conditions of orthogonality of the congruences. Both sets are included in the formula

$$\sum_1^n {}_i \lambda_{h|i} \lambda_k^i = \delta_h^k \qquad (h, k = 1, 2, \dots n); \quad . \quad . \quad (7)$$

if $h = k$, this is the usual relation between parameters and moments, and if $h \neq k$ it expresses the fact that the directions $\boldsymbol{\lambda}_h$ and $\boldsymbol{\lambda}_k$, i.e. the congruences (h) and (k), are orthogonal.

The equations (7) also express the essential fact that the n^2 parameters λ_k^i of a set of n orthogonal congruences are the reciprocal elements (in the algebraic sense) of the n^2 moments $\lambda_{h|i}$ of the same set of congruences, and vice versa (cf. Chapter IV, p. 74; Chapter VII, p. 206). In addition to (7) the equivalent formulæ

$$\sum_1^n {}_h \lambda_{h|i} \lambda_h^j = \delta_i^j \qquad (i, j = 1, 2, \dots n) \quad . \quad (7')$$

therefore hold. Multiplying these by a_{jk} and summing with respect to j we get the important formula

$$a_{ik} = \sum_1^n {}_h \lambda_{h|i} \lambda_{h|k} \qquad (i, k = 1, 2, \dots n) \quad . \quad (7'')$$

giving the coefficients of ds^2 in terms of the moments of any set of n orthogonal congruences. Analogously, multiplying (7′) by a^{ik}, summing with respect to i, and then putting i instead of j, we get

$$a^{ik} = \sum_1^n {}_h \lambda_h^i \lambda_h^k \qquad (i, k = 1, 2, \dots n). \quad . \quad (7''')$$

A vector $\mathbf{R}$ of our V_n is determined, as we know, by its covariant components R_i or its contravariant components R^i. Hence when a set of n congruences is fixed in V_n, the vector can also be determined by its n projections on the directions belonging to these congruences at the point where the vector is considered. By definition (Chapter V, p. 126), the projection of $\mathbf{R}$ on the direction $\boldsymbol{\lambda}_h$ is the invariant

$$c_h = \mathbf{R} \times \boldsymbol{\lambda}_h,$$

which can be expressed in either of the two equivalent forms

$$c_h = \sum_1^n{}_i R^i \lambda_{h|i}, \quad \ldots \ldots \quad (8)$$

$$c_h = \sum_1^n{}_i R_i \lambda_h^i. \quad \ldots \ldots \quad (9)$$

Thus the vector **R** is determined by the n invariants c_h. If we wish to deduce from these, in a given system of reference, the covariant or contravariant components, we need only solve the equations (8) or (9), which, together with (7′), give

$$R^i = \sum_1^n{}_h c_h \lambda_h^i, \quad \ldots \ldots \quad (8')$$

$$R_i = \sum_1^n{}_h c_h \lambda_{h|i}. \quad \ldots \ldots \quad (9')$$

If in particular the vector **R** is the gradient of an invariant f (i.e. if the components R_i are the derivatives f_i of f with respect to the variables x_i), then the invariant c_h represents the intrinsic derivative of f in the direction of the congruence (h). In fact, if s_h denotes the length of the arc of one of the lines of the congruence (h), measured from an arbitrary origin, then for a displacement ds_h along this direction the increment of f will be

$$df = \sum_1^n{}_i f_i \, dx_i,$$

where the dx's are the differentials corresponding to this displacement.

Dividing this quantity by ds_h we get by definition the derivative $\frac{\partial f}{\partial s_h}$ of f in the direction of the congruence (h); remembering that $\frac{dx_i}{ds_h} = \lambda_h^i$, we therefore have

$$\frac{\partial f}{\partial s_h} = \sum_1^n{}_i f_i \lambda_h^i, \quad \ldots \ldots \quad (10)$$

a formula corresponding to (9). Solving it, we get the formula

$$f_i = \sum_1^n{}_h \frac{\partial f}{\partial s_h} \lambda_{h|i}. \quad \ldots \ldots \quad (10')$$

corresponding to (9′); and lastly, changing to the reciprocal elements, we get also

$$f^i = \sum_1^n{}_h \frac{\partial f}{\partial s_h} \lambda^i_h, \quad . \quad . \quad . \quad . \quad . \qquad (10'')$$

which corresponds to (8′).

In general, it would be easy to show that when a set of n congruences is fixed a tensor of rank m can be determined by n^m *invariants*, instead of by that number of components, covariant, contravariant, or mixed, the proof being completely analogous to that given above for determining a vector by means of n invariants. This result simplifies the study of certain questions, so that we shall find it useful to carry somewhat further our investigations on sets of n congruences.

3. Geometrical definition of Ricci's coefficients of rotation.

We must now introduce a system of differential invariants which are closely connected with the set of n congruences. We shall reach the required result quickly by the following method.

Consider two very near points P and P' of V_n. At each of them the lines of the n congruences determine a *pyramid* (a generalization of the notion of the trihedron) whose directions are mutually orthogonal. If $\boldsymbol{\lambda}_1, \ldots \boldsymbol{\lambda}_n$ are the n directions at P, those at P' will be $\boldsymbol{\lambda}'_1 = \boldsymbol{\lambda}_1 + \delta'\boldsymbol{\lambda}_1, \ldots \boldsymbol{\lambda}'_n = \boldsymbol{\lambda}_n + \delta'\boldsymbol{\lambda}_n$, and we shall say that we pass from the first to the second by *local displacement*, i.e. by the law previously fixed which regulates the behaviour of the $n\,\infty^{(n-1)}$ lines of the set of n congruences. But the pyramid of directions can also be moved from P to P' by *parallel displacement*; we shall then get at P' n mutually orthogonal directions $\boldsymbol{\lambda}^*_1 = \boldsymbol{\lambda}_1 + \delta^*\boldsymbol{\lambda}_1, \ldots \boldsymbol{\lambda}^*_n = \boldsymbol{\lambda}_n + \delta^*\boldsymbol{\lambda}_n$, which will not in general coincide with those obtained by local displacement. We shall thus have at P' two pyramids infinitely near one another, since each is infinitely near the pyramid $\boldsymbol{\lambda}_1, \boldsymbol{\lambda}_2, \ldots \boldsymbol{\lambda}_n$. This means in particular that the ith direction of one makes an infinitesimal angle with the ith direction of the other, and an angle very nearly equal to $\frac{\pi}{2}$ with the remaining $n - 1$ directions of the other. We propose to examine these infinitesimal differences.

Consider two directions $\boldsymbol{\lambda}_h$, $\boldsymbol{\lambda}_k$ of the pyramid at P; these either coincide ($h = k$) or are orthogonal, so that we have

$$\cos\widehat{\boldsymbol{\lambda}_h\boldsymbol{\lambda}_k} = \delta_h^k.$$

Let them be displaced to P', the first by local and the second by parallel displacement, so that the first will coincide with $\boldsymbol{\lambda}'_h$ and the second with $\boldsymbol{\lambda}^*_k$. We shall calculate the resulting change in the cosine of the angle between them, i.e. the quantity

$$\delta \cos\widehat{\boldsymbol{\lambda}_h\boldsymbol{\lambda}_k} = \cos\widehat{\boldsymbol{\lambda}'_h\boldsymbol{\lambda}^*_k} - \cos\widehat{\boldsymbol{\lambda}_h\boldsymbol{\lambda}_k}.$$

This is an infinitesimal of the same order as the distance ds between P and P', and we shall therefore write it in the form $p_{hk}\, ds$; thus p_{hk} will give us a kind of measure of the rate at which the cosine in question changes for a displacement in the direction PP'. To calculate it, we start from the formula

$$\cos\widehat{\boldsymbol{\lambda}_h\boldsymbol{\lambda}_k} = \sum_1^n{}_i\, \lambda_{h|i}\, \lambda_k^i$$

and differentiate it, remembering that we have to operate on $\lambda_{h|i}$ with the symbol δ' (local displacement) and on λ_k^i with the symbol δ^* (parallel displacement). We shall get

$$p_{hk}\, ds = \delta \cos\widehat{\boldsymbol{\lambda}_h\boldsymbol{\lambda}_k} = \sum_1^n{}_i\, (\delta'\, \lambda_{h|i} \cdot \lambda_k^i + \lambda_{h|i} \cdot \delta^*\, \lambda_k^i). \quad . \quad (11)$$

We have also from the ordinary rule of the differential calculus

$$\delta'\, \lambda_{h|i} = \sum_1^n{}_j \frac{\partial \lambda_{h|i}}{\partial x_j}\, \delta x_j,$$

where δx_j denotes the increment of the co-ordinate x_j in passing from P to P', and from the law of parallelism

$$\delta^*\, \lambda_k^i = - \sum_1^n{}_{jl}\, \{jl, i\}\, \lambda_k^l\, \delta x_j.$$

Substituting in (11) we get

$$p_{hk}\, ds = \sum_1^n{}_{ij}\, \lambda_k^i \frac{\partial \lambda_{h|i}}{\partial x_j}\, \delta x_j - \sum_1^n{}_{ijl}\, \lambda_{h|i}\, \{jl, i\}\, \lambda_k^l\, \delta x_j.$$

In the second sum interchange the indices i and l, so as to get the same factor $\lambda_k^i \, \delta x_j$ as in the first. We can then write

$$p_{hk} \, ds = \sum_1^n {}_{ij} \left[\frac{\partial \lambda_{h|i}}{\partial x_j} - \sum_1^n {}_l \{ji, l\} \lambda_{h|l} \right] \lambda_k^i \, \delta x_j,$$

or, remembering formula (5) of p. 147,

$$p_{hk} \, ds = \sum_1^n {}_{ij} \lambda_{h|ij} \lambda_k^i \, \delta x_j. \quad . \quad . \quad . \quad (11')$$

Denoting the parameters of the direction PP' by

$$\xi^j = \frac{\delta x_j}{ds},$$

we have the formula

$$p_{hk} = \sum_1^n {}_{ij} \lambda_{h|ij} \lambda_k^i \, \xi^j, \quad . \quad . \quad . \quad . \quad (11'')$$

which holds for any direction whatever.

It is to be noted that p_{hk}, as given by the original definition (11), changes sign when the two indices are interchanged. This can be proved without difficulty, either from the final expression (11″), by going back to (7) and taking its covariant derivative; or more geometrically, by using the property that any $\cos \widehat{\boldsymbol{\lambda}_h \boldsymbol{\lambda}_k}$ is unchanged by either local or parallel displacement, so that the formulæ

$$\delta' \left(\sum_1^n {}_i \lambda_{h|i} \lambda_k^i \right) = 0, \qquad \delta^* \left(\sum_1^n {}_i \lambda_{h|i} \lambda_k^i \right) = 0$$

both hold.

Carrying out both differentiations and using the results to transform (11), we get

$$p_{hk} \, ds = - \sum_1^n {}_i (\lambda_{h|i} \cdot \delta' \lambda_k^i + \lambda_k^i \cdot \delta^* \lambda_{h|i}).$$

Further since $\cos \widehat{\boldsymbol{\lambda}_h \boldsymbol{\lambda}_k}$ can also be expressed in the form $\sum_1^n {}_i \lambda_h^i \lambda_{k|i}$, (11) is equivalent to

$$p_{hk} \, ds = \sum_1^n {}_i (\delta' \lambda_h^i \cdot \lambda_{k|i} + \lambda_h^i \cdot \delta^* \lambda_{k|i}).$$

Interchanging h and k, and adding to the previous equation, we get the required identity

$$p_{hk} + p_{kh} = 0 \qquad (h, k = 1, 2, \ldots n). \quad . \quad (12)$$

We shall now examine the case when the direction of displacement coincides with one of the directions belonging to the set of congruences, say the lth. We shall then have

$$\xi^j = \lambda_l^j,$$

and, denoting by γ_{hkl} the value of p_{hk} in this particular case (i.e. the rate of variation of $\cos\widehat{\boldsymbol{\lambda}_h\boldsymbol{\lambda}_k}$ for a displacement in the direction of $\boldsymbol{\lambda}_l$, in which $\boldsymbol{\lambda}_h$ is moved by local, and $\boldsymbol{\lambda}_k$ by parallel, displacement), we shall have from (11′)

$$\gamma_{hkl} = \sum_{1}^{n}{}_{ij}\, \lambda_{h|ij}\, \lambda_k^i\, \lambda_l^j \qquad (h, k, l = 1, 2, \ldots n). \quad . \quad (13)$$

The quantities γ were introduced by Ricci, and named by him the *coefficients of rotation* of the set of congruences. They have various important properties.

In the first place, they are *invariant*, as follows from (13) by the law of contraction. We have farther, as a particular case of (12),

$$\gamma_{hkl} + \gamma_{khl} = 0 \qquad (h, k, l = 1, 2, \ldots n), \quad . \quad (14)$$

which for $k = h$ reduces to

$$\gamma_{hhl} = 0. \quad . \quad . \quad . \quad . \quad . \quad . \quad (15)$$

We can also give a direct formal proof of (14), on the lines already suggested for the more general case of the p's. Starting from the identity (7), and taking the covariant derivative, we shall get (remembering Chapter VI, p. 152)

$$\sum_{1}^{n}{}_{i}\, \lambda_{h|ij}\, \lambda_k^i + \sum_{1}^{n}{}_{i}\, \lambda_{k|ij}\, \lambda_h^i = 0.$$

Multiplying by λ_l^j, and summing with respect to j from 1 to n, we get

$$\sum_{1}^{n}{}_{ij}\, \lambda_{h|ij}\, \lambda_k^i\, \lambda_l^j + \sum_{1}^{n}{}_{ij}\, \lambda_{k|ij}\, \lambda_h^i\, \lambda_l^j = 0$$

or

$$\gamma_{hkl} + \gamma_{khl} = 0.$$

The number of these invariants γ, which depend on three indices, is *a priori* n^3; but they are connected by the $\frac{1}{2}n^2(n+1)$ relations (14) of antisymmetry. Hence the number which are algebraically distinct is at most

$$n^3 - \frac{n^2(n+1)}{2} = \frac{n^2(n-1)}{2}.$$

The minuend n^3 is equal to the number of the derivatives $\lambda_{h|kl}$ of the quantities $\lambda_{h|k}$, and the subtrahend $\frac{1}{2}n^2(n+1)$ to the number of the relations given above as resulting from the differentiation of the equations (7) and connecting the n^3 derivatives. We can accordingly express the derivatives $\lambda_{h|kl}$ as functions of the quantities $\lambda_{h|k}$ and γ, by solving the equations (13). To do this, multiply (13) by $\lambda_{k|i'}\lambda_{l|j'}$, and sum with respect to k and l. We get

$$\sum_1^n{}_{kl}\gamma_{hkl}\lambda_{k|i'}\lambda_{l|j'} = \sum_1^n{}_{ij}\lambda_{h|ij}\sum_1^n{}_k\lambda^i_k\lambda_{k|i'}\sum_1^n{}_l\lambda^j_l\lambda_{l|j'}$$

$$= \sum_1^n{}_{ij}\lambda_{h|ij}\delta_i^{i'}\delta_j^{j'}$$

$$= \lambda_{h|i'j'};$$

or finally, replacing i' and j' by i and j,

$$\lambda_{h|ij} = \sum_1^n{}_{kl}\gamma_{hkl}\lambda_{k|i}\lambda_{l|j}. \quad . \quad . \quad . \quad . \qquad (16)$$

This result shows that in order to study the differential properties (i.e. the properties depending on the way in which the λ's vary) of the lines of the given congruences we need only consider the invariants γ, in terms of which all the derivatives of the λ's can be expressed.

The geometrical significance of the γ's, which we have already illustrated, is particularly expressive in the case of ordinary space. In this case the three congruences define at every point a triplet of orthogonal directions, and p_{23}, p_{31}, p_{12} are the components of a vector ω such that ωds is the *elementary rotation* of the triplet in the local displacement from P to P'.[1]

[1] See e.g. LEVI-CIVITA and AMALDI: *Lezioni di Meccanica Razionale*, Vol. I, p. 178; Bologna, Zanichelli, 1923. For the general case, the reader may be referred further to a paper by Signorina CARPANESE: "Parallelismo e curvatura in una varietà qualunque", in *Annali di Mat.*, Vol. XXVIII, 1919, pp. 147–169.

4. **Commutation formula for the second derivatives along the arcs.**

The invariants γ occur in another important formula, which we shall now establish.

We wish to compare the two second derivatives

$$\frac{\partial}{\partial s_k}\frac{\partial f}{\partial s_h} \quad \text{and} \quad \frac{\partial}{\partial s_h}\frac{\partial f}{\partial s_k};$$

we shall find that they are not equal, but are connected by a more complicated relation involving also the first derivatives and the γ's.

We have in the first place from (10), differentiating the invariant $\frac{\partial f}{\partial s_h}$ with respect to x_j and applying to the right-hand side the rule for differentiation given in Chapter VI, p. 152,

$$\frac{\partial}{\partial x_j}\frac{\partial f}{\partial s_h} = \sum_1^n{}_i f^i \lambda_{h|ij} + \sum_1^n{}_i \lambda_h^i f_{ij}.$$

We next replace f^i in the first term on the right by the expression given for it by (10″) (putting l instead of h for the index of summation), multiply both sides by λ_k^j, and sum with respect to j. We thus get

$$\sum_1^n{}_j \lambda_k^j \frac{\partial}{\partial x_j}\left(\frac{\partial f}{\partial s_h}\right) = \sum_1^n{}_{ijl} \frac{\partial f}{\partial s_l} \lambda_l^i \lambda_k^j \lambda_{h|ij} + \sum_1^n{}_{ij} f_{ij} \lambda_h^i \lambda_k^j.$$

By the definition (10) of the intrinsic derivative, the left-hand side of this equation is precisely $\frac{\partial}{\partial s_k}\frac{\partial f}{\partial s_h}$; the first term on the right, from the definition (13) of the invariants γ, reduces to

$$\sum_1^n{}_l \gamma_{hlk} \frac{\partial f}{\partial s_l} = -\sum_1^n{}_l \gamma_{lhk} \frac{\partial f}{\partial s_l}.$$

We therefore have

$$\frac{\partial}{\partial s_k}\frac{\partial f}{\partial s_h} = -\sum_1^n{}_l \gamma_{lhk} \frac{\partial f}{\partial s_l} + \sum_1^n{}_{ij} f_{ij} \lambda_h^i \lambda_k^j.$$

To get the other second derivative, we interchange h and k, which gives

$$\frac{\partial}{\partial s_h}\frac{\partial f}{\partial s_k} = -\sum_1^n{}_l\,\gamma_{lkh}\frac{\partial f}{\partial s_l} + \sum_1^n{}_{ij} f_{ij}\,\lambda^i_{k}\,\lambda^j_{h}.$$

Now take the difference of these two expressions. The second terms on the right cancel out, as on interchanging i and j they become identical. Hence

$$\frac{\partial}{\partial s_k}\frac{\partial f}{\partial s_h} - \frac{\partial}{\partial s_h}\frac{\partial f}{\partial s_k} = -\sum_1^n{}_l\frac{\partial f}{\partial s_l}(\gamma_{lhk} - \gamma_{lkh}). \quad . \quad (17)$$

This is the commutation formula required.

5. Case in which one of the congruences of the set is geodesic.

Suppose that one of the congruences of the set is geodesic (cf. § 1); without loss of generality we can always suppose that it is the nth. We propose to investigate the special characteristics of the coefficients of rotation γ in this case.

From formula (2′) we get the following relations for the elements of the direction $\boldsymbol{\lambda}_n$:

$$\sum_1^n{}_l\,\lambda_{n|il}\,\lambda^l_{n} = 0 \qquad (i = 1, 2, \ldots n). \quad . \quad (18)$$

We now multiply by λ^i_h and sum with respect to i; remembering the formula (13) defining the γ's, we get

$$\gamma_{nhn} = 0 \qquad (h = 1, 2, \ldots n); \quad . \quad . \quad (18')$$

this is equivalent to (18), as can be shown by multiplying by $\lambda_{h|i}$, summing with respect to h, and using (16).

The n equations (18′) are invariant not only for all possible changes of co-ordinates, but also for any change whatever of the $n - 1$ congruences (1), (2), ... $(n - 1)$, which together with (n) form an orthogonal set; in fact, to establish the equations (18) we made no special hypothesis as to the choice of these $n - 1$ congruences.

In particular, if the space is Euclidean, the equations (18′) are the intrinsic equations of a rectilinear congruence.

6. Geodesic curvature of one of the congruences of the set.

Returning to the case where the congruence (n) is general, we wish to show that the n invariants γ_{nhn} $(h = 1, 2, \ldots n)$ have a simple geometrical interpretation. It will be remembered that the left-hand side of (2′), which we denoted by p_k, is a covariant component of the geodesic curvature $\mathbf{p}$ (cf. Chapter V, p. 135). If the congruence considered is (n), we can therefore write

$$p_k = \sum_{1}^{n} {}_l\, \lambda_{n|kl}\, \lambda_n^l.$$

Now in accordance with § 2 the vector $\mathbf{p}$ can be represented by the n invariants

$$c_h = \sum_{1}^{n} {}_k\, p_k\, \lambda_h^k,$$

which give its orthogonal projections on the lines of the n congruences.

Carrying out this operation on the expression just given for p_k, we find, by (13),

$$c_h = \gamma_{nhn};$$

this shows that the invariants γ_{nhn} $(h = 1, 2, \ldots n)$ represent the orthogonal projections on the lines of the set of congruences of the vector which is the geodesic curvature of the congruence (n).

7. Case in which one of the congruences of the set is normal. Complete normality. Differential relations satisfied in every case by the γ's.

Suppose that (n) is a normal congruence. We know that the equivalent analytical condition is given by (6) where we take $X_i = \lambda_{n|i}$. We thus have

$$\left.\begin{aligned} \lambda_{n|k}(\lambda_{n|ij} - \lambda_{n|ji}) + \lambda_{n|i}(\lambda_{n|jk} - \lambda_{n|kj}) \\ + \lambda_{n|j}(\lambda_{n|ki} - \lambda_{n|ik}) = 0 \\ (i, j, k = 1, 2, \ldots n). \end{aligned}\right\} \quad (19)$$

We now multiply this equation by $\lambda_{n|k'}^k \lambda_{n|i'}^i$, where k' and i' are two new indices chosen among $1, 2, \ldots n - 1$, and sum with respect to i and k from 1 to n; remembering (13), we get

$$\lambda_{n|j}(\gamma_{nk'i'} - \gamma_{ni'k'}) = 0. \quad . \quad . \quad . \quad (19')$$

As j may have any value, we can always choose it so that $\lambda_{n|j} \neq 0$; thus we have

$$\gamma_{nki} = \gamma_{nik} \qquad (i, k = 1, 2, \ldots n - 1), \quad . \quad (20)$$

where we have written i, k instead of i', k'. Reciprocally, if the equations (20) are satisfied, the equations (19′) follow from them, and therefore also (19) as a necessary consequence. The equations (20) therefore constitute the required condition.

It is not without interest to find this condition by another method, starting from the remark that if the quantities $\lambda_{n|i}$ are to be proportional to the derivatives $\frac{\partial f}{\partial x_i}$ of a single function f, these derivatives can be substituted for them in the conditions of orthogonality

$$\sum_1^n{}_i \lambda_{n|i} \lambda_h^i = 0 \qquad (h = 1, 2, \ldots n - 1),$$

so that the hypothetical function f must satisfy the linear system of partial differential equations

$$\sum_1^n{}_i \lambda_h^i \frac{\partial f}{\partial x_i} = 0 \qquad (h = 1, 2, \ldots n - 1).$$

Reciprocally, if there exists a function f which satisfies these $n - 1$ equations, its derivatives must be proportional to the quantities $\lambda_{n|i}$.

Hence the conditions in question are the necessary and sufficient conditions that the given $n - 1$ equations may constitute a complete system (cf. Chapter III, § 9, p. 52).

To make the notation agree with that in Chapter III, we introduce the linear operators

$$X_h = \sum_1^n{}_i \lambda_h^i \frac{\partial}{\partial x_i} \qquad (h = 1, 2, \ldots n - 1),$$

noting that (10) shows that these operators are identical with the derivatives $\frac{\partial}{\partial s_h}$ with respect to the arcs. We thus have the system

$$X_h f = 0 \qquad (h = 1, 2, \ldots n - 1),$$

and we have to express the condition that, for $h, k = 1, 2, \ldots n - 1$, Poisson's parentheses

$$(X_h, X_k)f = X_h X_k f - X_k X_h f$$

are linear combinations of the terms $X_l f$.

Now, repeating the steps of the calculation in § 4, or better, borrowing from it the value already found for $\frac{\partial}{\partial s_h} \frac{\partial f}{\partial s_k}$, we have

$$X_h X_k f = \frac{\partial}{\partial s_h} \frac{\partial f}{\partial s_k} = - \sum_1^n {}_l \gamma_{lkh} \frac{\partial f}{\partial s_l} + \sum_1^n {}_{ij} f_{ij} \lambda_h^i \lambda_k^j.$$

Interchanging h and k, and subtracting, the second sum disappears. In the first, we must separate out the term corresponding to the value n of the index l, and put $X_l f$ again instead of $\frac{\partial f}{\partial s_l}$. We thus get

$$(X_h, X_k)f = \sum_1^{n-1} {}_l (\gamma_{lhk} - \gamma_{lkh}) X_l f + (\gamma_{nhk} - \gamma_{nkh}) \frac{\partial f}{\partial s_n}.$$

This must reduce to a linear combination of the quantities $X_l f$ $(l = 1, 2, \ldots n - 1)$.

As $\frac{\partial f}{\partial s_n}$ is independent of the X_l's, its coefficient in each of the parentheses included in the above expression (i.e. for $h, k = 1, 2, \ldots n - 1$) must vanish; this brings us back to (20).

It may be noted that if *all* the n congruences of the set are normal, the γ's with three distinct indices are all zero. In fact, choosing three distinct indices i, h, k, we have the following identities:

$$\gamma_{ihk} = \gamma_{ikh},$$
$$\gamma_{hki} = \gamma_{hik},$$
$$\gamma_{kih} = \gamma_{khi};$$

adding the first two and subtracting the third, and remembering that the γ's are antisymmetrical in the first two indices, we get

$$\gamma_{ihk} = - \gamma_{ihk},$$

or $$\gamma_{ihk} = 0 \qquad (i, h, k = 1, 2, \ldots n)$$

for every triplet of three distinct indices.

If we put

$$\gamma_{ij,hk} = \frac{\partial}{\partial s_k}\gamma_{ijh} - \frac{\partial}{\partial s_h}\gamma_{ijk} + \sum_1^n{}_l \{\gamma_{ijl}(\gamma_{lhk} - \gamma_{lkh}) + \gamma_{lik}\gamma_{ljh} - \gamma_{lih}\gamma_{ljk}\},$$

we get at once, by definition and the antisymmetry of the coefficients of rotation γ with respect to the first two indices,

$$\gamma_{ij,hk} = -\gamma_{ij,kh}, \qquad \gamma_{ij,hk} = -\gamma_{ji,hk}.$$

I add, but without giving a proof, that the cyclic identities

$$\gamma_{ij,hk} + \gamma_{ih,kj} + \gamma_{ik,jh} = 0$$

also hold; and from these it follows ultimately that

$$\gamma_{ij,hk} = \gamma_{hk,ij} \qquad (i, j, h, k = 1, 2, \ldots n).$$

Ricci discovered all these results as far back as 1895, basing his researches with regard to the four-index γ's on the analogous properties of Riemann's symbols of the first kind (cf. Chapter VII, p. 179). A particularly simple and direct proof has recently been given by Dei.[1]

8. **Canonical system with respect to a given congruence.**

In many questions a congruence of lines is either among the data of the problem, or is closely connected with them. In order to deal with these problems it is often useful to associate with the given congruence $n - 1$ others, forming with the given one a set of n mutually orthogonal congruences, so that the given congruence can be considered as the nth of this set. The choice of the $n - 1$ auxiliary congruences is *a priori* arbitrary; in many cases this arbitrariness may be taken advantage of to introduce some simplification. This is possible, as we shall now see; and the conclusion we shall reach is that given any congruence whatever, there is always *at least one* way of choosing the other $n - 1$ so that the relations

$$\gamma_{nkl} + \gamma_{nlk} = 0 \qquad (k \neq l;\ k, l = 1, 2, \ldots n - 1) \quad (21)$$

may be satisfied.

The system (or any one of the systems) of $n - 1$ congruences

[1] "Sulle relazioni differenziali che legano i coefficienti di rotazione del Ricci", in *Rend. della R. Acc. dei Lincei*, Vol. XXXII (first half-year, 1923), pp. 474–479.

which possesses this characteristic is called a *canonical system with respect to the given congruence.*

To prove that such a system exists, we associate with the given congruence a system—for the moment any whatever—of $n-1$ other orthogonal congruences, and fix our attention on a generic point P of the manifold; for shortness we shall denote by ϖ the pyramid of the $n-1$ directions $\boldsymbol{\lambda}_1, \boldsymbol{\lambda}_2, \ldots \boldsymbol{\lambda}_{n-1}$ drawn from P, orthogonal to $\boldsymbol{\lambda}_n$ and to one another. Suppose this pyramid rotated round the direction $\boldsymbol{\lambda}_n$, by which we mean that we pass from the pyramid ϖ to another ϖ' formed by $n-1$ other directions $\boldsymbol{\lambda}'_1, \boldsymbol{\lambda}'_2, \ldots \boldsymbol{\lambda}'_{n-1}$, also drawn from P, and orthogonal to $\boldsymbol{\lambda}_n$ and to one another. We wish, if possible, to determine the rotation so that after it has been effected the relations (21) may hold. For this we shall start from the relations connecting the $\boldsymbol{\lambda}'_h$'s with the $\boldsymbol{\lambda}_h$'s, which express analytically the rotation described.

Let a_{hk} $(h, k = 1, 2, \ldots n)$ be the cosine of the angle between the directions $\boldsymbol{\lambda}_h$ and $\boldsymbol{\lambda}'_k$. Naturally, if only one of the two indices h, k coincides with n, the corresponding a is zero $\left(\boldsymbol{\lambda}_n = \boldsymbol{\lambda}'_n\text{, and the corresponding angle is } \frac{\pi}{2}\right)$; while $a_{nn} = 1$. The formulæ for this are

$$a_{hn} = a_{nh} = 0 \qquad (h, k = 1, 2, \ldots n-1);$$
$$a_{nn} = 1.$$

We have in any case by definition

$$\sum_{1}^{n}{}_j \lambda_h^j \lambda'_{k|j} = a_{hk} \qquad (h, k = 1, 2, \ldots n),$$

and thence, multiplying by $\lambda_{h|i}$ and summing with respect to h,

$$\lambda'_{k|i} = \sum_{1}^{n}{}_h a_{hk} \lambda_{h|i}.$$

Limiting k to the values $1, 2, \ldots n-1$, for all of which $a_{nk} = 0$, we can take the sum on the right only to $n-1$, so that we have

$$\lambda'_{k|i} = \sum_{1}^{n-1}{}_h a_{hk} \lambda_{h|i} \qquad (k = 1, 2, \ldots n-1); \quad . \quad (22)$$

i.e. the moments of ϖ' are connected with those of ϖ by a linear substitution, as could have been anticipated, and the α_{hk}'s are the coefficients of this substitution. It is also to be anticipated that the substitution is orthogonal. To prove this, we take the equations (7″); putting $k = i$, they give

$$\sum_1^n{}_h (\lambda_{h|i})^2 = a_{ii} \qquad (i = 1, 2, \ldots n).$$

The coefficients a_{ii} on the right depend on the co-ordinates of reference, but not on the choice of the congruences associated with (n).

Since $\lambda_{h|n} = 0$ for $h \neq n$, it follows that *for any value of* i $(\neq n)$ the expression

$$\sum_1^{n-1}{}_h (\lambda_{h|i})^2$$

is invariant for rotations of the pyramid ϖ, and therefore the substitution defined by the α's is orthogonal. We have now to arrange this orthogonal substitution of order $n - 1$ in such a way that the relations (21) may be satisfied.

To do this we start from (16), from which we get as a particular case

$$\lambda_{n|ij} = \sum_1^n{}_{kl} \gamma_{nkl} \lambda_{k|i} \lambda_{l|j} \qquad (i, j = 1, 2, \ldots n - 1).$$

The terms of this sum in which $k = n$ vanish, by (15); those in which $l = n$ can be separated out by writing

$$\lambda_{n|ij} = \sum_1^{n-1}{}_{kl} \gamma_{nkl} \lambda_{k|i} \lambda_{l|j} + \lambda_{n|j} \sum_1^n{}_k \gamma_{nkn} \lambda_{k|i}.$$

The last sum can be suitably transformed by replacing γ_{nkn} by the expression given by (13); we then get successively

$$\begin{aligned}\sum_1^n{}_k \gamma_{nkn} \lambda_{k|i} &= \sum_1^n{}_{kpq} \lambda_{n|pq} \lambda_k^p \lambda_n^q \lambda_{k|i} \\ &= \sum_1^n{}_{pq} \lambda_{n|pq} \lambda_n^q \delta_i^p \\ &= \sum_1^n{}_q \lambda_{n|iq} \lambda_n^q.\end{aligned}$$

We can therefore write

$$\lambda_{n|ij} = \sum_{1}^{n-1}{}_{kl}\,\gamma_{nkl}\,\lambda_{k|i}\,\lambda_{l|j} + \lambda_{n|j}\sum_{1}^{n}{}_{q}\,\lambda_{n|iq}\,\lambda_{n}^{q}.$$

Now in this formula it is to be remarked that the left-hand side and the last term on the right depend on the parameters and moments of the direction $\boldsymbol{\lambda}_n$ alone, and do *not* depend on the other $n - 1$ associated directions; the same must therefore be true of the remaining part, i.e. of the sum

$$\sum_{1}^{n-1}{}_{kl}\,\gamma_{nkl}\,\lambda_{k|i}\,\lambda_{l|j} \qquad (i, j = 1, 2, \ldots n - 1). \quad . \quad (23)$$

We can therefore conclude that these expressions are *invariant* for any rotation whatever of the pyramid ϖ.

Of the $(n - 1)^2$ quadratic forms included in formula (23), which are obtained by choosing the indices i, j in every possible way, we are interested in any one in which $i = j$. Fixing the index i once for all, and putting for shortness

$$\lambda_{r|i} = z_r \qquad (r = 1, 2, \ldots n - 1),$$

the corresponding quadratic form is

$$\sum_{1}^{n-1}{}_{kl}\,\gamma_{nkl}\,z_k\,z_l = \tfrac{1}{2}\sum_{1}^{n-1}{}_{kl}\,(\gamma_{nkl} + \gamma_{nlk})\,z_k\,z_l. \quad . \quad (23')$$

In this the coefficient of the product $z_k z_l$ is $\gamma_{nkl} + \gamma_{nlk}$, i.e. the left-hand side of (21). If we wish to satisfy (21), we must make all the coefficients of the terms in $z_k z_l$, for which $k \neq l$, vanish by means of the orthogonal substitution (22), which we shall write in the form

$$z'_k = \sum_{1}^{n-1}{}_{k}\,a_{hk}\,z_h; \quad . \quad . \quad . \quad . \quad (22')$$

this is equivalent to *reducing the invariant quadratic form* (23′) *to the canonical form*

$$\sum_{1}^{n-1}{}_{k}\,\rho_k\,z_k^2 \quad . \quad . \quad . \quad . \quad . \quad . \quad . \quad (23'')$$

by an orthogonal substitution. This algebraic problem is always soluble. In the cases $n - 1 = 2$ or 3, it corresponds to the problem of finding the axes of a conic or a quadric, and is discussed in ordinary analytical geometry. In the general case the theory leads to the following result.

Consider the equation

$$\| \tfrac{1}{2}(\gamma_{nkl} + \gamma_{nlk}) - \delta_k^l \rho \| = 0, \quad . \quad . \quad . \quad (24)$$

which is of degree $n - 1$ in the unknown ρ, and is called a *secular equation.* Its $n - 1$ roots are always real (it is understood that we suppose the quantities γ_{nkl} real), and give the $n - 1$ coefficients ρ_k of the canonical form (23″).[1]

We can therefore always choose, at any point P, the pyramid ϖ and therefore the system of the $n - 1$ congruences (1), (2), . . . $(n - 1)$ so as to satisfy (21); i.e. there always exists *at least one* canonical system with respect to a given congruence. If the $n - 1$ roots of (24) are all different, the canonical system is uniquely determined; if they are all equal, any system of $n - 1$ congruences which are orthogonal to one another and to (n) satisfies (21) and may therefore be called canonical. In the general case where the number of different roots is p $(1 < p < n - 1)$, then $n - 1 - p$ coefficients of the orthogonal solution are arbitrary, and there are therefore ∞^{n-1-p} canonical systems.

9. Congruences of straight lines in Euclidean space. Geometrical significance of the canonical system.

In ordinary (i.e. Euclidean three-dimensional) space particular importance attaches to congruences of straight lines, which present themselves for consideration in various questions of geometrical optics; since the rays of a light pencil (in a homogeneous medium) form a rectilinear congruence.

We shall now discuss a geometrical property of these congruences, which will be seen to be connected with the discussion in the preceding section; or rather—since it involves no greater complication—we shall discuss congruences of lines in a Euclidean space of any number n of dimensions.

Consider a generic point P, and let r be the ray through P of the given rectilinear congruence; let X be the hyperplane (in ordinary space the plane) perpendicular to r at P. Take a displacement in X represented by the infinitesimal segment

[1] Compare Chapter VII, p. 205, where references are given.

$PP' = \epsilon$ in any direction; through P' will pass another ray r' of the congruence. In general, the two rays r, r' are skew; if for a particular direction of the displacement PP' it happens that they both lie in the same plane, i.e. that they meet or are parallel (more precisely, that the minimum distance between them is an infinitesimal of higher order than ϵ), this is called a *focal direction.* We shall now show that in general there exist $n - 1$ focal directions, all or some of which may be imaginary, coincident, or indeterminate; we shall then point out an important particular case in which these directions coincide with those of the canonical system.

Let PP' then be a focal direction; there will be a point C (which may be at an infinite distance) common to r and r'. Denote the length CP by $\frac{1}{\omega}$ (so that we shall have the particular case of the rays being parallel at the limit when $\omega = 0$), and let us take as axes of reference n orthogonal Cartesian axes y_ν $(\nu = 1, 2, \ldots n)$. Let $\lambda_{n|\nu}$ be the cosines of the direction n (i.e. its parameters or moments, since in Euclidean space $\lambda_{n|\nu} = \lambda_n^\nu$). The projection on the axis y_ν of the segment CP will then be given by $\frac{1}{\omega}\lambda_{n|\nu}$, and that of CP' will be

$$\frac{1}{\omega}\lambda_{n|\nu} + d\left(\frac{1}{\omega}\lambda_{n|\nu}\right),$$

while the projection of PP' is dy_ν. If then we express this last term as the difference of the other two (PP' being the third side of the triangle CPP'), we have

$$dy_\nu = d\left(\frac{1}{\omega}\lambda_{n|\nu}\right),$$

$$dy_\nu = \lambda_{n|\nu}\, d\frac{1}{\omega} + \frac{1}{\omega}d\lambda_{n|\nu}.$$

We now wish to use the methods of the absolute calculus. We shall therefore associate with the given congruence $n - 1$ other congruences, orthogonal to it and to each other, which we shall distinguish by the indices $1, 2, \ldots n - 1$. In addition to the projections of PP' on the axes we require also its projections on the set of n congruences so defined; for this we must multiply the last equation by λ_h^ν $(h = 1, 2, \ldots n)$ and sum with respect to ν. First, let $h = n$; the projection of PP' on the direction n is zero, as PP' by hypothesis belongs to the hyperplane X; hence the left-hand side is zero. Further, in consequence of the identity

$$\sum_1^n{}_\nu \lambda_n^\nu \lambda_{n|\nu} = 1,$$

it follows that
$$\sum_1^n{}_\nu \lambda_n^\nu d\lambda_{n|\nu} = 0;$$

hence finally we get
$$d\frac{1}{\omega} = 0,$$

which expresses the *a priori* evident fact that $CP = CP'$ (of course neglecting infinitesimals of higher order than the first). Putting h in turn equal to $1, 2, \ldots n - 1$, and denoting by ϵ_h the projection of PP' on the direction h, we find

$$\epsilon_h = \frac{1}{\omega} \sum_1^n{}_\nu \lambda_h^\nu d\lambda_{n|\nu} \qquad (h = 1, 2, \ldots n - 1).$$

We shall now expand $d\lambda_{n|\nu}$, remembering that since Christoffel's symbols are all zero, we can replace the ordinary by the covariant derivatives, and also that since $\epsilon_1, \epsilon_2, \ldots \epsilon_{n-1}$ are the projections of PP' on the directions of the set of congruences, and dy_k $(k = 1, 2, \ldots n)$ its projections on the axes, we therefore have

$$dy_k = \sum_1^{n-1}{}_j \epsilon_j \lambda_j^k.$$

The last formula thus becomes

$$\epsilon_h = \frac{1}{\omega} \sum_1^n{}_{\nu k} \lambda_h^\nu \lambda_{n|\nu k} \sum_1^{n-1}{}_j \epsilon_j \lambda_j^k,$$

or
$$\omega\epsilon_h = \sum_1^{n-1}{}_j \epsilon_j \sum_1^n{}_{\nu k} \lambda_{n|\nu k} \lambda_h^\nu \lambda_j^k.$$

Remembering the definition of the γ's we have the system of $n - 1$ equations

$$\omega\epsilon_h = \sum_1^{n-1}{}_j \epsilon_j \gamma_{nhj} \qquad (h = 1, 2, \ldots n - 1), \quad . \quad (25)$$

which we can also write

$$\sum_{1}^{n-1}{}_j (\gamma_{nhj} - \delta_h^j \omega)\epsilon_j = 0 \qquad (h = 1, 2, \ldots n-1). \quad (25')$$

This linear homogeneous system must determine the focal directions PP' (if they exist) in the hyperplane X, by giving their projections $\epsilon_1, \epsilon_2, \ldots \epsilon_{n-1}$ on the orthogonal directions $1, 2, \ldots n-1$ which we have associated with the ray r.

The necessary and sufficient condition that the system (25′) may have solutions ϵ which are not all zero, is that the determinant of the coefficients should vanish, i.e. that ω should satisfy the equation of degree $n-1$

$$\| \gamma_{nhj} - \delta_h^j \omega \| = 0 \qquad (h, j = 1, 2, \ldots n-1). \quad (26)$$

To every root ω corresponds at least one set of values of the ϵ's, i.e. at least one focal direction PP'. Hence in general there are $n-1$ of these directions, which, however, like the corresponding roots of (26), may be real or imaginary, distinct or coincident, or (in the case of multiple roots) may be capable of having an infinite number of determinations.

In fact, the properties of the secular equation, as noted in the preceding section, hold for *symmetrical* determinants of the type (24), while the left-hand side of (26) is not in general of this form. There is, however, an important category of congruences with this characteristic, which we shall now consider.

Normal congruences of rays.—If our congruence (n) is normal, then by (20)

$$\gamma_{nhj} = \gamma_{njh} \qquad (h, j = 1, 2, \ldots n-1).$$

We can therefore substitute $\frac{1}{2}(\gamma_{nhj} + \gamma_{njh})$ for γ_{nhj}, so that (26) at once becomes identical with (24), which defines the canonical directions. It follows that the *canonical and focal directions coincide.* Hence on the one hand we have the geometrical interpretation of the canonical directions; and on the other, from the properties noted at the end of the preceding section, we have the property that the focal directions are always real, and are in general determinate and mutually orthogonal; and further, that in the case of indeterminateness, when there is an infinite number of them, it is always possible (and in an infinite number of ways) to choose $n - 1$ of them which shall be mutually orthogonal.

As we are dealing with a normal congruence, there exists (by definition) a family of surfaces

$$f(x_1, x_2, \ldots x_n) = \text{constant},$$

which are cut orthogonally by the straight lines of the congruence; these lines therefore constitute the common normals to all the surfaces of the family. If we fix one of these surfaces, and associate with every point on it the $n - 1$ focal directions, we shall get $n - 1$ mutually orthogonal congruences of lines on the surface. These lines are called *lines of curvature*, by an obvious generalization from the lines so determined in the case of surfaces in ordinary space ($n = 3$). In fact, given such a surface, say σ, the normals to it form a normal congruence (since they cut σ and the surfaces parallel to σ orthogonally); and if we consider the two focal directions at every point of σ we arrive at precisely the ordinary definition of the lines of curvature as those lines of σ along which the normals to σ generate a developable ruled surface.

General Case.—If the congruence (n) under consideration is not normal, then in general, as we have seen, the focal and canonical directions at a generic point P of a ray r do not coincide. In order to find an interpretation of the canonical directions in this case, we should therefore have to examine in greater detail the behaviour of the rays of the congruence which are infinitely near r.

For $n = 3$ there is a classical discussion by Kummer,[1] giving a very illuminating interpretation of the canonical directions,[2] and pointing out in particular that the directions which bisect the angles between the canonical directions also bisect the angles between the focal directions (when the latter are real).

We shall leave the question at this point, merely pointing out to the reader the possibility of analogous interpretations for $n > 3$.

[1] See e.g. BIANCHI: *Lezioni di Geometria Differenziale*, Vol. I (third edition; Bologna, Zanichelli, 1922), Ch. X.

[2] Cf. T. LEVI-CIVITA: "Sulle congruenze di curve", in *Rend. della R. Acc. dei Lincei*, Vol. VIII (first half-year, 1899), pp. 239–46.

PART III

Physical Applications

CHAPTER XI

Evolution of Mechanics and Geometrical Optics; Their Relation to a Four-dimensional World according to Einstein

1. Hamilton's principle for a free particle.

We start from the equations of motion of a material particle in a conservative field. Let U be the potential for unit mass. The equations of motion, in Cartesian co-ordinates (referred to fixed axes) y_1, y_2, y_3, are

$$\ddot{y}_i = \frac{\partial U}{\partial y_i} \qquad (i = 1, 2, 3), \quad . \quad . \quad . \quad (1)$$

where as usual dots represent differentiation with respect to the time t. If we denote the square of the line element described by the moving particle in the small interval of time dt by

$$dl_0^2 = \sum_1^3{}_i \, dy_i^2$$

and if v is the velocity of the particle (in absolute value), then

$$v^2 = \frac{dl_0^2}{dt^2} = \sum_1^3{}_i \, \dot{y}_i^2.$$

Putting $$L = \tfrac{1}{2}v^2 + U$$

it is known that the equations (1) can be summed up in the equation of variation

$$\delta\int L dt = 0 \quad . \quad . \quad . \quad . \quad . \quad . \quad (2)$$

which expresses Hamilton's principle.

Let us fix our attention for a moment on (2). It implies an interval of integration (t_0, t_1), fixed arbitrarily in advance; and the vanishing of the left-hand side of (2) for variations δy_i of the y's, zero at the extremities but otherwise arbitrary, is equivalent to the equations (1) being satisfied in the same interval.

This case, in which t does not vary (i.e. $\delta t = 0$), is the simplest application of Hamilton's principle. Various generalizations, however, in which t also varies, either freely or subject to certain conditions, have become classical. We shall shortly have occasion to discuss one of these generalizations which concern the equivalence between the equations (1) and (2). Meanwhile we may note that if the co-ordinates are changed in any way, so that the Cartesians y_1, y_2, y_3 are replaced by any set of three curvilinear co-ordinates, or more generally by three Lagrangian parameters x_1, x_2, x_3, connected with y_1, y_2, y_3 by relations which may involve the time and which are regular and reversible in the field considered, namely,

$$(T_3): \qquad x_h = x_h(y_1, y_2, y_3, t) \qquad (h = 1, 2, 3),$$

or, solving with respect to y_i $(i = 1, 2, 3)$,

$$(T_3'): \qquad y_i = y_i(x_1, x_2, x_3, t) \qquad (i = 1, 2, 3);$$

then if we insert these expressions in L, it becomes a function $L(x \mid \dot{x} \mid t)$ of the arguments x_h, $\dot{x}_h$ $(h = 1, 2, 3)$, t, quadratic (in general not homogeneous) in the $\dot{x}$'s.

As we propose to consider L as an invariant, it follows that (2) will hold for the Lagrangian parameters x, and we have only to find its explicit form. Calculating the variation and integrating by parts in the usual way, we easily find

$$\delta\int L\,dt = -\int \sum_{1}^{3}{}_h \tau_h\,\delta x_h\,dt, \quad . \quad . \quad . \quad . \quad (3)$$

where for shortness we have put

$$\tau_h = \frac{d}{dt}\frac{\partial L}{\partial \dot{x}_h} - \frac{\partial L}{\partial x_h}$$

(Lagrangian binomials). The dynamical equations then take the form

$$\tau_h = 0 \qquad (h = 1, 2, 3) \quad . \quad . \quad . \quad (4)$$

(known as Lagrange's form); and it is to be noted that, in virtue of the invariance of the left-hand side of (3), the quantities τ_h constitute a co-variant tensor, as pointed out in a similar case in Chapter V, p. 110. It follows that the equations (4), i.e.

$$\frac{d}{dt}\frac{\partial L}{\partial \dot{x}_h} - \frac{\partial L}{\partial x_h} = 0 \qquad (h = 1, 2, 3), \quad . \quad . \quad (4')$$

are invariant (cf. Chapter V, p. 110) with respect to the transformations (T_3) which leave L invariant.

2. Time as a fourth co-ordinate. Space-time. World lines.

An obvious consequence of Lagrange's equations (4′) is the identity

$$\frac{d}{dt}\left\{L - \sum_1^3{}_i \frac{\partial L}{\partial \dot{x}_i}\dot{x}_i\right\} - \frac{\partial L}{\partial t} = 0.$$

Now suppose that in the interval (t_0, t_1) the independent variable t is also made to undergo a variation δt which is zero at the extremities and is otherwise arbitrary. Since the x_i's are unchanged by this, while the derivatives $\dot{x}_i = \frac{dx_i}{dt}$ undergo the increments

$$\delta \dot{x}_i = -\dot{x}_i \frac{d\delta t}{dt},$$

it will at once be seen that, by an obvious integration by parts, the contribution of the variation of t to $\delta\int L\,dt$, namely,

$$\int_{t_0}^{t_1} L\,\delta dt + \int_{t_0}^{t_1} dt\left\{\frac{\partial L}{\partial t}\delta t + \sum_1^3{}_i \frac{\partial L}{\partial \dot{x}_i}\delta\dot{x}_i\right\},$$

can be put in the form

$$\int_{t_0}^{t_1} dt\,\delta t\left\{\frac{\partial L}{\partial t} - \frac{d}{dt}\left(L - \sum_1^3{}_i \frac{\partial L}{\partial \dot{x}_i}\dot{x}_i\right)\right\}$$

which, as we have just pointed out, is zero in consequence of (4′).

It is therefore possible, in dealing with the Hamiltonian equation (2), to apply exactly the same treatment to the space-co-ordinates x_1, x_2, x_3 and the time t.

To simplify the argument, consider the four-dimensional manifold V_4 corresponding to four parameters x_i, t; the manifold, in which space and time are simultaneously represented, may be called space-time.

A set of three equations

$$x_i = x_i(t) \qquad (i = 1, 2, 3),$$

or, in terms of kinematics, a motion, corresponds to a curve belonging to V_4, and reciprocally. Such a curve is called a *world line*; it is an obvious generalization of the plane diagram (in which the abscissa is the time and the ordinate the space described) used to represent the circumstances of motion in a given trajectory. Adopting this expression, we can say that the integral curves of the equations (4′) are all those world lines of V_4, and only those, for which the variation of the integral $\int L\,dt$ vanishes, the extremities being fixed.

3. **General transformations of co-ordinates in space-time. Simultaneity.**

The most general transformation of parameters in V_4 obviously includes three equations of the type (T_3), which substitute for the Cartesian co-ordinates y_1, y_2, y_3 three independent combinations of them, x_1, x_2, x_3, also involving t; and a fourth equation which substitutes for the time t a further combination $x_0(y_1, y_2, y_3, t)$ (independent of the three preceding equations). This new parameter x_0 is sometimes called the *local time*, as it depends not only on the original time, but also on the point in question. A transformation (T_4) is thus represented by the formula:

$$(T_4): \qquad \begin{cases} x_0 = x_0(y_1, y_2, y_3, t), \\ (T_3). \end{cases}$$

An obvious but important property of such a transformation is the following. If two events are characterized by different values of y_1, y_2, y_3, but the same value of t, it will in general happen that after the transformation is effected, not only the space co-ordinates x_1, x_2, x_3 of the two events will be different, but also the time co-ordinates x_0. This implies that two events which appear simultaneous with reference to the system y_1, y_2, y_3, t are not in general simultaneous with reference to the system of the x's; simultaneity is therefore relative to the system of reference. This evidently does not happen when the first of the relations (T_4) is of the type $x_0 = x_0(t)$, or in particular $x_0 = t$, so that the (T_4) reduces to a (T_3). And it is precisely in order to avoid any conflict with the intuitive concept of (absolute) simultaneity that only transformations of the type (T_3) are considered in the classical physics. But a more acute criticism of this intuitive concept shows that, far from being a logical necessity, it has an empirical origin based on experimental results which can only be taken as a first approximation; it is therefore reasonable, in view of the speculative nature of our considerations, to admit the possibility of a more general conception of simultaneity.

4. Einstein's form for Hamilton's principle. Its invariant character under any transformation of co-ordinates.

So long as L is taken to be invariant, the form of the integral $\int L\,dt$ is evidently not invariant for a transformation (T_4), since in general dt is replaced by an expression linear in all four variables x. We might try to replace the base L by something more general; it would then be possible to reach the required result, but the method would be complicated and infertile, and the loss in simplicity both of concept and of form would be much greater than the gain in generality.

But it is not difficult to arrive at a significant form which shall be invariant for every (T_4) if we regard Hamilton's principle as an approximate result, the degree of approximation being of course sc high that in ordinary applications, astronomical as well as technical, the difference between it and the rigorous hypothetical principle shall be imperceptible. This will evidently be the case if the order of magnitude of the difference between the two, with respect to the values given by the ordinary theory, is not higher than the hundred-millionth (10^{-8}).

A concrete application of this criterion is as follows. Let c denote a constant velocity, large in comparison with the greatest velocity attained in the motions we propose to discuss. We shall consider quantities comparable with $\beta = \frac{v}{c}$ as small quantities of the first order, and we shall consider quantities of the second and higher orders as negligible in comparison with unity; we shall also suppose that the ratio $\frac{U}{c^2}$ is similarly negligible.

We note that this will in fact be the case if c is comparable with the velocity of light, not only for ordinary problems of terrestrial motion, but also in celestial mechanics. In order to see this, we need only suppose that v is a planetary velocity and U the Newtonian potential which determines it, so that by a well-known result U (in the field of motion of the planet) is of the same order of magnitude as v^2.

We may take 30 kilometres per second, corresponding to the earth's motion in its orbit, as the order of magnitude of v. In round numbers, $c = 300{,}000$ km./sec., so that we have $\frac{v}{c} = 10^{-4}$ (approx.), and therefore

$$\frac{v^2}{c^2} \text{ and } \frac{U}{c^2} = 10^{-8} \text{ (approx.)}.$$

We shall see farther on, in §§ 8, 9, and 16, that physical considerations lead us to take for c precisely the velocity of light.

Since δt must vanish at the limits of integration, we have

$$\delta \int_{t_0}^{t_1} dt = 0,$$

so that L can be replaced, as the integrand of (2), by

$$c^2 - L = c^2\left(1 - \tfrac{1}{2}\beta^2 - \frac{U}{c^2}\right).$$

The terms $-\frac{1}{2}\beta^2$, $-\frac{U}{c^2}$, though negligible in comparison with unity, are essential in order to prevent the equation of variation from reducing to an identity. Terms of higher order may however be neglected. We may therefore write

$$c^2 - L = c^2\sqrt{1 - \beta^2 - \frac{2U}{c^2}} = c\sqrt{c^2 - v^2 - 2U};$$

so that, omitting the constant factor c and writing $\frac{dl_0^2}{dt^2}$ instead of v^2, the equation of Hamilton's principle (which, as just pointed out, is equivalent to $\delta\int(c^2 - L)dt = 0$) can be replaced by

$$\delta\int\sqrt{c^2 - \frac{dl_0^2}{dt^2} - 2U}\,dt = 0,$$

or, putting
$$ds^2 = (c^2 - 2U)\,dt^2 - dl_0^2, \quad . \quad . \quad . \quad (5)$$

by
$$\delta\int ds = 0. \quad . \quad . \quad . \quad . \quad . \quad . \quad (6)$$

Since the value of dl_0^2 referred to Cartesian co-ordinates is $\sum_1^3 {}_i\, dy_i^2$, the ds^2 just introduced is a quaternary differential quadratic form; it is indefinite, since for real and infinitesimal values of dt, dy_1, dy_2, dy_3 it can have both positive and negative values. At the same time it is to be remembered that, *for the phenomena of motion at present under consideration, we have always* $ds^2 > 0$.

To show that this is so, note that, taking out the common factor c^2dt^2 and again replacing $\frac{dl_0^2}{dt^2}$ by v^2, we can write

$$ds^2 = c^2\,dt^2\left(1 - \frac{2U}{c^2} - \frac{v^2}{c^2}\right);$$

this proves the assertion, since the quantity in brackets is certainly positive when the quantitative relations stipulated at the beginning of our argument hold.

We may now note that if the ds^2 expressed by (5) is considered as the square of the line element of the manifold V_4 (which contains both space and time), then (6) represents the characteristic equation of geodesics of V_4 (cf. Chapter V, p. 130). It is true that the metric of this manifold is characterized by an indefinite quadratic form, but, as was pointed out on p. 142 of Chapter V, this does not introduce any real complication so long as we limit our considerations to lines wholly constituted of elements for which $ds^2 > 0$, as it is in the present case. We can therefore say that the proposed modification of Hamilton's principle imposes a metric limitation on the space-time manifold

V_4, and that the mechanical problem of the motion of a free particle under the action of forces derived from a potential has been transformed—with an alteration of the laws of dynamics which is quantitatively very small—into the purely geometrical problem of the determination of the geodesics of a certain four-dimensional metric manifold.

If for the arguments t, y_1, y_2, y_3 we substitute any four independent combinations of them whatever, x_0, x_1, x_2, x_3, by means of a substitution (T_4), ds^2 will lose the special form (5) and assume the general type of a quaternary quadratic,

$$ds^2 = \sum_{0}^{3}{}_{ik}\, g_{ik}\, dx_i\, dx_k, \quad . \quad . \quad . \quad . \quad (5')$$

whose ten co-efficients $g_{ik} (= g_{ki})$ will naturally be, in genéral, functions of the x's.

The essential point is that, ds^2 being invariant, (6) is also invariant for any choice whatever of co-ordinates in V_4. This constitutes a marked superiority of (6) over the original form of Hamilton's principle. From the conceptional point of view it is also to be noted that this change realizes Einstein's fundamental concept of *general relativity*, which requires that it shall be possible to express the laws of any physical phenomenon whatever in a form which is invariant for every possible choice of co-ordinates, both of space and of time, without the time having to hold the privileged position assigned to it in the classical theories.

5. Mass and energy: views suggested by the modification of the dynamical law.

We shall examine in detail the form taken by the dynamical equations of the free material particle when the classical Lagrangian function L is replaced by the function

$$L^* = -c\sqrt{c^2 - v^2 - 2U},$$

which we shall write briefly in the form

$$L^* = -c^2 K,$$

putting

$$K = \sqrt{1 - \beta^2 - \frac{2U}{c^2}}.$$

Substituting $-c^2K$ for L in Lagrange's equations, they become

$$\frac{d}{dt}\frac{\partial K}{\partial \dot{y}_i} - \frac{\partial K}{\partial y_i} = 0;$$

or since

$$\frac{\partial K}{\partial \dot{y}_i} = -\frac{\dot{y}_i}{c^2K} \quad . \quad . \quad . \quad . \quad . \quad . \quad (7)$$

we have also

$$\frac{d}{dt}\frac{\dot{y}_i}{K} = \frac{1}{K}\frac{\partial U}{\partial y_i} \qquad (i = 1, 2, 3).$$

Remembering that K differs by very little from 1, we see that quantitatively these equations differ by very little from the equations (1). Considering them from the point of view of form, and comparing them with the cardinal equation of classical dynamics

$$\frac{d\mathbf{Q}}{dt} = \mathbf{F}$$

(where $\mathbf{Q}$ is the momentum and $\mathbf{F}$ the force), we see that the momentum per unit mass of the old theory is replaced in the new by the vector whose components are $\frac{\dot{y}_i}{K}$. For a particle of mass m_0 and velocity $\mathbf{V}$ the vectorial expression for the momentum will therefore be

$$\mathbf{Q} = \frac{m_0\mathbf{V}}{K}.$$

If we wish to retain the formal property that the momentum is the product of the mass by the velocity, we must take as the mass not the constant m_0, an intrinsic property of the body in motion, but the quantity

$$m = \frac{m_0}{K},$$

which will be seen to depend on the velocity and the field of force. Neglecting the latter so as to fix the attention in the first place on the motion as it depends on the velocity, we reach the expression

$$m = \frac{m_0}{\sqrt{1 - \frac{v^2}{c^2}}},$$

from which it appears that m increases as the velocity increases and would tend to infinity if the velocity could reach the value c. In this sense we say that the typical velocity c, introduced to give an invariant form to Hamilton's principle, is a *limiting velocity*.

We now proceed to examine the concept of energy in the light of relativity mechanics.

In the classical mechanics, given a generic Lagrangian function $L(y \mid \dot{y})$ (where L does not explicitly contain the time t), the corresponding expression for the energy is

$$H = \sum_1^3 {}_i \frac{\partial L}{\partial \dot{y}_i} \dot{y}_i - L; \quad . \quad . \quad . \quad . \quad (8)$$

in the case where L can be broken up into a part $T(y \mid \dot{y})$ homogeneous of the second degree in the $\dot{y}$'s, and a part U independent of them, this becomes, by Euler's theorem,

$$H = T(y \mid \dot{y}) - U(y).$$

It is known that T can be interpreted as the kinetic and $-U$ as the potential energy. Since we have replaced the classical L by the expression

$$L^* = -c\sqrt{c^2 - v^2 - 2U} = -c^2 K,$$

we must now determine the new expression H^* for the energy per unit mass.

Applying (8) we get

$$\begin{aligned} H^* &= \sum_1^3 {}_i \frac{\partial L^*}{\partial \dot{y}_i} \dot{y}_i - L^* \\ &= -c^2 \left\{ \sum_1^3 {}_i \frac{\partial K}{\partial \dot{y}_i} \dot{y}_i - K \right\}; \end{aligned}$$

substituting from equations (7) and using the expression for K we get finally

$$H^* = \frac{c^2 - 2U}{K} = \frac{c^2 - 2U}{\sqrt{1 - \frac{v^2}{c^2} - \frac{2U}{c^2}}}. \quad . \quad . \quad (9)$$

We see therefore that the energy cannot be divided into a part due to motion and a part due to position. Further, for

$v = U = 0$, the energy does not vanish, but remains equal to c^2: a remarkable fact, the interpretation of which will be seen in a moment.

Expanding the radical in series we can write

$$H^* = c^2\left(1 - \frac{2U}{c^2}\right)\left(1 + \frac{v^2}{2c^2} + \frac{U}{c^2} + \ldots\right),$$

and therefore, retaining only terms of the second order,

$$\begin{aligned} H^* &= c^2\left(1 - \frac{U}{c^2} + \frac{1}{2}\frac{v^2}{c^2}\right) \\ &= c^2 - U + \tfrac{1}{2}v^2. \end{aligned}$$

To this degree of approximation, therefore, the energy is composed of a kinetic part expressed as usual by $\frac{1}{2}v^2$, a part due to position which is still given by $-U$, and in addition a constant part (i.e. a part independent of both position and velocity) equal to c^2; this last part is called the *intrinsic energy* of unit mass. A material particle of mass m_0 (at rest or moving under no forces) will thus have intrinsic energy m_0c^2. Now considerations of a different nature lead us to assign to this intrinsic energy a much more profound significance than that of a mere additive constant of conventional value; it is in fact taken to represent the effective atomic and molecular energy stored up in the body to the extent of 25 million kilowatt-hours for every gramme of matter. The possibility of the existence of this enormous quantity of latent energy is shown by phenomena of radioactivity: a sufficient example is the fact that any small mass of radium is capable of giving off for years and years, without perceptible modification, enough heat to raise an equal mass of water from 0° C. to boiling-point in every hour. The supply of heat would last for a very long time; more than 2500 years for radium, and for other radioactive elements a period comparable with geological epochs. While radioactivity is not a general property of all bodies, yet it demonstrates the fact that (at least in certain cases) matter contains an enormous store of energy, and in this form the assertion can be generalized so as to extend to every atom of ponderable matter.

Admitting the possibility of the existence of this intrinsic energy, the foregoing considerations result in our assigning to it the value m_0c^2. If instead we return to the expression (9) for the total energy, and suppose that the potential U is zero, we find for the total energy (kinetic and intrinsic) localized in a body whose mass when at rest is m_0, the expression

$$E = \frac{m_0c^2}{\sqrt{1 - \frac{v^2}{c^2}}};$$

and remembering the expression for the mass m as a function of the velocity we can also write this as

$$E = mc^2. \quad . \quad . \quad . \quad . \quad . \quad (10)$$

This result shows us that there is a proportional relation not only between the mass of the body when at rest and the intrinsic energy, but, more generally, between the mass and the total energy localized in the body. It also suggests the hypothesis that to any form of energy there must be assigned a mass connected with it by the relation (10); and, *vice versa*, that every mass m corresponds to a quantity of energy mc^2. This hypothesis is supported by other considerations, and leads to the view, of primary philosophical importance, that energy and matter may be considered as different manifestations of one single entity, which appears as ordinary matter when it is, so to speak, sufficiently concentrated, while it appears as energy in widely different forms when there are no condensation nuclei present.

6. **Einstein's form for the principle of inertia. Restricted relativity.**

The equations of motion in the original Newtonian form (1) imply, as is well known, a state of uniform motion when the forces are zero or, which comes to the same thing (except for a non-essential constant), for $U = 0$. Equation (2), which is rigorously equivalent to (1), therefore defines states of uniform motion for $U = 0$. This property also holds for the new Einsteinian form (6) of Hamilton's principle, though it is not rigorously equivalent to equations (2). Before proving this we may point out that, for $U = 0$, (5) gives

$$ds_0^2 = c^2\,dt^2 - dl_0^2. \quad . \quad . \quad . \quad . \quad . \quad (11)$$

By a mere change of the unit of measurement of time (the advantage of which will be seen shortly), i.e. by putting $ct = y_0$, this quadratic form becomes

$$ds_0^2 = dy_0^2 - dl_0^2,$$

and referring the space to orthogonal Cartesian co-ordinates,

$$ds_0^2 = dy_0^2 - dy_1^2 - dy_2^2 - dy_3^2. \quad . \quad . \quad (11')$$

This is analogous to the ordinary expression for the ds^2 of a Euclidean V_4 in orthogonal Cartesian co-ordinates, except for the signs of the co-efficients, which make it indefinite; in this case the index of inertia[1] is 3. Hence the V_4 with a metric of this kind is called *pseudo-Euclidean*; the system of co-ordinates y_0, y_1, y_2, y_3, which gives this form to ds^2, is called *pseudo-Cartesian* or *Galilean.*

It will sometimes be convenient to put the expression for a pseudo-Euclidean ds^2 back into the general form (5′); for this purpose we introduce the symbols

$$\delta_i'^k = \begin{cases} 1 \text{ for } i = k = 0, \\ -1 \text{ for } i = k \neq 0, \\ 0 \text{ for } i \neq k \end{cases}$$

(the notation being similar to that introduced in the note on p. 55 of Chapter III). We can then say that in pseudo-Cartesian co-ordinates the co-efficients of ds^2 are

$$g_{ik} = \delta_i'^k.$$

We then have also, as is easily verified,

$$g^{ik} = g_{ik} = \delta_i'^k. \quad . \quad . \quad . \quad . \quad . \quad (12)$$

Returning to the property enunciated at the beginning of this section, we note that, for $U = 0$, the expression for L^* becomes

$$L_0^* = -c\sqrt{c^2 - \sum_1^3{}_i \dot{y}_i^2},$$

[1] The index of inertia is the number of negative coefficients of a quadratic form when expressed (in any way) in the canonical form (i.e. so that it contains no product terms).

and that (6), which becomes

$$\delta \int ds_0 = 0, \quad . \quad . \quad . \quad . \quad . \quad (6')$$

can be written

$$\delta \int L_0^* \, dt = 0.$$

The corresponding Lagrangian equations, from the fact that L_0^* does not depend explicitly on the y's, at once give the three first integrals

$$\frac{\partial L^*}{\partial \dot{y}_i} = \text{constant} \qquad (i = 1, 2, 3),$$

whence there follows the constancy of all the $\dot{y}_i$'s (the principle of inertia). Now consider a particular, but very important, category of transformations (T_4) specified as follows. From the set of four co-ordinates (t, y_1, y_2, y_3) we pass to a new set $(\bar{t}, \bar{y}_1, \bar{y}_2, \bar{y}_3)$ for which the form (11) of ds_0^2 remains unchanged, this being understood in the sense that the transformation formulæ are to give identically

$$ds_0^2 = c^2 \, dt^2 - \sum_1^3{}_i \, dy_i^2 = c^2 \, d\bar{t}^2 - \sum_1^3{}_i \, d\bar{y}_i^2.$$

The equation (6′) then ensures that *in the new co-ordinates also, interpreting* $\bar{t}$ *as the time and* $\bar{y}_1, \bar{y}_2, \bar{y}_3$ *as Cartesian co-ordinates the motion will appear uniform* (*restricted relativity*).

Transformations of this kind were effectively constructed by Lorentz, so that they may be called Lorentz transformations; we shall denote them shortly by (Λ) and discuss them fully in Section 8. Meanwhile we may indicate the characteristic property, pointed out by Professor Marcolongo, that, if we put $\sqrt{-1}\, ct = y_4$, so that ds_0^2 takes the Euclidean form $-\sum_1^4{}_i \, dy_i^2$, a Lorentz transformation leaves unchanged the form $\sum_1^4{}_i \, dy_i^2$, and thus (here too apart from any question of imaginaries) is substantially identical with motion in a four-dimensional Euclidean space.

To close these remarks on the effective existence of these special transformations (Λ) we may note an important corollary. Every (Λ), as we have said, transforms a generic uniform motion into a new motion which is also uniform; but it is not possible to assert that the velocity is unaltered by the transformation. There is, however, at least one case in which this happens, namely, motion in which the velocity is that very large constant velocity c which we originally introduced in order to modify Hamilton's formula in a way which should be quantitatively imperceptible, but fertile in its results.

In fact, for a motion in which the velocity is c (with respect to the parameters t, y_1, y_2, y_3), we have obviously $c^2 = \frac{dl_0^2}{dt^2}$, and therefore

$$ds_0^2 = c^2\,dt^2 - dl_0^2 = 0.$$

In view of the invariance, not only of ds_0^2, but also of the special form $c^2\,dt^2 - \sum_1^3{}_i\, dy_i^2$ which we have given it, we have, on passing to the new variables $\bar{t}$, $\bar{y}_1$, $\bar{y}_2$, $\bar{y}_3$ by a Lorentz transformation,

$$c^2\,d\bar{t}^2 - \sum_1^3{}_i\, d\bar{y}_i^2 = 0$$

for the transformed motion as well as for the original one, and therefore the velocity is c.

7. The kinematics of rigid systems. Ordinary method of approach and possible variants.

In the foregoing sections we have been led to modify (very slightly in ordinary conditions) the dynamics of a material particle P, i.e. the relation between the motion and the disturbing force. Nothing however has been, or need be, modified as regards the kinematics, i.e. the description of the phenomenon of the change of position of a point P with respect to an assigned observer $\bar{\Sigma}$, or in other terms with respect to a Cartesian system, in a certain interval of time. For convenience (the reason for this choice will be clear in a moment) we shall denote these axes of reference by $\bar{O}\,\bar{y}_1\,\bar{y}_2\,\bar{y}_3$, and the time by $\bar{t}$.

The equations of motion of P,

$$\bar{y}_i = \bar{y}_i(\bar{t}) \qquad (i = 1, 2, 3), \quad . \quad . \quad . \quad (13)$$

the velocity **V** as a vector of components $\frac{d\bar{y}_i}{dt}$ $(i = 1, 2, 3)$, the acceleration, &c., will all be as in the ordinary case. In particular, there is uniform motion when **V** is constant, i.e. the $\bar{y}_i$'s are linear functions of $\bar{t}$. In this case, taking one of the axes, say that of the $\bar{y}_1$'s, parallel to **V**, the equations (13) can be put in the simplified form

$$\bar{y}_1 = \bar{y}_1^0 + vt, \quad \bar{y}_2 = \bar{y}_2^0, \quad \bar{y}_3 = \bar{y}_3^0, \quad . \quad . \quad (14)$$

where v obviously denotes the velocity in the scalar sense (the component of **V** along $\bar{y}_1$), and $\bar{y}_1^0$, $\bar{y}_2^0$, $\bar{y}_3^0$ are the initial values of the co-ordinates $\bar{y}_1$, $\bar{y}_2$, $\bar{y}_3$ of the moving point.

As is well known, there are in ordinary kinematics two ways of defining rigid motion and of investigating its problems; these are briefly as follows:

(1) A rigid system is defined as a system consisting of any number of points P, P', . . . of co-ordinates $\bar{y}_i$, $\bar{y}_i'$, . . . ($i = 1, 2, 3$), which move in such a way that their mutual distances apart remain unchanged; i.e. so that for any two points whatever of the system, P and P', and for any movement of these points, the relation

$$\sum_1^3{}_i (\bar{y}_i - \bar{y}_i')^2 = d^2$$

holds, the quantity d^2 on the right being constant (geometrical characteristics of the moving system).

In these relations and in their differential consequences are summed up all the properties concerning simultaneous positions, velocities, &c., of the various points of the system.

(2) The ground covered by the equations just given for the relations between pairs of points is, so to speak, divided into two parts, the first expressing the intrinsic circumstance (i.e. independent of the system of reference $\bar{\Sigma}$) that when the moving system changes its position with respect to $\bar{\Sigma}$ it keeps its configuration unchanged. This is equivalent to the possibility of placing at the points P, P', . . . an observer Σ rigidly attached to the body, who can be represented as usual by an orthogonal trihedron $O\,y_1\,y_2\,y_3$ with respect to which the position of each separate point of the moving system remains unchanged. In other words, the co-ordinates y_i, y_i', . . . of

these points with respect to these axes attached to the body do not vary with the time.

At this stage the argument usually is as follows. In order to determine the position, with respect to the original system of reference $\bar{\Sigma}$, of the whole moving system at a generic instant, it is only necessary to place the trihedron $O\,y_1\,y_2\,y_3$ (attached to the body) in its proper position with respect to $\bar{O}\,\bar{y}_1\,\bar{y}_2\,\bar{y}_3$. Thus *we again have to deal with transformation formulæ* (variable from moment to moment) *between two systems of orthogonal Cartesian axes*, and therefore of the type

$$\bar{y}_i = \sum_{1}^{3}{}_k\, \alpha_{ik}\, y_k + \phi_i(t) \qquad (i = 1, 2, 3), \quad . \quad (15)$$

where α_{ik} denotes the cosine (variable with the time, if the motion is not one of pure translation) of the angle between the fixed axis $\bar{O}\bar{y}_i$ and the moving axis Oy_k, and $\phi_i(t)$ is a function of the time (linear if the motion reduces to a uniform translation).

The proposition in italics, or the equivalent group of formulæ (15), constitutes the complement of what may be called the intrinsic rigidity of the body (the existence of the trihedron attached to the body); the combination of the two gives us once again the kinematics of a solid body in its classical form.

But if we analyse this complement a little further, we find that we can modify to some extent the ordinary idea of the motion of a solid body without giving up either intrinsic rigidity or the validity of Euclidean geometry.

We need only introduce the hypothesis (independent of both the geometry and the kinematics of the point) that the measures of the distances between the points P, P', . . . (and therefore also of angles) of our solid may differ according as they are made by an observer attached to P, P', . . . or by the fixed observer $\bar{\Sigma}$. While granting that the two observers may disagree as to the measures, it is to be borne in mind that, by hypothesis, the measurements are made by each of them in accordance with a Euclidean metric, and that (as in the classical scheme of things) the rigidity of the motion must always be respected, from the point of view of the fixed observer $\bar{\Sigma}$ as well as of the other. This requires that every distance apart of two points P, P', . . . of our system must remain unchanged in time, whether the distance is calculated by $\bar{\Sigma}$ or by the other. For this it is necessary and sufficient that the transformation formulæ between the $\bar{y}$'s and the y's,

$$\bar{y}_i = f_i\,(y_1, y_2, y_3, t) \qquad (i = 1, 2, 3), \quad . \quad . \quad (16)$$

where the f_i's are *a priori* unknown functions, should be such as to make

$$dl^2 = \sum_1^3{}_i\, d\bar{y}_i{}^2 = \sum_1^3{}_i \left(\frac{\partial f_i}{\partial y_1}\,dy_1 + \frac{\partial f_i}{\partial y_2}\,dy_2 + \frac{\partial f_i}{\partial y_3}\,dy_3\right)^2 \quad (17)$$

independent of t at every instant, whatever may be the differentials dy_i.

We get an obvious case in which this condition is satisfied if we suppose that the transformation formulæ are linear in the y_i's (though not necessarily with respect to t), and in particular that they are of the form

$$\bar{y}_i = \sum_1^3{}_k\, c_{ik}\, y_k + \phi_i(t), \quad . \quad . \quad . \quad . \quad (18)$$

where the c's are completely arbitrary constants, subject only to the qualitative condition that their determinant $\| c_{ik} \|$ does not vanish. It should be remembered that in the equations (15) the coefficients a_{ik} were in addition direction cosines (in some cases variable with the time) for two sets of orthogonal axes.

A transformation of the type (18) between the y's and the $\bar{y}$'s is, at every instant (i.e. for any assigned value of t), linear, and therefore homographic, or rather affine, so that straight lines are transformed by it into straight lines. From our point of view, this means that curves which appear to the observer Σ to be straight lines are so also for the observer $\bar{\Sigma}$, and inversely.

It would not be hard to show that, if we impose on a transformation (16) the double condition of making dl^2 or the right-hand side of (17) independent of t, and also of keeping geodesics unchanged, we necessarily reach, if not an affine transformation (18), at least the product (in the sense of a product of operators) of an affine transformation by a rigid motion in the ordinary sense of the term. By a suitable choice of the trihedra of reference, the passage between the y's and the $\bar{y}$'s can thus be effected by applying in succession the two following transformations:

(1) An affine transformation given in its canonical form, i.e. by means of equations of the type

$$y_1' = k_1 y_1, \quad y_2' = k_2 y_2, \quad y_3' = k_3 y_3, \quad . \quad (19)$$

where the k's are positive constants;

(2) a transformation (15) between the $\bar{y}$'s and the y''s.

We shall not spend time on the elementary considerations which lead to this conclusion, and shall merely point out that the coefficients k of the equations (19) determine the *deformation* consequent on the affine correspondence between the y's and the y''s, while in the second change, from the y''s to the $\bar{y}$'s, there is no further deformation.

It follows from this, taking (19) into account, and considering the two observers Σ and $\bar{\Sigma}$, that *a segment having the same direction as the axis* Oy_i, *and of length* l *with respect to the observer* Σ *attached to the body, will appear to the observer* $\bar{\Sigma}$ *as having the length* $k_i l$; hence the factor k_i is called the *coefficient of elongation.* The " elongation " of unit length is accordingly $k_i - 1$; this represents an expansion or contraction according as $k_i >$ or < 1. The formulæ (19) of course provide, more generally, information as to the alteration in length of segments (and therefore of vectors) in any direction whatever. If the coefficients α_i ($i = 1, 2, 3$) are the direction cosines with respect to the axes $Oy_1 y_2 y_3$ of a generic segment and l the length as it appears to the observer Σ, we obviously get, for the length $\bar{l}$ as estimated by $\bar{\Sigma}$,

$$\bar{l} = l\sqrt{k_1^2 \alpha_1^2 + k_2^2 \alpha_2^2 + k_3^2 \alpha_3^2}.$$

Returning for a moment to the ordinary equations (15) of rigid motion, we shall fix our attention in particular on the most elementary case (which will serve as a guide and a basis of comparison in the argument of the next section), that of uniform translatory motion. We can then take the trihedron of reference $\bar{O}\bar{y}_1 \bar{y}_2 \bar{y}_3$ with one of its axes, say $\bar{y}_1$, parallel to the direction (by hypothesis constant) of the velocity, and we shall take the trihedron $Oy_1 y_2 y_3$ attached to the body as coinciding with the fixed trihedron at the initial instant $t = 0$.

The motion being translatory, the axes attached to the body remain parallel to the corresponding fixed axes throughout; and if v is the velocity of translation, the formulæ determining the motion evidently reduce to

$$\bar{y}_1 = y_1 + vt, \quad \bar{y}_2 = y_2, \quad \bar{y}_3 = y_3. \quad . \quad . \quad (15')$$

We have thus again reached the typical equations (14) for each point (y_1, y_2, y_3, constant) of the body.

8. Römerian units. Study of Lorentz transformations.

The equations (15′), which define in the simplest form an ordinary uniform translation, can obviously be associated with the identity

$$\bar{t} = t.$$

We thus get a quaternary transformation between (y_1, y_2, y_3, t) and $(\bar{y}_1, \bar{y}_2, \bar{y}_3, \bar{t})$, which we shall denote by T.

We next observe that the most general representation of a uniform translation, with arbitrary choice of the two trihedra (one fixed, the other attached to the body), subject to the sole condition that the origins coincide initially, can be reduced to T together with two rotations independent of t. In fact, denoting as before the two trihedra of reference (fixed and moving with the body) by $\bar{\Sigma}$ and Σ, we shall denote by R a rigid rotation of Σ (round the origin O) which brings its axis Oy_1 into the direction parallel to the velocity of translation. Let R' be an analogous rotation (round O) of the trihedron $\bar{\Sigma}$; and $\bar{R}$ the inverse rotation. Then the transformation formulæ between (y_1, y_2, y_3, t) and $(\bar{y}_1, \bar{y}_2, \bar{y}_3, \bar{t})$ are represented by the symbolic product

$$\bar{R} T R.$$

Now consider the well-known kinematical deduction from the classical method of representing rigid motion, namely, that if we consider any velocity $c\mathbf{u}$ whatever with respect to Σ (c being the modulus and $\mathbf{u}$ the versor), this becomes $c\mathbf{u} + \mathbf{v}$ with respect to $\bar{\Sigma}$, if $\mathbf{v}$ is the vector representing the velocity of translation of Σ with respect to $\bar{\Sigma}$. This, as has been observed, is in contradiction with the results of experiment, at least as regards the velocity of light, for which c in cm. per second has the particular value $3 \cdot 10^{10}$, *which remains unaltered*, even when compounded with a uniform translation (Michelson-Morley experiment). The wish to restore concord between theory and experiment leads us to modify the equations (15′), and with them, if necessary, the equation $\bar{t} = t$, in such a way that the relation $d\bar{s}_0^2 = ds_0^2$ (not merely $d\bar{l}^2 = dl^2$) shall be rigorously satisfied; i.e. so that there shall be an identity between two quadratic forms involving not only the space co-ordinates but also the time.

Special transformations. We propose to try to modify these transformation formulæ (as usual very slightly, at least for small values of v) so as to make ds_0^2 invariant. For this purpose we shall have to replace t sometimes by the variable

$$y_0 = ct$$

and sometimes by the imaginary variable

$$y_4 = \iota ct \qquad (\iota = \sqrt{-1}).$$

With this change, putting also $\frac{v}{c} = \beta$, the equations (15′) and $\bar{t} = t$ become either

$$\bar{y}_1 = y_1 + \beta y_0, \quad \bar{y}_2 = y_2, \quad \bar{y}_3 = y_3, \quad \bar{y}_0 = y_0 \quad . \quad (20)$$

or

$$\bar{y}_1 = y_1 - \iota\beta y_4, \quad \bar{y}_2 = y_2, \quad \bar{y}_3 = y_3, \quad \bar{y}_4 = y_4. \quad (20')$$

The real variable y_0 introduced here is only the time measured by choosing as unit the time taken by light to traverse unit space. Thus the velocity of light is 1 and the dimensions of time become the same as those of length. The character of a primary magnitude ordinarily assigned to the time thus disappears, the unit of time being linked up with the unit of length by means of the phenomenon of the propagation of light. It will be convenient to apply the term "Römerian"[1] to measurements of time made in this way; we shall similarly use the term "Römerian velocities" (which are pure numbers) for velocities referred to the Römerian time y_0. It obviously follows from the equation $y_0 = ct$ that a Römerian velocity is only the corresponding ordinary velocity divided by c; in particular, the quantity

[1] From O. RÖMER (1644–1710), who was the first to discover and determine the velocity of light. His method was based on observation of the eclipses of Jupiter's satellites.

$\beta = \dfrac{v}{c}$ just introduced is only the Römerian velocity of translation.

In accordance with Marcolongo's remark quoted on p. 300, the transformations we are in search of must leave invariant the differential quadratic form ds_0^2, which we can write (introducing the imaginary variable y_4) in the form

$$-ds_0^2 = dy_1^2 + dy_2^2 + dy_3^2 + dy_4^2.$$

In order to obtain particular transformations satisfying this condition, we shall first consider linear homogeneous transformations. These will at once result, as we have already pointed out in the preceding section, in the condition (17) being satisfied, which interprets the transformation as equivalent to a rigid motion (if not in the ordinary sense, at least in the intrinsic sense there specified). As we are dealing with linear (and homogeneous) transformations, the invariance of the differential form $-ds_0^2$ implies that of the algebraic quadratic form

$$-q = y_1^2 + y_2^2 + y_3^2 + y_4^2,$$

and reciprocally.

Starting from the equations (20′) we shall examine whether we can reach the required result if we keep the co-ordinates y_2, y_3 invariant, i.e. if we suppose

$$\bar{y}_2 = y_2, \qquad \bar{y}_3 = y_3.$$

We have thus to find a linear transformation between the variables (y_1, y_4) and $(\bar{y}_1, \bar{y}_4)$ which will leave invariant the expression

$$y_1^2 + y_4^2.$$

Hence (apart from the question of imaginaries) we have to discuss a rigid rotation, round the origin of co-ordinates, in the plane y_1, y_4, and therefore of the form

$$\bar{y}_1 = y_1 \cos\phi - y_4 \sin\phi,$$
$$\bar{y}_4 = y_1 \sin\phi + y_4 \cos\phi.$$

If we introduce the real variables $y_0, \bar{y}_0$ instead of $y_4, \bar{y}_4$, it will be seen that the necessary and sufficient condition for the disappearance of imaginaries from the ultimate formulæ is that the coefficient of y_1 should be real and that of y_4 imaginary in the first equation, and *vice versa* in the second. To obtain this result, ϕ must be a pure imaginary; in fact, putting

$$\phi = \iota\psi \text{ (with } \psi \text{ real)},$$

we get

$$\cos\phi = \cos\iota\psi = \cosh\psi,$$
$$\sin\phi = \sin\iota\psi = \iota\sinh\psi,$$

where $\cosh\psi$ and $\sinh\psi$ as usual denote the hyperbolic cosine and sine.

Hence our transformation formulæ take the form

$$\left.\begin{aligned} \bar{y}_1 &= y_1 \cosh\psi + y_0 \sinh\psi \\ \bar{y}_2 &= y_2 \\ \bar{y}_3 &= y_3 \\ \bar{y}_0 &= y_1 \sinh\psi + y_0 \cosh\psi \end{aligned}\right\} \quad . \quad . \quad . \quad . \quad (21)$$

If we remember that in the equations (20) the pure number β is in ordinary cases fairly small, we see that in these cases the equations (21) differ quantitatively by very little from the equations (20), provided we suppose ψ sufficiently small for $\cosh\psi$ and $\sinh\psi$ not to differ by very much from 1 and 0 respectively.

But we get a precise kinematical interpretation of the parameter ψ on which the transformation (21) depends if for instance we fix our attention on the origin O of the moving axes, i.e. on the point whose co-ordinates are $y_1 = y_2 = y_3 = 0$, for a generic value of y_0, this last parameter denoting the time (Römerian time) as it appears to the observer Σ. For the fixed observer $\bar{\Sigma}$, with respect to whom $\bar{y}_0$ represents the time (likewise Römerian) and $\bar{y}_1, \bar{y}_2, \bar{y}_3$ the position, we have, corresponding to O and a generic value of y_0,

$$\bar{y}_1 = y_0 \sinh\psi, \qquad \bar{y}_0 = y_0 \cosh\psi,$$

while $\bar{y}_2$ and $\bar{y}_3$ vanish. Hence the motion of O is rectilinear, and the ratio

$$\frac{\bar{y}_1}{\bar{y}_0} = \tanh\psi,$$

which is obviously constant, is the (Römerian) velocity. Denoting this ratio by β, we have in the equation

$$\tanh\psi = \beta$$

the required kinematical significance of the parameter ψ. More generally, the same quantity β stands for the Römerian velocity of any other point P rigidly attached to Σ. In fact, if y_1, y_2, y_3 are constants and y_0 a generic value, the result of differentiating the equations (21) is to give

$$d\bar{y}_1 = \sinh\psi\, dy_0, \quad d\bar{y}_2 = d\bar{y}_3 = 0, \quad d\bar{y}_0 = \cosh\psi\, dy_0,$$

whence it follows that

$$\frac{d\bar{y}_1}{d\bar{y}_0} = \tanh\psi = \beta.$$

Q.E.D.

Applying the ordinary formulæ

$$\cosh = \frac{1}{\sqrt{1 - \tanh^2}}, \quad \sinh = \frac{\tanh}{\sqrt{1 - \tanh^2}},$$

we can put the equations (21) in the form commonly used (the special Lorentz transformation)

$$\left.\begin{aligned} \bar{y}_1 &= \frac{y_1 + \beta y_0}{\sqrt{1 - \beta^2}} \\ \bar{y}_2 &= y_2 \\ \bar{y}_3 &= y_3 \\ \bar{y}_0 &= \frac{y_0 + \beta y_1}{\sqrt{1 - \beta^2}} \end{aligned}\right\}, \quad . \quad . \quad . \quad . \quad (21')$$

or, using the ordinary instead of the Römerian unit of time (i.e. $\bar{t}$ and t instead of $\bar{y}_0$ and y_0, with $\bar{y}_0 = c\bar{t}$, $y_0 = ct$),

$$\left.\begin{aligned} \bar{y}_1 &= \frac{y_1 + vt}{\sqrt{1 - \frac{v^2}{c^2}}} \\ \bar{y}_2 &= y_2 \\ \bar{y}_3 &= y_3 \\ \bar{t} &= \frac{t + \frac{v}{c^2} y_1}{\sqrt{1 - \frac{v^2}{c^2}}} \end{aligned}\right\} \quad . \quad . \quad . \quad . \quad (21'')$$

It will be seen that the necessary condition for these formulæ to be real is $\beta^2 < 1$, or $v < c$; which once more demonstrates that the velocity of light c is a limiting velocity.

It can be easily verified that the formulæ obtained from (21′) or the equivalent (21″) by solving these equations with respect to y_1, y_2, y_3, and y_0 or t differ from the first set only by the change of v into $-v$ (and therefore of β into $-\beta$), and of course of the two sets of variables; precisely as happens for the equations (15′) and (20) which refer to an ordinary translation. If in particular we suppose β^2, i.e. $\frac{v^2}{c^2}$, (but not necessarily β), negligible in comparison with unity, the first three of the equations (21″) reduce to the formulæ (15′) of the ordinary translation, while the fourth gives rise to an additive term denoting the difference of time between the two observers Σ and $\bar{\Sigma}$, expressed by the equation

$$\bar{t} = t + \frac{v}{c^2} y_1.$$

It will be seen that the additional term $\frac{v}{c^2} y_1$ depends on the position of the point at which $\bar{\Sigma}$ has to apply his own measurements of the time; for this reason $\bar{t}$ is called the *local time*. It was associated by Lorentz with the ordinary uniform translations (15′) with the intention of explaining to a first approximation (i.e. neglecting β^2) the character of electromagnetic phenomena for bodies in motion; this requires explicitly that the relation $d\bar{s}_0{}^2 = ds_0{}^2$ should hold to the same order of approximation. Later on Lorentz himself discovered the equations (21″), which result in the rigorous invariance of $ds_0{}^2$. Einstein rediscovered them from the point of view of this invariance, which is the mathematical expression of his principle of relativity in its most elementary form.

Let us examine the formulæ (21″). They contain the best-known results (to some of which eminent students of relativity have assigned paradoxical consequences) of the kinematics of relativity. In the first place, the non-invariance of t, as noted above in § 3 in general for any (T_4), points to the necessity of abandoning the ordinary concept of simultaneity in the absolute sense. In fact, two instantaneous events, taking place at two different points of space, may correspond to the same value of t but not of $\bar{t}$ (a sufficient condition is that the y_1's should be different), and may therefore be simultaneous for one observer who uses Σ as his system of reference and not for another who uses $\bar{\Sigma}$. Hence the time ceases to be an absolute quantity and becomes relative to the system of reference and connected up with the space co-ordinates; it is in fact *local time*, to use the term already referred to as having been introduced by Lorentz in his researches on the electrodynamics of bodies in motion.

Suppose that two events take place at the same point P of the body (and therefore with the same y_1, y_2, y_3), but not at the same instant, being separated by an interval of time Δt (measured in the system Σ): for the observer $\bar{\Sigma}$ the interval will be $\Delta \bar{t}$, and the relation between the two is given at once by the fourth of the equations (21″), noting that y_1 is constant, so that

$$\Delta \bar{t} = \frac{\Delta t}{\sqrt{1 - \beta^2}}.$$

Hence for the observer who accompanies the point P where the phenomena take place, the interval of time Δt is shorter than for the fixed observer $\bar{\Sigma}$; i.e. we have a slowing down of the time with respect to $\bar{\Sigma}$'s measure, as if the unit of measurement had become $\dfrac{1}{\sqrt{1 - \beta^2}}$ times that used by $\bar{\Sigma}$.

Similarly two events which happen at what is for $\bar{\Sigma}$ the same point (i.e. with the same $\bar{y}_1, \bar{y}_2, \bar{y}_3$) but separated by an interval of time $\Delta \bar{t}$, will appear to Σ to be separated by a longer interval. This follows at once from the fact pointed out above that the inverse formulæ of (21′) and (21″) are found by changing v into $-v$ and therefore β into $-\beta$.

We shall now try to determine the difference, if any, in the estimates of lengths made by two observers Σ and $\bar{\Sigma}$, each at a specified instant of his own time. Suppose, for instance, we wish to carry over to the observer $\bar{\Sigma}$ measurements made by Σ. Substituting in the first three transformation formulæ of (21″) the value of t in terms of $\bar{t}$ given by the fourth, we get

$$\bar{y}_1 = \sqrt{1 - \frac{v^2}{c^2}}\, y_1 + v\bar{t}$$

$$\bar{y}_2 = y_2$$

$$\bar{y}_3 = y_3,$$

which may be considered as resulting from the product of the affine transformation

$$y_1' = \sqrt{1 - \frac{v^2}{c^2}}\, y_1, \quad y_2' = y_2, \quad y_3' = y_3$$

by the ordinary translation

$$\bar{y}_1 = y_1' + v\bar{t}, \quad \bar{y}_2 = y_2, \quad \bar{y}_3 = y_3.$$

In this form of the equations the change in length is at once obvious. We need only refer to the conclusions of the preceding section, noting that the elongation coefficients k_1, k_2, k_3 of the formulæ (19) are in this case represented by $\sqrt{1 - \beta^2}$, 1, 1. Hence if the fixed observer estimates distances at a generic instant $\bar{t}$, and if his results are compared with those of the observer attached to the moving body, who is also estimating the same distances at any instant whatever of his own time, then the former observes a *contraction, in the ratio* $\sqrt{1 - \beta^2} : 1$, *for longitudinal segments*, i.e. in the direction of motion, while there is *no change for transverse segments*, i.e. perpendicular to the velocity of translation.

The inverse formulæ, for the change from $\bar{\Sigma}$ to Σ, differ, as we have already said, only by the change of v into $-v$. Hence the same rules hold good; e.g. fixed segments in the direction of motion will appear to the moving observer as contracted in the ratio $\sqrt{1 - \beta^2} : 1$ in comparison with the measurement of them made by the observer $\bar{\Sigma}$; and so on.

General transformations. We propose lastly to prove a result analogous to one shown above to hold for the translations of classical kinematics, namely, that the most general Lorentz transformation (Λ) (i.e. a linear transformation between two sets of four variables y_i and $\bar{y}_i$ $(i = 1, 2, 3, 4)$ for which the quadric q remains invariant) can be represented in the symbolic form

$$\bar{R}\, \mathcal{L}\, R,$$

where R and $\bar{R}$ are ordinary orthogonal transformations (rotations) between (y_1, y_2, y_3) and $(\bar{y}_1, \bar{y}_2, \bar{y}_3)$, and $\mathcal{L}$ is a special Lorentz transformation of the type studied above.

The transformation (Λ) will be a quaternary orthogonal transformation of the type

$$\bar{y}_h = \sum_1^4{}_k \, a_{hk} \, y_k, \quad . \quad . \quad . \quad . \quad . \quad (22)$$

whose coefficients a_{hk} constitute an orthogonal matrix, i.e. such that

$$\sum_1^4{}_k \, a_{hk} \, a_{jk} = \delta_h^j, \quad . \quad . \quad . \quad . \quad . \quad (23)$$

$$\sum_1^4{}_k \, a_{kh} \, a_{kj} = \delta_h^j \quad . \quad . \quad . \quad . \quad . \quad (23')$$

$$(h, j = 1, 2, 3, 4).$$

In order that the variables $\bar{y}_1$, $\bar{y}_2$, $\bar{y}_3$, y_1, y_2, y_3, may be real, and $\bar{y}_4$, y_4 pure imaginaries, we must evidently have a_{hk} ($h, k < 4$) real, a_{h4} and a_{4h} ($h < 4$) pure imaginaries, and a_{44} real.

We shall of course interpret y_1, y_2, y_3 as Cartesian co-ordinates with respect to a trihedron K rigidly attached to Σ, and y_4 as the time variable; and similarly for the $\bar{y}$'s.

The directions of the trihedra K and $\overline{K}$ are *a priori* arbitrary; we shall now determine a rotation R for K and a rotation $\overline{R}$ for $\overline{K}$ such that we shall have

$$\Lambda = \overline{\mathrm{R}}\,\mathcal{L}\,\mathrm{R}.$$

To do this, we consider, with reference to $\overline{K}$, the vector whose components are a_{14}, a_{24}, a_{34}; let $\bar{\mathbf{i}}$ denote the relative versor. If we turn the trihedron $\overline{K}$ round in such a way that its axis $\bar{y}_1$ takes the direction of $\bar{\mathbf{i}}$, we shall have

$$a_{24} = a_{34} = 0;$$

we shall take this as the rotation $\overline{R}$.

Now from the identities (23) and the values just given it follows that

$$\sum_1^3{}_k \, a_{2k}^2 = \sum_1^3{}_k \, a_{3k}^2 = 1,$$

$$\sum_1^3{}_k \, a_{2k} \, a_{3k} = 0,$$

so that the two vectors determined (with respect to K) by the components a_{2k}, a_{3k} ($k = 1, 2, 3$) are of unit length and orthogonal. We shall call them $\mathbf{j}$, $\mathbf{k}$, and shall take as the rotation R

the rotation which turns the trihedron K so that its axes y_2, y_3 coincide in direction with the vectors **j**, **k**, so that we get

$$a_{21} = a_{23} = a_{31} = a_{32} = 0.$$

As a result of the two rotations R and $\bar{R}$ the form of the matrix of the a's comes to be

$$\left\| \begin{matrix} a_{11} & a_{12} & a_{13} & a_{14} \\ 0 & 1 & 0 & 0 \\ 0 & 0 & 1 & 0 \\ a_{41} & a_{42} & a_{43} & a_{44} \end{matrix} \right\|,$$

and, from the group properties of orthogonal substitutions, this matrix must also correspond to a substitution of this kind (since it is the result of the product of the original substitution by two rotations). A consequence of this is the vanishing of four other elements of the matrix: in fact, the conditions that the first line of the matrix shall be orthogonal with the second and third lines respectively are

$$a_{12} = a_{13} = 0,$$

and similarly, taking the fourth line with the second and third, we get

$$a_{42} = a_{43} = 0.$$

Thus we finally get the matrix in the form

$$\left\| \begin{matrix} a_{11} & 0 & 0 & a_{14} \\ 0 & 1 & 0 & 0 \\ 0 & 0 & 1 & 0 \\ a_{41} & 0 & 0 & a_{44} \end{matrix} \right\|,$$

which corresponds to a transformation of the type (21), i.e. to a special Lorentz transformation $\mathcal{L}$. We have thus shown that, through the two rotations R (for the trihedron K) and $\bar{R}$ (for $\bar{K}$), the general transformation (Λ) reduces to a special transformation $\mathcal{L}$.

So far we have considered only linear transformations. The question may be raised whether we should not get greater generality if this restriction were removed. In this connexion we shall merely say[1] that the linear transformations studied here are the only ones which, in addition to retaining the invariance of ds_0^2, make finite values of the $\bar{y}$'s correspond to finite values of the y's, and *vice versa.*

9. **Relative motion. Composition of velocities. Kinematical justification of a formula of Fresnel's.**

In order to show the relation between the various aspects of a single motion—let us say specifically the motion of an assigned point P—with reference to two different observers Σ and $\bar{\Sigma}$, it is only necessary to use the transformation formulæ between the corresponding co-ordinates. This holds both in ordinary kinematics and in relativity kinematics, with the reminder that for the latter the time y_0 is among the co-ordinates affected by the transformation.

Consider in particular a Lorentz translation, which, as we have seen in the preceding section, is defined by the formulæ (21″), suitable choice being made in advance of the two trihedra which represent the observers and are denoted by Σ and $\bar{\Sigma}$.

Now suppose that the motion (which we can call *relative*) of the point P in relation to Σ is given; i.e. that the expressions $y_i(y_0)$ $(i = 1, 2, 3)$ for its three space co-ordinates are known formally as functions of y_0 (the Römerian time). To obtain a representation of the absolute motion, i.e. the motion with reference to $\bar{\Sigma}$, it is obviously sufficient to find the expressions for the co-ordinates $\bar{y}_i$ $(i = 1, 2, 3)$ of the point P as functions of the new time variable $\bar{y}_0$. The transformation formulæ (21′) give the required result at once; in fact, if we insert in them for the y_i's the expressions $y_i(y_0)$ belonging to the moving point P, all the $\bar{y}$'s become known functions of y_0; and if we suppose this parameter found from the fourth equation

$$\bar{y}_0 = \frac{y_0 + \beta y_1}{\sqrt{1 - \beta^2}},$$

and substituted in the first three, we get the equations of absolute motion in their explicit form.

[1] For the proof, cf. C. MUNARI: "Sopra una espressiva interpretazione cinematica del Principio di Relatività", in *Rend. della R. Acc. dei Lincei*, Vol. XXIII (1914), p. 781.

The resulting relation between the absolute and relative velocities is especially interesting. The vector rule no longer holds that the absolute velocity = the relative velocity + the velocity of the moving origin (the latter, in the case of translations, of course reduces to the velocity of translation, whatever may be the instantaneous position of P). The relativity composition of velocities is a little more complicated. In order to see what happens in the clearest case, we shall consider a relative motion parallel to the translation $\mathbf{v}$. With this hypothesis the co-ordinates y_2 and y_3 of the point P are constant, and, from (21′), $\bar{y}_2$ and $\bar{y}_3$ are also constant, or, in other words, the motion with respect to $\bar{\Sigma}$ is also in the direction of the translation. Differentiating the first and fourth equations of (21′), we get

$$d\bar{y}_1 = \frac{dy_1 + \beta dy_0}{\sqrt{1-\beta^2}},$$

$$d\bar{y}_0 = \frac{dy_0 + \beta dy_1}{\sqrt{1-\beta^2}}.$$

Putting for the sake of shortness

$$\beta_a = \frac{d\bar{y}_1}{d\bar{y}_0}, \quad \beta_r = \frac{dy_1}{dy_0},$$

so that β_a and β_r are the velocities (scalar and Römerian) of the point P with respect to $\bar{\Sigma}$ and Σ respectively (absolute velocity and relative velocity), the foregoing formulæ, on dividing the first by the second, give

$$\beta_a = \frac{\beta_r + \beta}{1 + \beta\beta_r}; \quad . \quad . \quad . \quad . \quad . \quad (24)$$

this is what is called Einstein's law for the composition of velocities. Multiplying by c and remembering that $y_0 = ct$, $\bar{y}_0 = c\bar{t}$, we can evidently replace the Römerian velocities $\beta_a = \frac{d\bar{y}_1}{d\bar{y}_0}$, $\beta_r = \frac{dy_1}{dy_0}$, β, by the corresponding ordinary velocities $v_a = \frac{d\bar{y}_1}{d\bar{t}}$, $v_r = \frac{dy_1}{dt}$, v, and write

$$v_a = \frac{v_r + v}{1 + \beta\beta_r}. \quad . \quad . \quad . \quad . \quad (24')$$

If both the velocity v_r of P (with respect to Σ) and the velocity of translation v are small in comparison with c, the denominator differs from unity by a term of the second order; if we neglect this difference we get back to the fundamental relation of ordinary kinematics (which may be called Galilean)

$$v_a = v_r + v;$$

in view of the criterion we are applying, this result was of course to be expected. In general the equation (24) shows that, for $|\beta|$ and $|\beta_r|$ less than unity, $|\beta_a|$ also < 1; while, for $|\beta|$ or $|\beta_r|$ equal to unity, $|\beta_a|$ also $= 1$. To prove this, note that, whenever $|\beta| < 1$ and $|\beta_r| < 1$,

$$(1 + \beta\beta_r)^2 - (\beta + \beta_r)^2 = (1 - \beta^2)(1 - \beta_r^2)$$

is always positive, so that $\beta_a^2 = \left(\dfrac{\beta + \beta_r}{1 + \beta\beta_r}\right)^2 < 1$; while for $|\beta| = 1$, or $|\beta_r| = 1$, $\beta_a^2 = 1$; which proves the required result. We thus find once more the limiting character of the velocity c of light: however near v_r may be to c, provided it is less than c ($\beta_r < 1$), if it is compounded with another velocity of translation v, less than c, but as nearly equal to c as we please ($|\beta| < 1$), the result will always be less than c, or in other words $|\beta_a|$ always < 1. *Vice versa*, the velocity c for Σ remains c for any $\bar{\Sigma}$, whatever may be the velocity of the (Lorentz) translation with which the two observers are moving with respect to one another.

Within the scale of velocities of ponderable bodies (velocities small compared with c), the relation (24′) reduces sensibly to the Galilean formula $v_a = v_r + v$, as we have already said. But when the phenomenon of motion under consideration is the propagation of light in a transparent medium, so that the velocity has an order of magnitude comparable with that of c, then the divergence between the Einsteinian and the Galilean kinematics becomes striking, and lends itself to experimental verification.

Einstein has in fact drawn from this a magnificent argument in support of the theory of relativity. He deduced logically (by a purely kinematical proof [1]) from (24′) a formula of Fresnel's

[1] Even before Einstein, Lorentz had given a theoretical justification of Fresnel's formula, based on his celebrated electron theory of the electromagnetic phenomena of bodies in motion. Einstein's explanation is plainly more attractive.

concerning the movement of light waves through transparent media in translatory motion; a formula which was experimentally confirmed for the first time by Fizeau (1851), whose experiments were repeated with improved methods by Michelson and Morley and by Zeeman.

The argument is briefly as follows. In a medium of refractive index μ, it is known that light is propagated with velocity $\frac{c}{\mu}$, if the medium is at rest. Suppose instead that the medium has a velocity v in the direction of propagation of light (in the same or the opposite sense). Ordinary kinematics would lead us to expect that the velocity of propagation (with respect to the observer) would become $\frac{c}{\mu} \pm v$; Fizeau and the others, however, by delicate experiments on interference phenomena, found that the amount to be added to $\frac{c}{\mu}$ (or subtracted from it) is not the whole of v, but v multiplied by the coefficient (<1) $1 - \frac{1}{\mu^2}$, so that the velocity of propagation is

$$\frac{c}{\mu} \pm v\left(1 - \frac{1}{\mu^2}\right). \quad . \quad . \quad . \quad . \quad . \quad . \quad (25)$$

The factor $1 - \frac{1}{\mu^2}$ is known as *Fresnel's convection coefficient.*

The expressions (25) are evidently not in agreement with the Galilean kinematics. But they are in excellent agreement with the Einsteinian kinematics. In fact, let us consider, to fix ideas, the case in which the motion of the medium is in the same sense as the propagation of light, so that we take the $+$ sign in (25). Then (24′) holds when we put $\frac{c}{\mu}$ for v_r, and therefore $\frac{1}{\mu}$ for β_r. Hence it gives

$$v_a = \frac{\frac{c}{\mu} + v}{1 + \frac{\beta}{\mu}},$$

or, neglecting terms of the second order in $\beta\left(= \frac{v}{c}\right)$,

$$v_a = \left(\frac{c}{\mu} + v\right)\left(1 - \frac{\beta}{\mu}\right) = \frac{c}{\mu} + \left(1 - \frac{1}{\mu^2}\right)v - \frac{v\beta}{\mu}.$$

The last term is also of the second order with respect to the first, so that we are left finally with Fresnel's formula

$$v_a = \frac{c}{\mu} + \left(1 - \frac{1}{\mu^2}\right) v.$$

10. **Further generalization of the metric of V_4, still coinciding to a first approximation with ordinary dynamics.**

We now propose to see whether it is possible to assign to the V_4 other metrics slightly different from that characterized by (5), but such that the dynamical principle underlying them is still equivalent, to a first approximation, to Hamilton's principle.

We return to the general form (5′) of ds^2, and observe first that the particular form (5) just considered is a case of (5′) obtained by identifying the time co-ordinate x_0 with ct and the space co-ordinates x_1, x_2, x_3 with the Cartesian co-ordinates y_1, y_2, y_3, and putting

$$g_{00} = 1 - \frac{2U}{c^2},\ g_{0i} = 0,\ g_{ik} = -\delta_i^k \quad (i, k = 1, 2, 3). \quad (26)$$

If now we wish to consider a metric whose coefficients g differ by very little from the values (26), we can put

$$g_{00} = 1 - 2\phi,\ g_{0i} = -\gamma_i,\ g_{ik} = -\delta_i^k - \gamma_{ik}, \quad (26')$$

where

$$\phi = \frac{U}{c^2} + \psi,$$

with the understanding that the quantities γ (which as regards dimensions are pure numbers) are of the second order $\left(\text{like } \frac{U}{c^2}\right)$ or higher order with respect to $\frac{v}{c}$, while ψ (which also has the dimensions of a number) is to be considered as of at least the third order.

With these values for the coefficients, and taking the variables y_1, y_2, y_3 to be co-ordinates differing very little from Cartesians, we can write ds^2 in the form

$$ds^2 = (1 - 2\phi)dy_0^2 - 2dy_0 \sum_{1}^{3}{}_i \gamma_i dy_i - \sum_{1}^{3}{}_{ik} (\delta_i^k + \gamma_{ik}) dy_i dy_k. \quad (27)$$

If we denote derivation with respect to y_0 by a dash, and put

$$\left.\begin{aligned} \beta^2 &= \frac{v^2}{c^2} = \sum_1^3{}_i\, y_i'^2 \\ T_1 &= \sum_1^3{}_i\, \gamma_i y_i' \\ T_2 &= \tfrac{1}{2}\sum_1^3{}_{ik}\, \gamma_{ik} y_i' y_k' \end{aligned}\right\} \qquad . \quad . \quad . \quad . \quad (28)$$

we shall have

$$\frac{ds^2}{dy_0^{\,2}} = 1 - 2\phi - 2T_1 - 2T_2 - \beta^2.$$

It is to be noted that since $y_i' = \frac{1}{c}\frac{dy_i}{dt}$ is of the first order, it follows that T_1 is of the third and T_2 of the fourth order at least.

To shorten the work, we shall introduce the quadrinomial

$$\Gamma = \phi + T_1 + T_2 + \tfrac{1}{2}\beta^2,$$

observing that it is composed of terms of the second order, which can be written $\frac{1}{c^2}(\frac{1}{2}v^2 + U)$, plus terms of higher order. We then have

$$\frac{ds^2}{dy_0^{\,2}} = 1 - 2\Gamma,$$

and we can extract the square root, neglecting powers of Γ higher than the second (i.e. terms of order higher than the fourth). To this degree of approximation we get

$$\frac{ds}{dy_0} = 1 - \Gamma - \tfrac{1}{2}\Gamma^2,$$

i.e. rewriting $c\,dt$ for dy_0, and multiplying by c^2dt,

$$c\,ds = c^2dt - c^2dt(\Gamma + \tfrac{1}{2}\Gamma^2).$$

Substituting this expression in the variational equation of dynamics

$$\delta\int c\,ds = 0,$$

and remembering that δt vanishes at the limits of integration, we see that this equation reduces to

$$\delta\int c^2(\Gamma + \tfrac{1}{2}\Gamma^2)dt = 0,$$

and that the corresponding Lagrangian function is therefore

$$L = c^2(\Gamma + \tfrac{1}{2}\Gamma^2),$$

or expanding, and neglecting terms of order higher than the fourth (in the sense just defined, i.e. ignoring the presence of the factor c^2),

$$L = \tfrac{1}{2}v^2 + U + c^2T_1 + c^2T_2 + c^2\psi + \frac{c^2}{2}\left(\frac{\tfrac{1}{2}v^2 - U}{c^2}\right)^2 + v^2\frac{U}{c^2}. \quad (29)$$

The first two terms of the expression on the right (reduced to zero dimensions, i.e. divided by c^2) are of the second order; they constitute the Lagrangian function of the classical mechanics, from which we began our investigations.

The successive terms of (29) (reduced to zero dimensions in the same way) are of higher order: hence they will represent small corrections to be applied to the equations of motion. The metric (27) which we have here assumed still gives, therefore, to a first approximation, the same laws as are deduced from Hamilton's classical principle. Besides the potential U, it contains the ten functions ψ, γ_i, γ_{ik} of the four variables y (position and time); these are small, as we agreed, and as we have repeatedly had to remember in making the various transformations, but are *a priori* arbitrary. We shall see farther on how the law of universal gravitation and a criterion provided by the tensor calculus lead to the determination of these ten functions (from ten differential equations), and so to an explanation of some slight divergences which have been observed between the results predicted by the Newtonian mechanics and the true motion of the heavenly bodies. This more exact correspondence between theory and observation provides a physical justification of Einstein's new method of approach, which further incontestably represents an enormous speculative advance through its characteristic of securing invariance for all transformations of the co-ordinates, not only of space, but also of time.

11. An important particular case. Corresponding trajectories and their identity with those of an ordinary mechanical problem.

We shall now apply the expression (29) for L to a special case, the interest of which will be seen in next chapter (§ 8, p. 394). Suppose that we have

$$T_1 = 0,$$
$$T_2 = \tfrac{1}{2}\chi\frac{v^2}{c^2}$$

(either exactly, or neglecting terms of order higher than the third and fourth respectively), where χ is a function of the y's of at least the second order. Suppose further that ψ and χ, like U, do not explicitly depend on the time. We shall meet later on a characteristic example in which this condition is satisfied.

The expression (29) can now be written

$$L = \tfrac{1}{2}v^2\left(1 + \frac{2U}{c^2} + \chi\right) + U + c^2\psi + \frac{c^2}{2}\left(\frac{\tfrac{1}{2}v^2 - U}{c^2}\right)^2. \quad (30)$$

It is to be noted that $\left(\frac{\tfrac{1}{2}v^2 - U}{c^2}\right)^2$ in the last term is of the fourth order, while the principal part of L is of the second order; hence, in this last term, we may calculate $\frac{\tfrac{1}{2}v^2 - U}{c^2}$ only to a first approximation. But we know that to a first approximation the classical mechanics holds, and that therefore the integral of *vis viva* exists in the form

$$\tfrac{1}{2}v^2 - U = E_0 = \text{constant};$$

hence the last term on the right of (30) can be replaced by the constant $\tfrac{1}{2}E_0^2/c^2$, or even suppressed, since a constant contributes nothing to the variational equation.

The remaining terms of L can be separated into two groups, according as they do or do not depend on the velocity, by putting

$$\boldsymbol{T} = \tfrac{1}{2}v^2\left(1 + \frac{2U}{c^2} + \chi\right)$$
$$\boldsymbol{U} = U + c^2\psi$$

and therefore

$$L = \boldsymbol{T} + \boldsymbol{U}.$$

This form of the Lagrangian function corresponds exactly to the form found in the classical mechanics (for a system with three degrees of freedom, if not for a material particle), if we consider $\boldsymbol{T}$ as corresponding to the *vis viva* and $\boldsymbol{U}$ to the potential. Further, it is known [1] that, whenever (as in this case) $\boldsymbol{T}$ is a quadratic form in the quantities $\dot{y}_i = \frac{dy_i}{dt}$, not explicitly containing t, and $\boldsymbol{U}$ is a function only of the Lagrangian co-ordinates y, then the differential equations arising from

$$\delta\int(\boldsymbol{T}+\boldsymbol{U})dt = 0$$

admit of the integral (of *vis viva*)

$$\boldsymbol{T}-\boldsymbol{U} = E,$$

where E is a constant (the total energy), and the trajectories corresponding to a given value of E are identical with the geodesics of a manifold such that the square of its line element is defined by

$$ds^2 = 2(\boldsymbol{U}+E)\boldsymbol{T}\,dt^2.$$

(Principle of Stationary Action.) Applying all this to our case we shall have the integral of *vis viva* in the form

$$\tfrac{1}{2}v^2\left(1+\frac{2U}{c^2}+\chi\right)-\left(U+c^2\psi\right) = E,$$

and, for any value of E fixed in advance, we can assert that the trajectories coincide with the geodesics of the manifold

$$ds_1^{\,2} = \left(U+c^2\psi+E\right)\left(1+\frac{2U}{c^2}+\chi\right)dl_0^{\,2},$$

where
$$dl_0^{\,2} = \sum_1^3{}_i\, dy_i^{\,2},$$

or with the trajectories of the motion, in ordinary space, of a material particle with total energy zero, and acted on by forces derived from the potential

$$U_1 = \left(U+c^2\psi+E\right)\left(1+\frac{2U}{c^2}+\chi\right),$$

[1] Cf. for example LEVI-CIVITA and AMALDI: *Lezioni di meccanica razionale*, Vol. II, Chapter XI, No. 16 (Bologna, Zanichelli; in the press); or WHITTAKER: *Analytical Dynamics*, 2nd edition, Chapter IX (Cambridge University Press, 1917).

which can also be written (neglecting constant terms and terms of higher order) as

$$U_1 = U + c^2\psi + \frac{2U^2}{c^2} + U\chi + E\left(\frac{2U}{c^2} + \chi\right). \qquad (31)$$

12. **Qualitative characteristics of relativity metrics. Geodesic principle for the dynamics of a material particle. Stationary and, in particular, statical line elements.**

In accordance with the remarks at the end of § 10, the metric of the space-time manifold V_4 in the region round a generic point must be regarded in concrete cases in close connexion with the physical phenomena which take place in space and time, particularly in the neighbourhood of the point and instant considered. The quantitative dependence will be duly established in next chapter. At any rate, in ordinary cases, as has been seen, we can never go far from a pseudo-Euclidean metric. This leads to the condition that in the real world of physics the metric of V_4 is to have the same qualitative properties as those belonging to the pseudo-Euclidean metrics. In particular, the index of inertia must be 3, which implies (as could be proved) that in every set of four orthogonal directions drawn from a generic point, three are spacelike ($ds^2 < 0$) and one is timelike ($ds^2 > 0$).

By " relativity metric " we shall from now onwards mean an indefinite metric subject to these qualitative restrictions.

In a V_4 with a definite metric there is no qualitative distinction to be made between the various lines in it, while in a relativity V_4, as we have already pointed out, we have at every point three kinds of direction, according as $ds^2 <$ or $>$ or $=$ 0, and, corresponding to these, three kinds of line—spacelike, timelike, and lines of zero length. Naturally the classification is much more complicated for manifolds of two or three dimensions immersed in a V_4 with an indefinite metric; and the same choice of the variables of reference (which is geometrically equivalent to the choice of co-ordinate hypersurfaces) would in general require preliminary close study of the local behaviour from this point of view.

We shall avoid any discussion of this kind, and shall impose some limits on the arbitrariness of the choice of co-ordinates by taking as a model what happens in the case of a pseudo-Euclidean ds^2 referred to ordinary time t (or a linear function y_0

of t) and three space co-ordinates x_1, x_2, x_3, which are entirely arbitrary. Of the four co-ordinate lines one (y_0) will then be timelike and the others spacelike; further, on any hypersurface $y_0 =$ constant we have

$$(ds^2)_{y_0 = const.} = -dl^2,$$

where dl^2 is a positive definite differential quadric, so that we can say that a purely spacelike metric, like that of ordinary geometry, holds in every timelike section of the space-time. We shall constantly refer the relativity manifold V_4 to co-ordinates x_0, x_1, x_2, x_3 for which this qualitative property holds.

Granting these various preliminaries we reach the following *geodesic principle*—derived from the particular cases in sections 4 and 10 by an obvious generalization—which, in Einstein's work, appears as a fundamental law of the dynamics of a material particle in clearly specified physical conditions (i.e. for an assigned ds^2):

The world lines of a generic free material particle are identical with the geodesics of the corresponding ds^2, and more precisely with the timelike geodesics. In other words, these world lines satisfy the variational equation

$$\delta\int ds = 0,$$

making at the same time $ds^2 > 0$.

Among the relativity metrics special interest attaches to those in which it is possible to choose a system of reference such that the ten coefficients g_{ik} shall all be independent of the timelike parameter x_0; metrics of this kind are called *stationary* (in relation to the particular system of reference chosen). The justification for this name is obvious if it is remembered that in physics a phenomenon which takes place in a continuous medium, and is determined by a certain number of parameters which are functions of position and of the time (e.g. the motion of a fluid) is called *stationary* if these parameters do not depend explicitly on the time.

In particular, a stationary metric will be called *statical* when the coefficients g_{0i} ($i = 1, 2, 3$) of the three product terms in dx_0 vanish, i.e. when in the expression

$$L^2 = g_{00} + 2\sum_1^3{}_i\, g_{0i}\, x_i' + \sum_1^3{}_{ik}\, g_{ik}\, x_i'\, x_k' \quad . \quad . \quad (32)$$

(in which dashes denote differentiation with respect to x_0) the terms of the first degree in x_i' are missing. The justification for the name is somewhat more indirect, and will appear from the following considerations.

It is known, and can in any case be verified at once, that when L is an even function of the x''s (as in the case we are considering) the Lagrangian equations (4′) define a *reversible* motion, i.e. such that if $P = P(t)$ represents the motion starting from a certain initial position P_0 with an initial velocity v_0, then on changing t into $-t$ (i.e. considering the motion defined by $P = P(-t)$) we have the solution corresponding to the same initial position and the same initial velocity but in the reverse direction. Further, in the classical mechanics it is known that the motion of a particle is reversible whenever the field of force is invariable with respect to the time, i.e. when the field is *statical* (in the ordinary sense of the word). Hence the application of the term *statical* to a relativity metric whose geodesics are reversible with respect to the timelike variable x_0.

In the statical case it is usual to put

$$g_{00} = V^2, \quad g_{ik} = -a_{ik}$$

so that (32) becomes

$$L^2 = V^2 - \sum_{ik}^{3}{}_{1} a_{ik} x_i' x_k'; \quad . \quad . \quad . \quad . \quad (32')$$

this coefficient V^2 has an important mechanical meaning, which we shall now explain.

If at a given instant the velocity of the moving point vanishes, i.e. if each $x_i' = 0$ (case of initial motion starting from rest), we have in particular from (4′) and (32′)

$$\sum_{k}{}_{1}^{3} a_{ik} x_k'' = -\tfrac{1}{2}\frac{\partial V^2}{\partial x_i} \qquad (i = 1, 2, 3)$$

(the two dashes of course denoting double differentiation with respect to x_0); these define the quantities x_i'' as functions of position. The terms on the right,

$$X_i = -\tfrac{1}{2}\frac{\partial V^2}{\partial x_i},$$

being derivatives of a single function $-\frac{1}{2}V^2$, evidently constitute a covariant system (for any transformations whatever of the space co-ordinates). Hence the X_i's constitute the covariant components of a spacelike vector $\mathbf{F} = \text{grad}\,(-\frac{1}{2}V^2)$. The contravariant components

$$X^i = \sum_{1}^{3}{}_k\, a^{ik} X_k$$

of this vector, from the preceding formulæ, are identical with the initial accelerations. Hence the vector $\mathbf{F}$ obviously provides the statical measure of the force (per unit mass) of the field (the initial acceleration of a free material particle, or, if preferred, the force per unit mass which must be overcome to maintain the particle at rest).

Consider, beside the point P of co-ordinates x_i, a neighbouring point P' of co-ordinates $x_i + dx_i$ and the (invariant) trinomial

$$\sum_{1}^{3}{}_i\, X_i\, dx_i = -\tfrac{1}{2}dV^2.$$

Defining, as is natural, the virtual work of $\mathbf{F}$ for the displacement PP' as the product of the displacement by the orthogonal projection of the force (just as in ordinary Euclidean space), the preceding identity shows that $-\frac{1}{2}V^2$ constitutes the potential function of the force acting in the field in statical conditions.

As has been seen just above, in the statical case the force in the field can be very simply expressed by means of the single coefficient $g_{00} = V^2$. In more general conditions, the whole of the mechanics of the point is summed up in Einstein's geodesic principle, or, as an alternative form, in the consequent Lagrangian equations (4′); an analogous argument can also be developed for the initial motion, and the expression for the force in the field (at a generic point and instant) as a function of the g's deduced from it, but the results are by no means so simple and expressive as in the statical case. To put it briefly, the concepts of mass, force, and energy are all contained in the four-dimensional metric, but, at least in general, the task of distinguishing between them and associating them with the coefficients of ds^2 seems to be neither easy nor fruitful of further results.

13. Versors in a V_4 with pseudo-Euclidean metric.

An important fact in connexion with a versor (unit vector) in the space-time manifold V_4 is that it can always be made to correspond to a vector in three dimensions. This follows from the fact that it has four parameters (or moments), only three of which are independent, in virtue of the quadratic identity expressing that the length of the vector is unity (cf. Chap. V, p. 91). The interpretation of a vector of this kind as a velocity gives particularly interesting results.

Let us consider—limiting the case to a pseudo-Euclidean V_4—a generic motion defining y_1, y_2, y_3 as functions of y_0, and giving rise to a world line in V_4. If, as we shall first suppose, the velocity of the motion $< c$, we shall get a timelike line, in which the corresponding

$$ds^2 = dy_0^2 - \sum_{1}^{3}{}_i\, dy_i^2 = dy_0^2 (1 - \beta^2)$$

is positive. If, on the other hand, the velocity $> c$ (i.e. $\beta > 1$), ds^2 is negative, and we shall have a spacelike versor (cf. Chap. V, p. 142. In either case, denoting as usual the components $\frac{dy_i}{dy_0}$ of the Römerian velocity by β_i and the direction cosines of this velocity by $\alpha_i = \frac{\beta_i}{\beta}$, we obviously get the expressions

$$\xi^0 = \frac{dy_0}{|\, ds_0 \,|} = \frac{1}{\sqrt{1-\beta^2}},$$

$$\xi^i = \frac{dy_i}{dy_0}\frac{dy_0}{|\, ds_0 \,|} = \frac{\beta_i}{\sqrt{1-\beta^2}} = \frac{\beta}{\sqrt{1-\beta^2}}\,\alpha_i \ (i = 1, 2, 3),$$

for the parameters of the world line (i.e. of the versor $\boldsymbol{\xi}$ tangential to it).

Given the three components of an ordinary vector $\boldsymbol{\beta}$, these formulæ determine the four parameters ξ^i of a four-dimensional unit vector (versor) $\boldsymbol{\xi}$, and *vice versa*. Given β, the versor (in the ordinary sense) $\boldsymbol{\alpha}$ belonging to it is of course fixed without ambiguity (provided $\beta \neq 0$). It will sometimes be convenient to describe $\boldsymbol{\alpha}$ as the versor *reduced* from the four-dimensional versor $\boldsymbol{\xi}$. For $\beta = 0$ the versor $\boldsymbol{\xi}$ has its components ξ^1, ξ^2, ξ^3

all zero, and is accordingly called *purely timelike.* If instead we consider the case of a very large velocity in a direction $\boldsymbol{\alpha}$ (i.e. if we make β tend to infinity, while the ratios between the β^i's remain determinate), then we have $\xi^0 = 0$, while the other components ξ^i reduce to the direction cosines α^i of the reduced versor. In this case the four-dimensional versor $\boldsymbol{\xi}$ is called *purely spacelike*; it is tangential to the three-dimensional manifold (space) $x_0 =$ constant, or rather coincides with the versor $\boldsymbol{\alpha}$ belonging to this manifold.

All this can easily be extended to the case of a V_4 of any metric whatever, referred to any co-ordinates x_0, x_1, x_2, x_3, the first timelike and the other three spacelike, and characterized by the form

$$ds^2 = \sum_0^3{}_{ik}\, g_{ik}\, dx_i\, dx_k = dx_0{}^2 L^2,$$

where L^2 denotes, as in § 12, the expression

$$V^2 + 2 \sum_0^3{}_i\, g_{0i}\, x_i' - \beta^2,$$

and, as usual, $\beta = \dfrac{dl}{dx_0}$.

Given a generic versor of parameters

$$\xi^i = \frac{dx_i}{|\,ds\,|} \qquad (i = 0, 1, 2, 3)$$

we shall have

$$\xi^0 = \frac{dx_0}{|\,ds\,|} = \frac{1}{L},$$

$$\xi^i = \frac{dx_i}{|\,ds\,|} = \frac{dx_i}{dx_0}\,\frac{dx_0}{|\,ds\,|} = \frac{x_i'}{L} = \frac{\beta}{L}\,\frac{dx_i}{dl}.$$

14. **Digression on geodesics of zero length.**

Let τ denote a parameter of any kind such that the co-ordinates x can be considered functions of it, and put

$$T = \tfrac{1}{2}\frac{ds^2}{d\tau^2} = \tfrac{1}{2}\sum_0^3{}_{ik}\, g_{ik}\, \dot{x}_i\, \dot{x}_k$$

(dots denoting differentiation with respect to τ). Consider the equations of motion of a material system as summed up in the variational equation

$$\delta\int 2T\,d\tau = 0. \quad . \quad . \quad . \quad . \quad . \quad . \quad (33)$$

We know from ordinary mechanics that if τ denotes the time and T the *vis viva* of a material system, then the Lagrangian equations implicit in (33), i.e.

$$\frac{d}{d\tau}\frac{\partial T}{\partial \dot{x}_i} - \frac{\partial T}{\partial x_i} = 0 \qquad (i = 0, 1, 2, 3), \quad . \quad (34)$$

define the spontaneous motion of the system and have as a first integral the equation

$$T = E = \text{constant}.$$

Whenever the value of the constant E is different from zero, then by using the equation $T = E$ it is easy to eliminate the parameter τ from (33) and obtain from it a variational equation capable of defining the trajectories. We have in fact, from the definition of T,

$$\sqrt{2T}\,d\tau = ds,$$

so that the expression for the *action*, i.e. the integral $\int 2T d\tau$ which occurs in (33), can be written

$$\int\sqrt{2E}\,\sqrt{2T}\,d\tau = \sqrt{2E}\int\sqrt{2T}\,d\tau = \sqrt{2E}\int ds.$$

Hence, for $E \neq 0$, the variational equation (33), by elimination of the parameter τ, gives the equation

$$\delta\int ds = 0, \quad . \quad . \quad . \quad . \quad . \quad . \quad (35)$$

which is the characteristic equation of the geodesics in the V_4 whose line element is ds. From this equation we can deduce, as in § 24, p. 131, the differential equations

$$\ddot{x}_i + \sum_{0}^{3}{}_{jl}\,\{jl, i\}\,\dot{x}_j\,\dot{x}_l = 0 \quad . \quad . \quad . \quad . \quad (36)$$

of the geodesics, where dots denote differentiation with respect to s. The same equations would also be obtained, but with τ instead of s, by writing out (34) in full and solving for the $\ddot{x}$'s.

To sum up, for $E \neq 0$, it is a matter of indifference whether we define the geodesics of V_4 as trajectories derived from the variational equation (33), or by means of the typical property (35).

We now propose to examine separately the case $E = 0$; since $T = E$ and $ds^2 = 2T\, d\tau^2$, this is equivalent to $ds = 0$ along the whole of the line in question, which therefore in this case takes the name of *geodesic of zero length.* (Such lines are of course real only if ds^2 is an indefinite form.) In this case (35) is no longer suitable for defining geodesics; the method just referred to and used in § 24, p. 131, to obtain the differential equations also breaks down, since it assumes s as the independent variable, and therefore excludes the possibility of ds being identically $= 0$. The equations (34), however, keep their significance, and therefore offer a means of defining geodesics of zero length by a process of passing to the limit (in conditions of complete analytical regularity) from ordinary geodesics. We shall thus apply the term "geodesics of zero length" to the lines represented by solutions of the Lagrangian system (34) for the value zero of the constant E.

The differential equations (36) of ordinary geodesics give x_0, x_1, x_2, x_3 directly as functions of a parameter τ (or in particular of s). We can suppose the parameter eliminated after integration, giving, for example, x_1, x_2, x_3 as functions of x_0. But it is also possible to eliminate the parameter beforehand, by obtaining from (35) three differential equations which define x_1, x_2, x_3 as functions of x_0. To do this, we introduce x_0 as the independent variable in (35), so that it takes the form

$$\delta \int L dx_0 = 0, \quad . \quad . \quad . \quad . \quad . \quad . \quad (35')$$

where, as in § 12, we have put

$$L^2 = g_{00} + 2\sum_1^3{}_i g_{0i}\, x'_i + \sum_1^3{}_{ik} g_{ik}\, x'_i\, x'_k.$$

From (35') we deduce, by the ordinary method, the three required Lagrangian equations, which are

$$\frac{d}{dx_0}\frac{\partial L}{\partial x'_i} - \frac{\partial L}{\partial x_i} = 0 \qquad (i = 1, 2, 3).$$

These are completely equivalent to (35'), since, as was seen in § 2, the fourth equation (to be obtained by making x_0 vary and equating to zero the coefficient of δx_0) is a necessary consequence of these three.

These equations, like (35) above, lose their significance in the case of geodesics of zero length ($E = 0$). An analogous reduction can be found in this case too, but it is preferable to follow another method and leave the pure Lagrangian form. This method is as follows.

We start from (33) instead of (35), and note that from the definition of T and L we have obviously

$$2T = L^2 \frac{dx_0^2}{d\tau^2}.$$

Suppose that x_0 is not constant along the geodesic (or arc of geodesic) of zero length under consideration.[1] In the integral (33) (corresponding to a generic geodesic of the kind in question) we can then assume x_0 instead of τ as the independent variable, so that

$$\delta \int L^2 \frac{dx_0}{d\tau} dx_0 = 0.$$

The parameter τ is a function, *a priori* unknown, of x_0, such that $\frac{dx_0}{d\tau}$ remains finite and not zero. We can therefore put

$$\frac{d\tau}{dx_0} = \frac{1}{\Lambda(x_0)} = e^{\int \lambda(x_0) dx_0},$$

where Λ and λ are also finite and not zero. Then the preceding variational formula becomes

$$\delta \int L^2 \Lambda dx_0 = 0,$$

from which we get for the geodesics the equations

$$\frac{d}{dx_0} \frac{\partial (L^2\Lambda)}{\partial x_i'} - \frac{\partial (L^2\Lambda)}{\partial x_i} = 0 \qquad (i = 1, 2, 3);$$

[1] From the limitations introduced in § 12, this condition can always be satisfied in the real field. In fact, if we put $dx_0 = 0$ in ds^2, there remains a definite negative form, which cannot vanish along an actual line, i.e. when dx_1, dx_2, dx_3, do not vanish simultaneously.

expanding and dividing by Λ, these can be written in the form

$$\frac{d}{dx_0}\frac{\partial L^2}{\partial x_i'} - \frac{\partial L^2}{\partial x_i} = \lambda \frac{\partial L^2}{\partial x_i'}.$$

The parameter λ can be at once eliminated from these. Denoting for shortness the Lagrangian binomial on the left-hand side by τ_i, we get finally the two equations

$$\frac{\tau_1}{\dfrac{\partial L^2}{\partial x_1'}} = \frac{\tau_2}{\dfrac{\partial L^2}{\partial x_2'}} = \frac{\tau_3}{\dfrac{\partial L^2}{\partial x_3'}}$$

which are to be taken together with the equation

$$L^2 = 0.$$

15. Some elementary theorems of geometrical optics.

It is known that in a transparent homogeneous medium light is propagated in a straight line with constant velocity if no disturbing influence is at work. In the case of an isotropic medium —the only one we shall consider—the velocity is always the same in all directions and therefore is a constant characteristic of the medium. *In vacuo* (cf. § 4) the velocity is, in round numbers,

$$c = 3 \times 10^{10} \text{ cm./sec.}$$

or 300,000 kilometres per second.

If instead we have a heterogeneous medium, in which the refractive index μ (which is defined as the reciprocal of the velocity of propagation) varies from point to point, then the rays are in general not rectilinear but are bent in accordance with a law which depends on the way in which μ varies, i.e. on the function $\mu(x, y, z)$. This law can be put in a compact and useful form in the following way.[1] If the initial point P_0 and the final point P_1 of the path of a ray of light are fixed, the time taken by the ray to go from P_0 to P_1 along a line s will obviously be expressed by the integral

$$t = \int_{P_0}^{P_1} \mu \, ds,$$

since μ, as we have just said, is the reciprocal of the velocity.

[1] Cf. for example Levi-Civita and Amaldi: *op. cit.*, Chap. XI, No. 18.

Now the line actually followed by the light is the one which makes this integral a minimum, and therefore satisfies the condition

$$\delta t = 0.$$

This variational equation, which sums up the whole of geometrical optics, is known as *Fermat's principle.*

16. **Geometrical optics according to Einstein and the meaning of the constant** c.

In constructing a geometrical scheme to represent light rays the existence is assumed of an absolute frame of reference, exactly as is done in the Newtonian mechanics. In order to help the imagination, the system of reference is supposed to be provided by a hypothetical medium at rest—the so-called cosmic ether—which constitutes as it were a background or support for all optical phenomena. In space free from ponderable matter light is propagated in a straight line with constant velocity c with respect to the ether, or, which is the same thing, with respect to fixed axes, where " fixed " axes mean axes at rest with respect to the ether. Hence c is the velocity of light as it appears to a generic observer O, at rest with respect to the ether.

Consider a solid C moving with velocity u (a pure translatory motion) and a pencil of parallel rays of light which are being propagated in the same sense as the motion of C.

With respect to the observer O, the luminous phenomenon is diagrammatically represented, as we have just noted, as a particular uniform motion with velocity c.

According to ordinary kinematics, the analogous velocity with respect to an observer O' rigidly attached to C is $c - u$.

Now within the range of velocities which can be realized by material bodies the ratio $\frac{u}{c}$, and still more its square $\frac{u^2}{c^2}$ (only the latter of which can be submitted to effective experimental control) are small; we can, however, take it as definitely established that the velocity of propagation is still c with respect to O' also. This follows from the classical Michelson-Morley experiment, subsequently repeated by other physicists, and recently on new bases by Professor Majorana.

In order to explain this experimental result, it is evidently sufficient that the phenomenon which appears to macroscopic methods of measurement as the translation of a body C with velocity u should, with more refined methods of measurement, be a transformation (Λ). The study of these transformations has in fact shown that any ordinary uniform translation is almost indistinguishable from a (Λ), the difference being of the order of one ten-millionth, provided that $\frac{u}{c} \leqslant 10^{-4}$.

The classical laws of geometrical optics (that the propagation of light is rectilinear, uniform, and with velocity c), and the famous experiments referred to above, will therefore still hold if we suppose that for the propagation of light, as for the motion of a material particle under no forces, the equation

$$\delta \int ds_0 = 0$$

holds, with the condition

$$ds_0^2 = 0$$

(equations of uniform motion with velocity c); and if, on the other hand, we consider the phenomenon of the translation of solid bodies as very slightly different from the description of ordinary kinematics, so that it corresponds to a transformation (Λ).

Hence these special kinds of motion which correspond to the propagation of light in the ether, in the absence of disturbing influences, are dependent on the form

$$ds_0^2 = c^2\, dt^2 - dl_0^2, \quad . \quad . \quad . \quad . \quad (37)$$

in which the constant c has a specific numerical value.

For ordinary motion, with velocities which are at most planetary, and under the action of conservative forces—e.g. in the presence of assigned masses—the same part is played by the form

$$ds^2 = (c^2 - 2U)\, dt^2 - dl_0^2 \quad . \quad . \quad . \quad (37')$$

in which on the one hand the constant c is subject only to the qualitative restriction of being sufficiently large, and on the other the influence of the masses modifies to some extent the coefficient of dt^2. If we aim at attaining unity of conception of physical phenomena, we shall obviously be constrained, *cæteris paribus*, to adopt a single differential form ds^2 as the determining form both for the motion of material particles and for the behaviour of light rays, serving as a basis for both cases. We must therefore assign to the constant c, in the general dynamical case, the same specific value as belongs to it in the particular optical phenomenon. In the absence of disturbing influences, in particular of masses at a perceptible distance, so that $U = 0$, the ds^2 of mechanics then becomes identical with the ds^2 of optics (the limiting case).

Further, since in the case $U = 0$ (i.e. in the absence of masses at a perceptible distance) the intervention of ds^2 has led to geometrical optics being summarized in two laws which appear as limiting cases of dynamical laws, we are led to hope for the extension of the same criterion also to the case in which masses exist ($U \neq 0$).

The propagation of light will therefore be governed in any case by the following postulates:

(1) The geodesic principle (as for material motion),

$$\delta \int ds = 0; \quad . \quad . \quad . \quad . \quad . \quad (38)$$

(2) $ds^2 = 0$, which is equivalent to saying that the motions in question have the square of the velocity $\dfrac{dl_0^2}{dt_0^2}$ equal to

$$c^2 - 2U = c^2\left(1 - \frac{2U}{c^2}\right).$$

The velocity V is thus slightly less than c; neglecting terms which are in fact absolutely negligible, it is given by

$$V = c\left(1 - \frac{U}{c^2}\right).$$

These two postulates can be summed up in a single illuminating geometrical assertion:

In the metric we have assigned to V_4, *the world lines of light are geodesics of zero length.*

It is to be noted that this assertion has an invariant form, and is therefore suitable for defining the behaviour of light rays, even if these are referred to a system of any co-ordinates x_0, x_1, x_2, x_3 whatever, instead of to the particular system t, y_1, y_2, y_3. The assertion lends itself to an obvious generalization, since it is natural to extend its scope so that it shall continue to hold even when the ds^2 which characterizes the metric of the V_4, though satisfying the qualitative restrictions of § 12, is not reducible to the particular form (37).

17. **Interpretation in geometrical optics of the condition** $ds^2 = 0$.

Given any direction in the four-dimensional space (t, x_1, x_2, x_3), i.e. any system of increments (dt, dx_1, dx_2, dx_3), we can obviously make a vector (velocity) $\mathbf{v}$ correspond to it in the physical space whose line element is given by

$$dl^2 = -\sum_1^3{}_{ik}\, g_{ik}\, dx_i\, dx_k = \sum_1^3{}_{ik}\, a_{ik}\, dx_i\, dx_k, \quad . \quad (39)$$

or more precisely in the Euclidean space tangential to the given space at the generic point from which the specified increments are drawn.

We shall take the ratios

$$\frac{dx_i}{dt} = \dot{x}_i \qquad (i = 1, 2, 3)$$

for the contravariant system of this vector with respect to the metric (39). Writing these in the form

$$\frac{dx_i}{dl}\,\frac{dl}{dt},$$

we see from the presence of the factor $\frac{dx_i}{dl}$, which is the direction parameter, that the positive factor $\frac{dl}{dt}$ measures the length of the vector. Referring back to the equation (39), we have for the square of this length

$$v^2 = \frac{dl^2}{dt^2} = \sum_1^3{}_{ik}\, a_{ik}\, \dot{x}_i\, \dot{x}_k.$$

Another vector $\mathbf{w}$, a function solely of position and time (of position alone in stationary conditions), can be made to correspond to the set of three coefficients g_{0i}, which are covariant with respect to any transformations whatever of the space co-ordinates alone, by taking these three quantities for the covariant system of the vector. Then, denoting as usual the coefficients of the form reciprocal to (39) by a^{ik} and putting

$$w^2 = \sum_{1}^{3}{}_{ik}\, a^{ik} g_{0i} g_{0k},$$

we get w for the length and (for $w > 0$) the ratios $\frac{g_{0i}}{w}$ for the moments (the system reciprocal to the parameters) of the direction of this vector. It is to be noted that if the spacelike co-ordinates x have the dimensions of a length, the coefficients a_{ik} of dl^2, and therefore their reciprocals a^{ik}, are pure numbers, while the coefficients g_{0i} of the product terms in t have the dimensions of a velocity. Hence the vector **w**, like **v**, can be interpreted as a velocity. It will be obvious that this conclusion still holds even if the dimensions of the co-ordinates x_1, x_2, x_3 are left indeterminate.

If ϕ denotes the angle between **v** and **w**, both for the moment supposed not zero, we have for the metric (39)

$$\cos\phi = \sum_{1}^{3}{}_{i} \frac{g_{0i}}{w} \frac{\dot{x}_i}{v},$$

and therefore identically

$$vw \cos\phi = \sum_{1}^{3}{}_{i}\, g_{0i}\, \dot{x}_i, \quad . \quad . \quad . \quad . \quad (40)$$

which holds even if v or w vanishes.

Using (39) and (40), the expression for ds^2 can now be written in the form

$$ds^2 = dt^2 (V^2 + 2vw \cos\phi - v^2),$$

putting $V^2 = g_{00}$.

This makes it evident that the condition $ds^2 = 0$, characteristic of the propagation of light, defines its velocity v as a function of the position and direction of the ray, as well as of the time, in the general case in which the coefficients of ds^2, and with them V, w, and ϕ, depend on t.

Representing the ratios $\frac{v}{V}$ and $\frac{w}{V}$ (both positive and pure numbers) by β and p, we have for β the equation of the second degree

$$\beta^2 - 2p \cos\phi\, \beta - 1 = 0; \quad . \quad . \quad . \quad (41)$$

the product of the roots being -1, it follows that one is positive and the other negative. By definition v is necessarily positive, so that it is uniquely determined by (41).

When all the product terms in dt vanish (the statical case), $w = 0$; hence $\beta = 1$, and v coincides with V. In general $p > 0$, and the difference between v and V (for a specified position and time) depends on the direction of the ray, i.e. on the angle ϕ which it makes with $\mathbf{w}$. We also have $v = V$ for every ray perpendicular to $\mathbf{w}$. It is obvious from (41) that the maximum and minimum values of β correspond to $\phi = 0$ and $\phi = \pi$. This is equivalent to saying that the maximum velocity of propagation

$$V(\sqrt{1+p^2}+p)$$

is along $\mathbf{w}$, and the minimum velocity

$$V(\sqrt{1+p^2}-p)$$

is in the same direction but in the opposite sense.

Except in the statical case, it will be seen that the propagation of light in physical space is not only non-symmetrical for opposite senses but is completely irreversible.

18. **Fermat's principle in stationary relativity metrics.**

We saw in § 14 how the difficulty involved in the variational principle $\delta\int ds = 0$ for $ds^2 = 0$ can be evaded in finding the explicit form of the differential equations of the propagation of light. It is not without interest to note that for every stationary ds^2 the behaviour of the light rays can also be defined by associating Fermat's principle of the minimum time with the equation $ds^2 = 0$, i.e. by assuming

$$\delta\int dx_0 = 0, \quad . \quad . \quad . \quad . \quad . \quad . \quad (42)$$

with the condition that dx_0 is to be connected with x_0, the space co-ordinates, and their differentials by $ds^2 = 0$. Naturally while in the four-dimensional geodesic principle expressed by (38) not only dx_1, dx_2, dx_3 but also dx_0 are to be zero at the extremities of the interval of integration, in (42) this condition must not apply to dx_0, as it would reduce (42) to a mere identity.

We now propose to establish the equivalence, for every stationary metric, of the two principles of geometrical optics: (*a*) the four-dimensional geodesic principle, and (*b*) the principle of minimum time.

To do this we must consider the geodesics of zero length as derived by the method of limits from timelike geodesics ($ds^2 > 0$). For the latter we put as usual

$$\left.\begin{aligned} x_i' &= \frac{dx_i}{dx_0} \qquad (i = 1, 2, 3) \\ \beta^2 &= \frac{dl^2}{dx_0^{\,2}} = \sum_{1}^{3}{}_{ik}\, a_{ik}\, x_i'\, x_k' \\ L^2 &= \frac{ds^2}{dx_0^{\,2}} = V^2 + 2\sum_{1}^{3}{}_i\, g_{0i}\, x_i' - \sum_{1}^{3}{}_{ik}\, a_{ik}\, x_i'\, x_k' \end{aligned}\right\} \quad (43)$$

where the function L has finite partial derivatives, since ds^2, and therefore L, is not to vanish.

The equation (38) can be written

$$\delta\int L\, dx_0 = 0. \quad . \quad . \quad . \quad . \quad . \quad . \quad (44)$$

Taking the variation with respect to the co-ordinates x_1, x_2, x_3, we get by the classical procedure the Lagrangian equations

$$\frac{d}{dx_0}\frac{\partial L}{\partial x_i'} - \frac{\partial L}{\partial x_i} = 0 \quad (i = 1, 2, 3); \quad . \quad . \quad (45)$$

while the variation with respect to x_0 gives

$$\frac{d}{dx_0}\left(\sum_{1}^{3}{}_i \frac{\partial L}{\partial x_i'} x_i' - L\right) + \frac{\partial L}{\partial x_0} = 0,$$

which is a necessary consequence of the equations (45).

On the hypothesis, characteristic of the stationary case, that L does not explicitly contain x_0, we get the integral

$$L - \sum_{1}^{3}{}_i \frac{\partial L}{\partial x_i'} x_i' = E, \quad . \quad . \quad . \quad . \quad (46)$$

where the constant E represents the total energy of the moving point.

Multiplying by L, the left-hand side may be written in the form

$$\tfrac{1}{2}L^2 + \tfrac{1}{2}\left(L^2 - \sum_1^3{}_i \frac{\partial L^2}{\partial x_i'} x_i'\right).$$

It follows from the third of the equations (43) that L^2 is a polynomial of the second degree in x_1', x_2', x_3'; it is there already divided up into three homogeneous sets of terms of degree 0, 1, 2 respectively. By Euler's theorem on homogeneous functions, the linear term disappears from the difference $L^2 - \sum_1^3{}_i \frac{\partial L^2}{\partial x_i'} x_i'$, which reduces to $V^2 + \beta^2$. Hence (46) multiplied by L gives

$$\tfrac{1}{2}L^2 + \tfrac{1}{2}(V^2 + \beta^2) = EL.$$

The left-hand side is essentially positive when L tends to zero, being in fact, for $L = 0$, $\geqslant \tfrac{1}{2}V^2$ (which is to be taken as having a lower limit which is not zero in the field considered). The product EL can therefore be considered as *a function of the* x*'s and* x′*'s which is always regular, and not zero, when* L *tends to zero*; in the latter hypothesis the constant E obviously tends to infinity.

Further, for all motions with the same total energy E, the principle (44), in which we suppose that δx_0 vanishes at the extremities of the interval of integration, can be replaced by an analogous one which has the advantage over the first of not requiring this condition to be satisfied. In fact, for δx_0 zero at the extremities, we have $\delta\int dx_0 = 0$, and in consequence (44) is equivalent to

$$\delta\int (L - E)\,dx_0 = 0$$

or, for $E \neq 0$, to

$$\delta\int\left(1 - \frac{L}{E}\right)dx_0 = 0;$$

and in this last equation we can drop the condition that δx_0 vanishes at the extremities, since if we transfer the δ under the integral sign and apply it to dx_0 (both explicit, and implicit in the x''s) we get

$$-\frac{1}{E}\int \delta dx_0\left(L - E - \sum_1^3{}_i \frac{\partial L}{\partial x_i'} x_i'\right),$$

which vanishes in virtue of (46).

It is therefore established that, for an assigned non-zero value of E, the equations of motion can be expressed by means of the formula

$$\delta\int\left(1-\frac{L}{E}\right)dx_0 = 0 \quad . \quad . \quad . \quad . \quad (47)$$

without the necessity of imposing any condition as to δx_0.

The function under the integral sign can be written $1 - \frac{L^2}{EL}$, from which it appears, remembering what was said above about the behaviour of EL, that this function is regular and tends to unity if L tends to zero. Now this is precisely the hypothesis which corresponds to the transition from material motion to the limiting case of the propagation of light. Since the function is regular, the order of the operations $\delta\int$ and passage to the limit may be interchanged, so that (47) gives Fermat's principle

$$\delta\int dx_0 = 0.$$

Fermat's principle can be put in a purely geometrical form, referred to the spacelike metric with line element dl, if we give dx_0 the value found from $ds^2 = 0$ in terms of x_0, x_1, x_2, x_3, dx_1, dx_2, dx_3, and insert this in the formula just above. The result is particularly easy to interpret in the statical case ($g_{0i} = 0$, $i = 1, 2, 3$), in which we have evidently $dx_0 = \frac{dl}{V}$, and Fermat's principle takes the form

$$\delta\int\frac{dl}{V} = 0.$$

This shows that the light rays coincide with the geodesics of the three-dimensional space with line element $\frac{dl}{V}$; alternatively, referring to the physical space dl^2 and again applying the theorem of least action (cf. § 11), we can say that they coincide with a pencil of trajectories corresponding to the potential $\frac{1}{2V^2}$ and total energy 0.

19. The stress tensor and its divergence in the classical theory.

Let there be given a continuous medium, and in it a surface element (facet) $d\sigma$; one side of this facet is supposed chosen as the positive side, and one sense of the normal direction is associated with it. We shall agree that this sense is the one which corresponds to the passage from the negative to the positive side, and shall denote its versor by **n**. The resultant of the molecular actions which the particles on the negative side of the element exert on those on the positive side is ordinarily called the *stress* [1] relative to the positive side of the element considered.[2] In normal cases—the only ones we propose to consider—this resultant is of the same order of magnitude as $d\sigma$, and is represented by $\mathbf{\Phi}_n d\sigma$, where $\mathbf{\Phi}_n$ is the *specific stress* on the positive side of the surface element normal to **n**.

Referred to orthogonal Cartesian axes $Oy_1y_2y_3$, the three components of the vector $\mathbf{\Phi}_n$ will obviously be denoted by $\mathbf{\Phi}_{ni}$ $(i = 1, 2, 3)$. To characterize the distribution of the stresses at a single point P, we introduce the three stresses $\mathbf{\Phi}_1$, $\mathbf{\Phi}_2$, $\mathbf{\Phi}_3$ which act on the facets at P parallel to the co-ordinate planes, or, more precisely, the facets whose normal versors are in the positive directions of the co-ordinate axes. Their components are denoted in order by

$$\begin{array}{lll} \Phi_{11}, & \Phi_{12}, & \Phi_{13}\,; \\ \Phi_{21}, & \Phi_{22}, & \Phi_{23}\,; \\ \Phi_{31}, & \Phi_{32}, & \Phi_{33}\,; \end{array}$$

it follows from the postulates of ordinary mechanics that the matrix formed by these terms is symmetrical or that

$$\Phi_{32} = \Phi_{23}, \quad \Phi_{13} = \Phi_{31}, \quad \Phi_{21} = \Phi_{12},$$

so that there are really six of these quantities Φ_{ik} $(i, k = 1, 2, 3)$.

[1] Cf., for example, A. E. H. Love, *Mathematical Theory of Elasticity*, third edition, Chap. II; Cambridge University Press, 1920.

[2] Some authors, Love in particular, invert the respective rôles of the two sides of the facet in their definitions, and therefore, by the principle of reaction, change the sense of the vector described as the stress. The sign of its components will be changed accordingly, and the inequality will be inverted which determines whether a given stress is of the nature of a pressure or a pull with respect to the element considered.

Putting n^k for the components of the versor $\mathbf{n}$ (its direction cosines) we get the fundamental formula

$$\mathbf{\Phi}_n = \sum_{1}^{3}{}_k \mathbf{\Phi}_k n^k, \quad . \quad . \quad . \quad . \quad . \quad (48)$$

and hence for the three components in the direction of the co-ordinate axes

$$\Phi_{n|i} = \sum_{1}^{3}{}_k \Phi_{ik} n^k.$$

If $\boldsymbol{\xi}$ is a generic direction of direction cosines ξ^i, the scalar product $\mathbf{\Phi}_n \times \boldsymbol{\xi}$, i.e. the component of the stress $\mathbf{\Phi}_n$ along $\boldsymbol{\xi}$, can naturally be written in the form

$$\mathbf{\Phi}_n \times \boldsymbol{\xi} = \sum_{1}^{3}{}_{ik} \Phi_{ik} n^k \xi^i.$$

From the symmetry of the Φ_{ik}'s, it follows that in the sum just written down Φ_{ik} can be replaced by Φ_{ki}; the sum is therefore, by (48), equivalent to the scalar product $\mathbf{\Phi}_\xi \times \mathbf{n}$. Hence we have the relation of reciprocity, expressed by the equation

$$\mathbf{\Phi}_n \times \boldsymbol{\xi} = \mathbf{\Phi}_\xi \times \mathbf{n}.$$

For $\boldsymbol{\xi} = \mathbf{n}$ we have in particular what is called the *normal stress*, i.e. the component along the normal to the facet of the stress with respect to the facet itself. In accordance with the conventions we have adopted, the stress will be of the nature of a push or a pull according as this normal component is positive or negative. From the remarks above, the necessary criterion is provided by the sign (for $\boldsymbol{\xi} = \mathbf{n}$) of the expression

$$\sum_{1}^{3}{}_{ik} \Phi_{ik} \xi^i \xi^k.$$

To make the notation uniform, we shall write $\boldsymbol{\xi}'$ instead of $\mathbf{n}$. Consider the bilinear form

$$\Phi = \sum_{1}^{3}{}_{ik} \Phi_{ik} \xi^i \xi'^k, \quad . \quad . \quad . \quad . \quad . \quad (49)$$

which represents either the component along $\boldsymbol{\xi}$ of the specific stress on the facet normal to $\boldsymbol{\xi}'$, or the component along $\boldsymbol{\xi}'$ of the specific stress on the facet normal to $\boldsymbol{\xi}$.

If now we replace the y's by any curvilinear co-ordinates x whatever (the geometrical nature of the space characterized by $\sum_{i=1}^{3} dy_i^2$ being of course regarded as invariant), then the parameters ξ^i, ξ'^i of the directions $\boldsymbol{\xi}$, $\boldsymbol{\xi}'$ constitute, as we know, two contravariant systems, reducing to the direction cosines in Cartesian co-ordinates, while the scalar quantity Φ just defined will behave as an invariant on account of its intrinsic meaning. It follows (cf. Chapter IV, p. 70) that the coefficients of the bilinear form Φ (referred to these parameters as arguments) will constitute a symmetrical covariant double system which is called the *stress tensor*. Extending the notation adopted in the case of Cartesian co-ordinates we shall denote it by Φ_{ik}. This tensor will of course have the contravariant components $\Phi^{ik} = \Phi^{ki}$ and the mixed components Φ_i^k, which can be obtained in the ordinary way by composition with the coefficients of the fundamental form.

The stress tensor depends in general on the position of the point considered; the components Φ_{ik}, referred to generic co-ordinates x, can therefore in any case be thought of as functions of the co-ordinates, and therefore as having derivatives—ordinary, covariant, and contravariant. As we saw in Chapter VI, p. 153, from a given double tensor X_{ik} we can always obtain a vector $\mathbf{Y}$ intrinsically related to it, which we called its divergence, and whose covariant components are defined for $n = 3$ by

$$Y_i = \sum_{kl=1}^{3} a^{kl} X_{ik|l}. \qquad (50)$$

Now the divergence of the stress tensor has an important mechanical interpretation, which can be found at once by using Cartesian co-ordinates. We know in fact that the molecular forces applied to a given particle by all the surrounding particles have for their resultant a vector $\boldsymbol{\chi}$, whose components per unit volume, in orthogonal Cartesian co-ordinates, are given by

$$\chi_i = -\sum_{k=1}^{3} \frac{\partial \Phi_{ik}}{\partial y_k}. \qquad (51)$$

Noting that in this system of reference the divergence of Φ_{ik} is expressed by precisely the sum on the right of (51), and remembering that the covariant components of a vector are identical in this case with the ordinary components, we see at once that the vector $\boldsymbol{\chi}$ is the divergence of the stress tensor with its sign changed. Applying the formula (50) we can therefore write

$$\chi_i = -\sum_{1}^{3}{}_{kl}\, a^{kl}\, \Phi_{ik|l}. \quad . \quad . \quad . \quad . \quad (51')$$

20. **The fundamental equations of the mechanics of continuous systems, referred to fixed axes; transformations of them in general co-ordinates (space co-ordinates).**

It is known that, when no hypothesis is made as to the nature of the medium, and when therefore the stresses are not particularized, the fundamental equations of the mechanics of a continuous system reduce to the dynamical equation

$$\rho \mathbf{f} = \rho \mathbf{F} + \boldsymbol{\chi} \quad . \quad . \quad . \quad . \quad . \quad (52)$$

(where ρ is the density, $\mathbf{f}$ the acceleration, $\mathbf{F}$ the force per unit mass, and $\boldsymbol{\chi}$ the vector defined in the preceding section), together with the equation of continuity

$$\frac{\partial \rho}{\partial t} + \operatorname{div}(\rho \mathbf{v}) = 0 \quad . \quad . \quad . \quad . \quad (53)$$

($\mathbf{v}$ being the velocity), which can also be written

$$\frac{d\rho}{dt} + \rho \operatorname{div}(\mathbf{v}) = 0, \quad . \quad . \quad . \quad . \quad (53')$$

where the symbol $\frac{d\rho}{dt}$ denotes a " proper " derivative, i.e. one which considers ρ as depending on t in such a way that as t varies ρ refers always to one and the same particle of matter.

If now we wish to find the explicit form of these two equations with reference to any co-ordinates x whatever, connected with the y's by formulæ which do not involve the time, all we need do is to obtain the expressions for the covariant (or contravariant) components of the vector $\mathbf{f}$, since those of $\boldsymbol{\chi}$ are already known from the preceding section (cf. formula (51')) and the invariant expression of $\operatorname{div}(\rho \mathbf{v})$ is known from p. 153, Chapter VI; the force $\mathbf{F}$ will naturally be supposed given by means of its covariant (or contravariant) components.

The acceleration $\mathbf{f}$ is defined by

$$\mathbf{f} = \frac{d\mathbf{v}}{dt}$$

where the (proper) derivative is supposed to be calculated with respect to an observer (system of axes or, more generally, co-ordinate net) fixed in the mechanical sense of the word.

Referred to co-ordinates y this relation is equivalent to the three scalar relations

$$f_i = \frac{dv_i}{dt} = \frac{\partial v_i}{\partial t} + \sum_1^3{}_k \frac{\partial v_i}{\partial y_k} v_k \qquad (i = 1, 2, 3). \quad . \quad (54)$$

If now, with reference to any co-ordinates x whatever connected with the y's by relations which do not involve the time, we consider the simple system

$$\frac{\partial v_i}{\partial t} + \sum_1^3{}_k v_{i|k} v^k, \quad . \quad . \quad . \quad . \quad . \quad (54')$$

it is easy to see that this is covariant. In fact, on the one hand the quantities $\frac{\partial v_i}{\partial t}$ (t being a parameter not involved in the transformations) are covariant like the v_i's, and on the other the quantities $\sum_1^3{}_k v_{i|k} v^k$ are covariant from the law of contraction of tensors. Noting once more that in orthogonal Cartesian co-ordinates the covariant derivatives reduce to the ordinary derivatives, and also the covariant and contravariant components of a vector to the ordinary components, we see that in these co-ordinates the expressions (54′) are identical with those on the right of (54), i.e. with the (covariant) components f_i of $\mathbf{f}$. This identity will still hold with reference to the x's, and we can write

$$f_i = \frac{\partial v_i}{\partial t} + \sum_1^3{}_k v_{i|k} v^k.$$

We can now find the explicit form of the equations (52) and (53) with reference to the co-ordinates x. The first will give the covariant equations

$$\rho\left(\frac{\partial v_i}{\partial t} + \sum_1^3{}_k v_{i|k} v^k\right) = \rho F_i - \sum_1^3{}_{kl} a^{kl} \Phi_{ik|l} \quad . \quad (55)$$

and the second the invariant equation

$$\frac{\partial \rho}{\partial t} + \sum_{1}^{3}{}_i (\rho v_i)^i = 0, \quad . \quad . \quad . \quad . \quad . \quad (56)$$

or

$$\frac{\partial \rho}{\partial t} + \sum_{1}^{3}{}_i (\rho v^i)_i = 0. \quad . \quad . \quad . \quad . \quad . \quad (56')$$

21. **Galilean systems of reference.**

Among the purely spacelike transformations a particularly simple group consists of those which give the change from a system of fixed (in the mechanical sense of the word) Cartesian axes to a system of Cartesian axes in uniform translatory motion with respect to the first set; the latter system is called *Galilean.* The definitions of force, specific stress on a generic surface element, and divergence (whether of a vector or a tensor) are not changed in a transformation of this kind, but the velocity $\mathbf{v}$ of a generic point is altered by the addition of a constant quantity represented by the velocity of translation $\boldsymbol{\tau}$; this addition, however, evidently does not alter the acceleration (i.e. the proper derivative of $\mathbf{v}$). It follows that such a transformation leaves unchanged the dynamical equation (52), and also the equation of continuity; the latter is evident from the form (53′), which, in addition to div($\mathbf{v}$) (which, as just pointed out, is invariant), contains the proper derivative of ρ, which from its intrinsic meaning is obviously independent of the axes of reference.

Furthermore, all the laws of the classical mechanics are known to be unaltered if the axes of reference are supposed to be in uniform translatory motion.

22. **Equivalent form for the system (52) and (53).**

In the general equations of motion of a continuous system the force per unit mass $\mathbf{F}$ occurs explicitly. From the formal point of view we can always, and in an infinite number of ways, consider $\mathbf{F}$ as the divergence of a suitable tensor; its components can then be supposed amalgamated with the Φ_{ik}'s, so that we can at once put $\mathbf{F} = 0$ in the equation (52).

From the point of view of application this is not always convenient, and in many cases the direct method is preferable; but from the speculative point of view this process of submerging the force per unit mass in the stress is not only legitimate, but in accordance with the physical standpoint which refuses to admit action at a distance, asserting that every disturbance is transmitted by mediate action. In virtue of these considerations we shall put $\mathbf{F} = 0$ in the vector equation (52).

We now propose to transform, without altering their content, the three scalar equations included in (52) and the equation of continuity (53), in such a way as to replace these four equations by a set of four substantially identical equations.[1]

Referring to orthogonal Cartesian axes y, we project the equation (52) (in which we have now put $\mathbf{F} = 0$) on the axis y_i, using (55), we get

$$\rho\left(\frac{\partial v_i}{\partial t} + \sum_1^3{}_k \frac{\partial v_i}{\partial y_k} v_k\right) = -\sum_1^3{}_k \frac{\partial \Phi_{ik}}{\partial y_k}, \quad . \quad . \quad . \quad (57)$$

while the equation of continuity (53) or (53′) takes the well-known form

$$\frac{\partial \rho}{\partial t} + \sum_1^3{}_k \frac{\partial(\rho v_k)}{\partial y_k} = 0. \quad . \quad . \quad . \quad . \quad (58)$$

Adding (58) multiplied by v_i to (57) we get

$$\frac{\partial(\rho v_i)}{\partial t} + \sum_1^3{}_k \frac{\partial(\rho v_i v_k)}{\partial y_k} = -\sum_1^3{}_k \frac{\partial \Phi_{ik}}{\partial y_k},$$

which can be written

$$\frac{\partial(\rho v_i)}{\partial t} + \sum_1^3{}_k \frac{\partial}{\partial y_k}(\rho v_i v_k + \Phi_{ik}) = 0. \quad . \quad . \quad (57')$$

It will now be seen that the quantity on the left of (57′) and (58) is in all four cases the sum of partial derivatives with respect to the independent variables t, y_1, y_2, y_3. It follows from § 5 that, since ρ denotes the material density, $\epsilon = c^2\rho$ can be interpreted as the energy density; further, it may be seen in a moment that the vector $\rho\mathbf{v}$ (the momentum density) represents the flux of matter (per unit of surface and of time), and therefore the flux of energy will be $c^2\rho\mathbf{v} = \epsilon\mathbf{v}$.

[1] Cf. particularly G. D. Mattioli, *Rend. Acc. Lincei*, Series V, Vol. XXIII (second half-year, 1914), pp. 328–334, 427–432.

Now to give greater uniformity to the equations (58) and (57′), and to use in them the quantities whose physical interpretation has just been noted, we must replace t and the v_i's by their Römerian expressions $y_0 = ct$, $\beta_i = \frac{v_i}{c}$, and put

$$T_{00} = \epsilon = c^2\rho, \quad \ldots \quad (59)$$

$$T_{0i} = T_{i0} = -\epsilon\beta_i = -c\rho v_i, \quad \ldots \quad (60)$$

$$T_{ik} = T_{ki} = \Phi_{ik} + \epsilon\beta_i\beta_k = \Phi_{ik} + \rho v_i v_k, \quad . \quad (61)$$

$$(i, k = 1, 2, 3).$$

The result is that the four equations (58) and (57′) are all included in the single equation

$$\frac{\partial T_{i0}}{\partial y_0} - \sum_{1}^{3}{}_k \frac{\partial T_{ik}}{\partial y_k} = 0 \quad \ldots \quad (62)$$

by giving i in turn the values, 0, 1, 2, 3.

From the equations (59), (60), (61), we can see the interpretation of the various T's. T_{00} represents the energy density; T_{0i} $(i = 1, 2, 3)$ the components with their sign changed of the relative Römerian flux; the T_{ik}'s $(i, k = 1, 2, 3)$ in statical conditions $(v_i = 0)$, reduce to the ordinary stress components, from which they differ in general by the additive terms $\epsilon\beta_i\beta_k = \rho v_i v_k$ (which, however, in ordinary circumstances are unimportant compared with the other terms). To distinguish when necessary the T_{ik}'s from the ordinary stress Φ_{ik}, we shall call them the *kinetic stress*.

23. **Einsteinian modification of the equations of motion of a continuous system in a particular case.**

The original equations (52) and (53), and therefore the equivalent set (62), are invariant when the axes of reference undergo an ordinary uniform translation. In the earlier stages of the argument we set out to give the dynamics of a material particle a form which should be invariant for a generic transformation (T_4), and we were induced to use Hamilton's principle in order to modify the equations of motion slightly. It followed from this operation that when there are no external forces the equations so modified keep their algebraic form unaltered, not only for ordinary translations but also for the Lorentz transformations which we studied in detail in § 8.

Now the dynamics of a continuous system must clearly include as a limiting case (corresponding to a medium of density everywhere zero except in one very small region) the mechanics of a single material particle. This at once shows that it is absolutely necessary that the postulates introduced for the mechanics of a continuous system should be brought into harmony with the modifications accepted above in the mechanics of the material particle. The form of the equations (62), when there are no external forces, must therefore remain unchanged for any Lorentz transformation. If in accordance with (59), (60), and (61) we take for the T_{ik}'s the expressions

$$T_{00} = \epsilon, \quad T_{0i} = -\epsilon\beta_i, \quad T_{ik} = \epsilon\beta_i\beta_k, \quad . \quad (63)$$

this condition is not rigorously satisfied, though, as we have just pointed out, there is invariance for ordinary translations; but it is easy to show that the required invariance for Lorentz transformations can be obtained by a modification, which, as usual, is very slight in the conditions ordinarily realized.

To do this, we take the four-dimensional form

$$ds_0^2 = dy_0^2 - dl_0^2$$

used above in discussing the dynamics of a particle, where as usual $dl_0^2 = \sum_1^3{}_i \, dy_i^2$.

Denoting by dy_i ($i = 0, 1, 2, 3$) the increments of the coordinates of the generic material element of the system under consideration, and by dl_0 and ds_0 the corresponding elements of the (spacelike) trajectory and the world line, we have by definition

$$\beta_i = \frac{dy_i}{dy_0}, \quad . \quad . \quad . \quad . \quad . \quad . \quad (64)$$

whence

$$\beta^2 = \frac{dl_0^2}{dy_0^2} \quad . \quad . \quad . \quad . \quad . \quad . \quad (64')$$

and

$$ds_0^2 = dy_0^2 \, (1 - \beta^2). \quad . \quad . \quad . \quad . \quad (64'')$$

The parameters of the world line are

$$\lambda^i = \frac{dy_i}{ds_0}$$

(where we have suppressed the sign of absolute value, since in dealing with the motion of a material particle we must have $\beta^2 < 1$, or $ds_0{}^2 > 0$); they can be expressed in terms of the β_i's, using (64), (64'), and (64''), in the form

$$\lambda^0 = \frac{dy_0}{ds_0} = \frac{1}{\sqrt{1-\beta^2}},$$

$$\lambda^i = \frac{dy_i}{ds_0} = \frac{\beta_i}{\sqrt{1-\beta^2}} \qquad (i = 1, 2, 3).$$

From these, taking account of the general formula

$$\lambda_i = \sum_0^3{}_k \, g_{ik} \lambda^k$$

and of the values of $g^0_{ik} = g^{0ik}$ corresponding to $ds_0{}^2$ (cf. § 6, formula (12)), we get the moments

$$\lambda_0 = \frac{1}{\sqrt{1-\beta^2}},$$

$$\lambda_i = -\frac{\beta_i}{\sqrt{1-\beta^2}}.$$

If we take the values of the monomials

$$\epsilon \lambda_i \lambda_k$$

as given by these formulæ, and compare them with the expressions (63) for the T_{ik}'s, we see at once that the difference between each of them and the corresponding T_{ik} is of the second order.

We shall now show that if in the equations (62) we replace the values (63) of the T_{ik}'s by the very slightly different values

$$T_{ik} = \epsilon \lambda_i \lambda_k \qquad (i, k = 0, 1, 2, 3), \quad . \quad . \quad (65)$$

the equations will behave in the required manner for Lorentz transformations; and we shall be able to deduce the criterion to be applied for transforming the equations in the more general case.

Note first that, if ds^2 is taken as the fundamental form, the terms on the right of (65), and therefore the T_{ik}'s, constitute a covariant double system. Further, taking into account once more the particular values g^0_{ik} of the coefficients of ds^2 expressed in terms of the co-ordinates y, it will be clear that the covariant derivatives of the T_{ik}'s are identical with the ordinary derivatives, and that the terms on the left-hand side of (62) can be written in the form

$$\sum_{0}^{3}{}_{kl}\, g^{kl} \frac{\partial T_{ik}}{\partial y_l} = \sum_{0}^{3}{}_{kl}\, g^{kl}\, T_{ik|l} = \sum_{0}^{3}{}_{k}\, T_{ik}^{|k},$$

and are therefore identical with the covariant components of the divergence of the tensor T_{ik} (cf. Chapter VI, p. 153). These equations therefore collectively express the fact that the divergence of this tensor vanishes—a property which is invariant for any transformations whatever of both the space and the time co-ordinates. Remembering finally that a Lorentz transformation leaves unchanged the form of ds^2, we can now assert that the equations (62), with the values of T_{ik} given by (65), will still hold after the application of any Lorentz transformation.

Q. E. D.

24. General case. Introduction of the energy tensor, and meaning of its components in general co-ordinates.

When there are no stresses, the result we have arrived at is that we assign to the T_{ik}'s corresponding to the motion of a generic continuous system the tensor value given by

$$T_{ik} = \epsilon \lambda_i \lambda_k,$$

where ϵ is the energy density and the λ_i's are the moments of the world line of the material element. Further, given any distribution of stresses, referred to Cartesian co-ordinates, then in order to transform the equations of motion into any spacelike co-ordinates (leaving the time unchanged) we have traced out an argument based on the invariance of the bilinear form $\Phi = \sum_{1}^{3}{}_{ik}\, \Phi_{ik}\, \xi^i\, \xi'^k$ (which showed us the three-dimensional tensor character of the Φ_{ik}'s when we pass to generic co-ordinates), on the vector character of the velocity, and on the invariance of the density.

We now propose to consider more generally transformations (T_4) of space and time (i.e. of the set y_0, y_1, y_2, y_3 into a new set x_0, x_1, x_2, x_3), keeping the results already obtained in the two particular cases just referred to (cf. § 23 and § 20). A sufficient condition is that the T_{ik}'s (defined physically with reference to a particular system of co-ordinates) shall have the character of a tensor for any transformations whatever. The tensor so introduced is called the *energy tensor*.

This is equivalent to asserting the invariance of a bilinear form in four variables

$$B = \sum_{0}^{3}{}_{ik} T_{ik}\, \xi^i \xi'^k$$

having for its coefficients the quantities T_{ik} and for its arguments the parameters

$$\xi^i = \frac{dx_i}{ds}, \quad \xi'^i = \frac{d'x_i}{ds'}$$

of two arbitrary four-dimensional versors ξ, ξ'. It will be seen at once that this postulate covers the two particular cases already discussed. In fact, when there are no stresses the tensor character of the T_{ik}'s follows from the expressions (65) adopted for them, while for transformations (T_3) which leave the timelike co-ordinate unchanged, the invariance of B involves that of the form Φ, as will be seen from the following argument.

When we pass from a system x to a system $\bar{x}$, it follows from the invariance of B that (with obvious meanings for the notation used)

$$\sum_{0}^{3}{}_{ik} T_{ik}\, dx_i\, d'x_k = \sum_{0}^{3}{}_{ik} \overline{T}_{ik}\, d\bar{x}_i\, d'\bar{x}_k.$$

In the case of a (T_3), we shall have $dx_0 = d\bar{x}_0$, $d'x_0 = d'\bar{x}_0$, and therefore

$$T_{00}\, dx_0 d'x_0 + dx_0 \sum_{1}^{3}{}_k T_{0k} d'x_k + d'x_0 \sum_{1}^{3}{}_k T_{0k} dx_k + \sum_{1}^{3}{}_{ik} T_{ik} dx_i d'x_k$$

$$= \overline{T}_{00} dx_0 d'x_0 + dx_0 \sum_{1}^{3}{}_k \overline{T}_{0k} d'\bar{x}_k + d'x_0 \sum_{1}^{3}{}_k \overline{T}_{0k} d\bar{x}_k + \sum_{1}^{3}{}_{ik} \overline{T}_{ik} d\bar{x}_i d'\bar{x}_k;$$

and as this must hold whatever the differentials dx_0, $d'x_0$ may be, it follows that

$$T_{00} = \overline{T}_{00},$$

$$\sum_1^3{}_k T_{0k}\, dx_k = \sum_1^3{}_k \overline{T}_{0k}\, d\bar{x}_k,$$

$$\sum_1^3{}_{ik} T_{ik}\, dx_i\, d'x_k = \sum_1^3{}_{ik} \overline{T}_{ik}\, d\bar{x}_i\, d'\bar{x}_k.$$

As the differentials dx_i, $d'x_i$ are arbitrary, these relations express the fact that T_{00} is an invariant (the energy density), that the T_{0k}'s are the components of a vector (the flux of the energy with its sense changed), and the T_{ik}'s those of a covariant double tensor (the kinetic stress).

Q. E. D.

We now propose to examine, with reference to pseudo-Cartesian co-ordinates y, the physical significance of the form B when the directions ξ, ξ' are chosen in a particular way.

Suppose first that both the directions are purely timelike, i.e. that

$$dx_i = d'x_i = 0 \qquad (i = 1, 2, 3)$$

and therefore $\qquad ds^2 = dy_0{}^2, \quad ds'^2 = d'y_0{}^2.$

Then the only parameters which are not zero are ξ^0 and ξ'^0, which are equal to 1, and there remains

$$B = T_{00};$$

i.e. in this case B represents the energy density.

Now suppose that ξ, ξ' are purely spacelike, i.e. that $dy_0 = d'y_0 = 0$, and consequently

$$ds^2 = -\, dl_0^2, \quad ds'^2 = -\, dl_0'^2.$$

Then there remains

$$B = \sum_1^3{}_{ik} T_{ik} \frac{dy_i}{dl_0} \frac{d'y_k}{dl_0'},$$

i.e. B reduces to the linear invariant of the kinetic stress.

Lastly, suppose that $\boldsymbol{\xi}$ is purely spacelike and $\boldsymbol{\xi}'$ purely timelike; i.e. that $dy_0 = 0$, $d'y_i = 0$ $(i = 1, 2, 3)$, and therefore

$$ds^2 = -dl_0^2, \quad ds'^2 = d'y_0^2.$$

Then
$$B = \sum_1^3{}_i\, T_{0i} \frac{dy_i}{dl_0}$$

and is therefore identical with the flux of the energy in the direction $\boldsymbol{\xi}$ with its sign changed.

We can now determine the physical significance of the T_{ik}'s with reference to any system of co-ordinates whatever. This follows easily from the invariance of the form B if we allow that the physical significance of this form in the particular cases noted above remains the same in any other system of reference. The different cases are in detail:

(a) The energy density at a generic instant and point x_0, x_1, x_2, x_3 will be what B becomes for $\boldsymbol{\xi}$, $\boldsymbol{\xi}'$ purely timelike, i.e. for

$$\xi^0 = \xi'^0 = \frac{1}{\sqrt{g_{00}}}, \quad \xi^i = \xi'^i = 0 \qquad (i > 0);$$

i.e. it will be
$$\frac{T_{00}}{g_{00}}.$$

(b) The flux of the energy along a specified (spacelike) direction $\boldsymbol{\alpha}$ of parameters $\alpha^i = \dfrac{dx_i}{dl}$ will be what $-B$ becomes when we put in it

$$\left.\begin{aligned} \xi^i &= \frac{dx_i}{|ds|} = \frac{dx_i}{dl} = \alpha^i, \quad \xi^0 = 0 \\ \xi'^i &= 0, \quad \xi'^0 = \frac{1}{\sqrt{g_{00}}} = \frac{1}{V} \end{aligned}\right\} (i = 1, 2, 3),$$

i.e. it will be
$$-\frac{1}{V} \sum_1^3{}_i\, T_{i0}\, \alpha^i.$$

If in particular the direction $\boldsymbol{\alpha}$ coincides with one of the co-ordinate directions, say x_h, we have

$$\alpha^h = \frac{1}{\sqrt{-g_{hh}}},$$

and the other α^i's are zero; hence the flux of the energy in that direction is given by

$$-\frac{1}{V}\frac{T_{h0}}{\sqrt{-g_{hh}}}.$$

(*c*) The component in a direction $\boldsymbol{\alpha}$ of the kinetic stress relative to a facet normal to a direction $\boldsymbol{\alpha}'$ will be what B becomes when we put

$$\xi^i = \frac{dx_i}{|\,ds\,|} = \frac{dx_i}{dl} = \alpha^i, \quad \xi^0 = 0,$$

$$\xi'^i = \frac{d'x_i}{|\,ds'\,|} = \frac{d'x_i}{dl'} = \alpha'^i, \quad \xi'^0 = 0;$$

i.e. it will be

$$\sum_{ik}^{3}{}_{1} T_{ik}\,\alpha^i\,\alpha'^k.$$

If in particular the direction $\boldsymbol{\alpha}$ coincides with that of one of the co-ordinate lines, say x_r, and $\boldsymbol{\alpha}'$ with that of another, say x_s, we shall have

$$\alpha_r = \frac{1}{\sqrt{-g_{rr}}}, \quad \alpha'_s = \frac{1}{\sqrt{-g_{ss}}},$$

and all the other α's will be zero. Hence the component in the direction x_r of the kinetic stress relative to a facet normal to x_s will be

$$\frac{T_{rs}}{\sqrt{g_{rr}\,g_{ss}}}.$$

Before concluding this section we wish to make one last remark. We have seen that when there are no stresses (the case of discrete particles of matter) the energy tensor takes the particularly simple form

$$T_{ik} = \epsilon\lambda_i\,\lambda_k. \quad . \quad . \quad . \quad . \quad . \quad . \quad (65)$$

Another important particular case is when the energy tensor has the form

$$T_{ik} = \epsilon\lambda_i\,\lambda_k - p g_{ik}, \quad . \quad . \quad . \quad . \quad . \quad (66)$$

where p is any invariant function of the position and the time. In order to see the physical significance of this expression, consider a specified point of V_4, and take a system of co-ordinates which are, at least locally, pseudo-Cartesian, which we know is always possible. Then the g_{ik}'s take the values $\delta_i'^c$, while if we make the direction x_0 coincide with that of the world line, the λ_i's all become zero, except λ_0, which is 1.

In these conditions we shall have

$$T_{00} = \epsilon - p,$$
$$T_{ik} = 0 \qquad (i \neq k),$$
$$T_{ii} = p \qquad (i > 0).$$

The last two formulæ tell us that on every facet there is exerted a stress normal to it and independent of its direction: the scalar quantity p measures the value of this stress per unit of surface. The medium under consideration therefore behaves like a perfect fluid (a fluid incapable of transmitting a shearing stress), and p represents its pressure. It is hardly necessary to point out that if p is negative it represents a uniform pull in all directions —which, within certain limits, is known to be a possible condition even in a real liquid.

25. **Relativistic form of the equations of motion of a continuous system.**

In the particular case of no forces, we saw in § 23 how the general equations of motion of a continuous system can be put in the form

$$\sum_{0}^{3}{}_k T_{ik}^{|k} = 0 \qquad (i = 0, 1, 2, 3), \quad . \quad . \quad (67)$$

where the T_{ik}'s are regarded as elements of a tensor, and that this equation holds in general co-ordinates x whatever may be the transformations (involving both space and time) imposed on the original co-ordinates y. The proof of this consists in the invariant character of the equations (67) (which express the vanishing of the divergence of the tensor T_{ik}), together with the fact that in the original co-ordinates y the equations (67) reduce to the form (62), and that the quantities T_{ik} become identical (neglecting terms of the second order, if not rigorously) with the expressions (63) which are their values in the classical mechanics. All this holds without change even if we drop the particular hypothesis suggested by the law of transformation of the T_{ik}'s when the transformations (T_4) are applied; that the forces are zero. It is only necessary to retain the tensorial character of the T_{ik}'s in every case, as we have already agreed; which, in the particular case where stresses Φ_{ik} are present, means that their experimental values are determined, say, with reference to the co-ordinates which formed the starting-point of the investigation.

The equations (67) thus hold so long as the metric considered is pseudo-Euclidean, and for any co-ordinates of reference whatever. But the invariant expression for the laws of motion, which is seen to hold under this hypothesis, can be at once extended to the general case of any metric whatever, in virtue of the observation made earlier in this book (cf. Chapter VI, p. 164) that in a first-order region every metric behaves as if it had constant coefficients, and is therefore Euclidean in the proper sense in the case of a definite ds^2, and pseudo-Euclidean in the cases which concern relativity mechanics (cf. §§ 6 and 12). In fact, the equations (67) contain only contravariant derivatives of the T_{ik}'s, or, in other words, combinations of their ordinary first derivatives with the g_{ik}'s and their first derivatives; the argument thus does not go beyond the consideration of a first-order region round the generic point which is being studied.

26. A particular class of motions of a continuous system.

In the classical mechanics the equation (52) of the motion of a continuous medium, when there are no forces and no molecular action (a discrete system), evidently reduces to

$$\mathbf{f} = 0,$$

with which is to be associated the equation of continuity. It follows that the vector equation is satisfied at once by the uniform rectilinear motion of single particles, the density being then determined by the equation of continuity. This is conceptually evident; in order to translate it into a formula, we assign to any material particle, initially at P_0, a velocity $\mathbf{v}(P_0)$ which is a function (*a priori* arbitrary) of the position P_0: the geometrical equation of motion is then evidently

$$P(t) = P_0 + \mathbf{v}(P_0)t, \quad . \quad . \quad . \quad . \quad (68)$$

which shows that the solution depends in substance on three arbitrary functions of three arguments each.

If we wish to find an explicit expression for the law of variation of the density it is perhaps preferable to go back to the molecular equation of continuity instead of to the equation (53) which is its local form. It is a well-known result that if we introduce the functional determinant D of the actual co-ordinates $y(t)$ with respect to the initial co-ordinates y_0 we get

$$\rho D = \rho_0,$$

where ρ_0 is the initial value of ρ, and is *a priori* arbitrary just as is the initial distribution of the velocity. Projecting the equation (68) on the axes, and denoting the components of $\mathbf{v}$ by $v_i(y_1^0, y_2^0, y_3^0)$, we get

$$y_i = y_i^0 + v_i t,$$

whence

$$D = \left\| \delta_i^k + \frac{\partial v_i}{\partial y_k^0} t \right\|.$$

It follows from this that D is a polynomial of the third degree in t, which reduces to unity for $t = 0$. Naturally (supposing that the v_i's and their first derivatives are finite and continuous) the motion remains regular so long as D does not vanish; the smallest positive root (if such exists) of the equation of the third degree $D = 0$ determines the amplitude of the interval of regularity, &c.

A particular case worth noting is when the density remains constant for each particle (incompressible systems). In this case $\frac{d\rho}{dt} = 0$, and the equation of continuity, in the original Eulerian form (53′), gives

$$\operatorname{div}(\mathbf{v}) = 0. \qquad (69)$$

This implies in particular that the divergence vanishes at the initial instant, and therefore gives as a necessary condition for the constancy of the density that the field of the initial velocities must be solenoidal, i.e. that $\operatorname{div} \mathbf{v}(P_0) = 0$. This condition is not however sufficient. In fact, if the density is to remain constant it is necessary and sufficient that $D = 1$ at any instant t; the expansion of D as a polynomial of the third degree in t shows that this imposes three conditions, corresponding to the vanishing of the coefficients of t, t^2, t^3, and (69) expresses only the first of these conditions. Further, if these conditions are satisfied initially, ρ remains constant for every particle (i.e. $\frac{d\rho}{dt} = 0$), which ensures that the equation (69) is satisfied at every instant, or in other words that the field of the velocities is always solenoidal.[1]

We have dealt at some length with this class of elementary solutions, because the results can easily be generalized for any V_4 whatever. If $\lambda_i(x_0, x_1, x_2, x_3)$ denote the moments of a generic congruence of lines in the V_4, we know (cf. Chapter X, p. 274) that the necessary and sufficient condition for the congruence to be geodesic is that the curvature vector, or, what is equivalent, its covariant components, shall vanish, i.e. that

$$\sum_{0}^{3}{}_l \, \lambda_{i|l} \lambda^l = 0 \qquad (i = 0, 1, 2, 3). \quad . \quad . \quad (70)$$

We now propose to show that in a V_4 with any metric whatever we get solutions of the equations (67) by taking for world lines the lines of any geodesic congruence whatever, or, in other words, by supposing that the λ's satisfy the equations (70) and by assigning a suitable value to the density ρ, and through it to the quantity ϵ which appears in the expression (65) for the energy tensor of a discrete system (i.e. a system with no molecular action).

Take the general equations (67), which we shall write in the form

$$\sum_{0}^{3}{}_{kl} \, g^{kl} \, T_{ik|l} = 0 \qquad (i = 0, 1, 2, 3)$$

and in them give T_{ik} the value $\epsilon \lambda_i \lambda_k$. We shall have

$$T_{ik|l} = \epsilon_l \lambda_i \lambda_k + \epsilon \lambda_{i|l} \lambda_k + \epsilon \lambda_i \lambda_{k|l}$$

and therefore by substitution

$$\lambda_i \sum_{0}^{3}{}_{kl} \, g^{kl} \epsilon_l \lambda_k + \epsilon \sum_{0}^{3}{}_k \, \lambda_i^{|k} \lambda_k + \epsilon \, \lambda_i \sum_{0}^{3}{}_k \, \lambda_k^{|k} = 0 \qquad (i = 0, 1, 2, 3).$$

[1] Cf. Cisotti: "Moti di un liquido che lasciano inalterata la distribuzione locale delle pressioni", in *Rend. della R. Acc. dei Lincei*, Series V, Vol. XIX (first half-year, 1910), pp. 373–376. The observation is there limited to the case of permanent motion.

The second term vanishes in virtue of (70), and therefore the four equations reduce to the single condition

$$\sum_{0}^{3}{}_{kl}\, g^{kl}\, \epsilon_l\, \lambda_k + \epsilon \sum_{0}^{3}{}_{k}\, \lambda_k^{|k} = 0. \quad . \quad . \quad . \quad (71)$$

If we choose ϵ so that this condition is satisfied, the equations of motion will all be satisfied also.

Q. E. D.

The equation (71) defining ϵ can be put in a somewhat more expressive form by using the results (cf. Chapter X, p. 267) that

$$\sum_{0}^{3}{}_{kl}\, g^{kl}\, \epsilon_l\, \lambda_k = \sum_{0}^{3}{}_{l}\, \epsilon_l\, \lambda^l = \frac{d\epsilon}{ds},$$

where s denotes the arc of the world line, and noting that

$$\sum_{0}^{3}{}_{k}\, \lambda_k^{|k} = \operatorname{div}\boldsymbol{\lambda}.$$

Hence (71) can be written

$$\frac{d\epsilon}{ds} + \epsilon \operatorname{div}\boldsymbol{\lambda} = 0 \quad . \quad . \quad . \quad . \quad (71')$$

which is precisely the form of the equation of continuity.

If in particular we consider a solenoidal geodesic congruence ($\operatorname{div}\boldsymbol{\lambda} = 0$), the last equation becomes

$$\frac{d\epsilon}{ds} = 0,$$

whence $\epsilon =$ constant along any world line; i.e. the density of a particle remains constant throughout the motion.

27. **Experimental determination of the coefficients of an Einsteinian ds^2.**

We shall close this chapter by some remarks of a general character on the experimental determination of the coefficients g_{ik}.

We suppose ourselves fixed in determinate physical conditions, so that, as already noted in § 10 and § 12, we must also regard as determinate the Einsteinian

$$ds^2 = \sum_{0}^{3}{}_{ik} g_{ik}\, dx_i\, dx_k \quad . \quad . \quad . \quad . \quad (72)$$

of the field which we wish to explore by means of suitable experiments. It is of course understood that we admit the validity of the fundamental postulates of general relativity, and more precisely:

(*a*) (cf. § 16) the propagation of light always takes place in such a way that

$$ds^2 = 0 \quad . \quad . \quad . \quad . \quad . \quad . \quad . \quad (73)$$

along every world line;

(*b*) (cf. § 12) the world lines of the motion of a material particle in a field of force for which ds^2 can be expressed by (72) are timelike geodesics for this ds^2.

We propose to show that (*a*) suffices to determine the ratios of the coefficients g_{ik}, or, which comes to the same thing, gives ds^2 except for a factor which can in turn be found from (*b*). Of the four parameters, x_0 will as usual denote the time, in the sense of the *conventional time*, measured at any single point by a clock which may be of any kind and even incorrect. However the timelike parameter x_0 is chosen, the mere fact that it is timelike implies, according to the Einsteinian theory, that ds^2 will always be greater than 0 if x_0 alone varies, x_1, x_2, x_3 remaining constant. But, when $dx_1 = dx_2 = dx_3 = 0$, ds^2 reduces to $g_{00} dx_0{}^2$, so that the coefficient g_{00} necessarily > 0, and we can therefore put

$$g_{00} = c^2 e^{2\nu}, \quad . \quad . \quad . \quad . \quad . \quad (74)$$

where c is a positive constant (introduced for the sake of homogeneity) and ν, like g_{00}, is an unknown function of x_0, x_1, x_2, x_3 (a pure number, i.e. of zero dimensions).

We shall now choose any instant x_0 we please, and three values x_1, x_2, x_3 of the space co-ordinates, i.e. a point P; we propose in the first place to determine the ratios of the g's at P and at the instant x_0.

For this we shall use light signals between P and very near points in the surrounding physical space, which is by hypothesis (at any given moment) in one-to-one correspondence with the sets of three co-ordinates x_1, x_2, x_3. In consequence, surfaces and lines in this physical space represented by equations between x_1, x_2, x_3 at the moment x_0 are perfectly determinate: in particular the lines x_1 (given by the equations x_2 = constant, x_3 = constant) on which only x_1 varies, the lines x_2, &c.

We shall choose two points Q and Q' very near P, on the same line x_1 as P. Suppose that Q and Q' correspond to increments (to be treated as infinitesimals) dx_1 and $-dx_1$ of the co-ordinate x_1; dx_2, dx_3 are zero in both cases since the displacement is along a line x_1.

Suppose that two light rays start from P at the instant x_0, one towards Q, the other towards Q'. Let $x_0 + dx_0$ be the instant when the first ray arrives at Q; $x_0 + d'x_0$ the instant (not in general the same as the first) when the second ray arrives at Q'. Using the expression (72) for ds^2 and the condition $ds^2 = 0$ for the propagation of light, we shall have in passing from P to Q

$$g_{00}\, dx_0^2 + 2g_{01}\, dx_0\, dx_1 + g_{11} dx_1^2 = 0, \quad . \quad (75)$$

and in passing from P to Q'

$$g_{00}\, d'x_0^2 - 2g_{01}\, d'x_0\, dx_1 + g_{11}\, dx_1^2 = 0. \quad . \quad (76)$$

These two equations, in which dx_1, dx_0, $d'x_0$ are known (the first chosen as we please, the other two found by experiment), obviously give the ratios $\frac{g_{01}}{g_{00}}, \frac{g_{11}}{g_{00}}$. It is to be noted that if the elementary times of propagation dx_0, $d'x_0$ (found by observation) are equal, then (75) and (76) give by subtraction $g_{01} = 0$. Reciprocally, if $g_{01} = 0$, the two intervals of time must be equal. Hence the elementary propagation of light in the direction of a line x_1 is a reversible phenomenon if and only if $g_{01} = 0$.

In the same way, considering the other two co-ordinate lines x_2 and x_3, we can determine the four ratios

$$\frac{g_{02}}{g_{00}}, \frac{g_{22}}{g_{00}}; \quad \frac{g_{03}}{g_{00}}, \frac{g_{33}}{g_{00}}.$$

To obtain the other three ratios

$$\frac{g_{23}}{g_{00}}, \frac{g_{31}}{g_{00}}, \frac{g_{12}}{g_{00}}$$

we must make further experiments of the same type, but with the point Q in a direction other than those of the co-ordinate lines.

Thus to determine $\frac{g_{23}}{g_{00}}$, we can use a line on the surface through P, $x_1 =$ constant, which is neither x_2 nor x_3; e.g.

$$x_3 - x_2 = \text{constant}.$$

We then have, in passing from P to a very near point Q on this line, the increments

$$0, \quad dx_2, \quad dx_2,$$

with dx_2 arbitrary.

If we make a light ray start from P at the instant x_0 towards this point Q, and if dx_0 denotes the small time of propagation, we get from (72) divided by g_{00}

$$dx_0{}^2 + 2\frac{g_{02}}{g_{00}}dx_0\,dx_2 + 2\frac{g_{03}}{g_{00}}dx_0\,dx_2$$

$$+ \frac{g_{22}}{g_{00}}dx_2{}^2 + \frac{g_{33}}{g_{00}}dx_2{}^2 + 2\frac{g_{23}}{g_{00}}dx_2{}^2 = 0, \quad . \quad (77)$$

whence we get the ratio $\frac{g_{23}}{g_{00}}$, all the other quantities in this equation being known or already determined. In a similar way we can find $\frac{g_{31}}{g_{00}}$ and $\frac{g_{12}}{g_{00}}$.

It is not inapposite to add that from other experiments of the same type we can get any number (in fact an infinite number) of further equations between the ratios of the g's. The consistency of these results, in so far as this is borne out by the further experiments, affords a very significant control of the validity of the Einsteinian hypothesis so far as concerns the postulate (a).

The ratios

$$g'_{ik} = \frac{g_{ik}}{g_{00}} \qquad (i, k = 0, 1, 2, 3) \quad . \quad . \quad (78)$$

being thus determined, if we put

$$ds^2 = g_{00}\,ds'^2 = c^2e^{2\nu}ds'^2 \quad . \quad . \quad . \quad (79)$$

(using (74)), it follows that the individual coefficients of the differential form

$$ds'^2 = \frac{ds^2}{g_{00}}$$

are all known, and therefore the form itself is completely determined.

From (72) and (78), separating out the terms which contain the suffix 0, we get ds'^2 in the form

$$ds'^2 = dx_0{}^2 + \sum_1^3{}_i\, g'_{0i}\, dx_0\, dx_i + \sum_1^3{}_{ik}\, g'_{ik}\, dx_i\, dx_k. \quad . \quad (80)$$

At this point we find that we have to determine the function ν by gravitational experiments, and more precisely by experiments on the motion of material particles in the field in which the expression (79) holds for ds^2.

The equations of motion are included in the variational equation

$$\delta\int ds = 0. \quad . \quad . \quad . \quad . \quad . \quad (81)$$

Now suppose that the time x_0 is taken as the independent variable along the trajectory. Let $\dot{x}_i$ ($i = 1, 2, 3$) denote the derivatives $\frac{dx_i}{dx_0}$; and using (80), put

$$\left.\begin{aligned}\frac{ds'}{dx_0} &= \sqrt{1 + \sum_1^3{}_i\, g'_{0i}\, \dot{x}_i + \sum_1^3{}_{ik}\, g'_{ik}\, \dot{x}_i\, \dot{x}_k} \\ &= L\,(x_0 \mid x_1, x_2, x_3 \mid \dot{x}_1, \dot{x}_2, \dot{x}_3).\end{aligned}\right\} \quad . \quad . \quad (82)$$

Then, remembering (79), the variational equation (81) can be written in the form

$$\delta\int (e^{\nu} L)\, dx_0 = 0. \quad . \quad . \quad . \quad . \quad (81')$$

This is equivalent to the three Lagrangian equations

$$\frac{d}{dx_0}\frac{\partial(e^{\nu}L)}{\partial\dot{x}_i} - \frac{\partial(e^{\nu}L)}{\partial x_i} = 0 \qquad (i = 1, 2, 3).$$

Noting that ν does not depend on the $\dot{x}$'s, and putting for the sake of brevity

$$\frac{\partial L}{\partial \dot{x}_i} = \alpha_i, \quad \frac{d\alpha_i}{dx_0} - \frac{\partial L}{\partial x_i} = \beta_i \quad . \quad . \quad . \quad (83)$$

it follows that

$$a_i \frac{d\nu}{dx_0} - L\frac{\partial \nu}{\partial x_i} + \beta_i = 0. \quad . \quad . \quad . \quad . \qquad (84)$$

It is further to be noted that direct observation of the motion enables us to determine how the co-ordinates x_i vary as functions of the time x_0, so that we must consider the functions $x_i(x_0)$, and therefore also the derivatives $\dot{x}_i$ and $\ddot{x}_i$, known for every material particle left to itself in or projected into the field of force we are considering. It follows that the quantities α_i, β_i defined by (83) are also known. Since

$$\frac{d\nu}{dx_0} = \frac{\partial \nu}{\partial x_0} + \sum_1^3{}_i \frac{\partial \nu}{\partial x_i} \dot{x}_i,$$

it follows that ultimately the equations (84) are three linear equations in the four partial derivatives of the unknown function ν. If we fix a generic point P and an instant x_0, any arbitrary choice of the velocity of the body under experiment (i.e. of the three numerical values to be assigned to $\dot{x}_1$, $\dot{x}_2$, $\dot{x}_3$) will give three equations in the four derivatives

$$\frac{\partial \nu}{\partial x_0}, \frac{\partial \nu}{\partial x_1}, \frac{\partial \nu}{\partial x_2}, \frac{\partial \nu}{\partial x_3}$$

referred to the given position and time. The equations are therefore more than sufficient to determine the numerical values of these derivatives, in the sense that by making a larger number of experiments we can not only determine the four unknowns, but also test the accuracy of the results as many times over as we wish.

The derivatives of ν, at every point in a certain field and at every instant in a certain interval, being known, ν itself is determined except for an additive constant; hence, from (74), g_{00} is known except for a constant multiplier, which we may suppose absorbed into the factor of homogeneity c^2, so that c^2 remains arbitrary. The presence of this constant in the expression for g_{00}, and hence, by (79), in ds^2, seems to be in the nature of things, corresponding in substance to the choice, which remains arbitrary, of the unit chosen to measure ds^2, the space-time interval.

CHAPTER XII

The Gravitational Equations and General Relativity

1. Qualitative properties of the coefficients of ds^2.

It follows from the results in the preceding chapter (p. 325) that when the variables of reference y_0, y_1, y_2, y_3 are such that they can be interpreted, without sensible error, the first as absolute time, and the others as Cartesian co-ordinates, then the coefficients g_{ik} of the Einsteinian ds^2 of space-time, in conditions corresponding to the motion of the celestial bodies (in particular, of the bodies forming the planetary system), differ by very little from $\delta_i'^k$, the difference being of at least the second order, in the sense explained above. More precisely we can say that:

(*a*) The coefficient g_{00} differs from $1 - \dfrac{2U}{c^2}$ by terms of order higher than the second (cf. p. 320 in the preceding chapter), where U represents the ordinary Newtonian potential of the field considered.

(*b*) The coefficients g_{0i} $(i > 0)$ are of order higher than the second. If in fact they were only of the second order, it follows from p. 339 in the preceding chapter that the difference between the velocities of propagation of light in the various directions round a point would also have to be of the second order; this, however, is physically inadmissible, as a difference of this magnitude could be detected by means of optical experiments.

(*c*) The other coefficients g_{ik} $(i, k > 0)$ differ from $\delta_i'^k$ by terms of the second or higher order.

Now let us consider the absolute motion of a generic material particle P, e.g. a small planet. Let $u(P, P')$ be the Newtonian potential of the attraction exerted on it by any particle P' of the other attracting bodies, which we shall suppose to be of fairly large mass compared with P, as is in fact the case in the typical examples offered by astronomy. The disturbing effects of P on the motion of P' being supposed negligible, the dependence of u on the space co-ordinates y_1, y_2, y_3 involves the co-ordinates of P, while its dependence on the Römerian time y_0 involves the co-ordinates of the attracting body P'. If δP is a generic displacement (of components δy_i) of the point P, and

$$\delta u = \sum_1^3 {}_i \frac{\partial u}{\partial y_i} \delta y_i$$

is the corresponding increment of u, we have

$$\delta u = \mathbf{F} \times \delta P,$$

where $\mathbf{F}$ is the force exerted on P by P'. Further, if we consider a small interval of time dy_0 and denote by dP' the displacement of P' during that interval, and by du the increment of u, we get similarly, applying the principle of reaction,

$$du = -\mathbf{F} \times dP'.$$

After this it is easy to determine the order of magnitude of the timelike derivative of u in relation to the spacelike derivatives. In fact, from the first formula, putting $\delta P = \mathbf{n}\delta l$ (where $\mathbf{n}$ is the versor of a generic direction) we get the well-known result that the derivative of u in this direction has the value $\mathbf{F} \times \mathbf{n}$, and is therefore of the same order of magnitude as the intensity F of the force; while from the second formula, on dividing by dy_0, it follows that

$$\frac{du}{dy_0} = -\mathbf{F} \times \frac{dP'}{dy_0} = -\mathbf{F} \times \frac{1}{c}\frac{dP'}{dt},$$

which shows that the order of magnitude of this derivative is that of βF (with the usual meaning of β). Hence, in the supposed conditions, the timelike derivative of u is of the first order in relation to the spacelike derivatives. The same result holds without change for g_{00}, which, as we have just said, is $1 - \frac{2U}{c^2}$ (neglecting terms of order higher than the second), U being a sum of terms of the type just considered.

Taking the case of g_{00} as typical, we shall assume, in ordinary astronomical conditions, that:

(d) The derivatives of the coefficients g_{ik} with respect to y_0 are of higher order by at least one unit than the analogous derivatives with respect to the other y's.

We can sum up all this in the statement that if we are content with approximate results (meaning that we stop short at terms of the second order), everything happens as if the coefficients g_{i0} were zero, and the other g_{ik}'s independent of y_0. This is equivalent to the statement that, to the given order of approximation and in ordinary astronomical conditions, *every ds^2 behaves as if it were statical* (cf. Chapter XI, p. 326).

2. The tensor G_{ik} and its divergence. The gravitational tensor.

We have already noted (cf. Chapter VII, p. 200) that for any V_n whatever we can construct from the Riemannian tensor the symmetrical double tensor

$$G_{ik} = \sum_{1}^{n}{}_{jh}\, a^{jh}\,(ij,\, hk), \quad . \quad . \quad . \quad . \quad (1)$$

and its linear invariant

$$G = \sum_{1}^{n}{}_{ik}\, a^{ik}\, G_{ik}. \quad . \quad . \quad . \quad . \quad . \quad (2)$$

This definition naturally holds also for an indefinite metric: in particular therefore for the ds^2 of relativity ($n = 4$), in which case the tensor under discussion is called the *Einstein tensor*; its components are

$$G_{ik} = \sum_{0}^{3}{}_{jh}\, g^{jh}\,(ij,\, hk), \quad . \quad . \quad . \quad . \quad (1')$$

and its linear invariant therefore takes the form

$$G = \sum_{0}^{3}{}_{ik}\, g^{ik}\, G_{ik} = \sum_{0}^{3}{}_{ijhk}\, g^{ik}\, g^{jh}\,(ij,\, hk). \quad . \quad . \quad (2')$$

We may note incidentally that for a V_2 the tensor G_{ik} is related to the fundamental tensor g_{ik} and to the Gaussian curvature by the formula [1]

$$G_{ik} = -\,K g_{ik} \qquad (i,\, k = 1,\, 2);. \quad . \quad . \quad (3)$$

[1] In fact, for $n = 2$, it follows from the definitions of K (p. 194, formula (28)) and of the ϵ-systems (Chap. VI, p. 158) that $(ij,\, hk) = K\epsilon_{ij}\,\epsilon_{hk}$, as can at once be verified, remembering that the symbol $(ij,\, hk)$ either reduces to $(12,\, 12) = Ka$, or vanishes. Further, with the same definition of ϵ, we have also the identities $\sum_{1}^{2}{}_{jh}\, g^{jh}\,\epsilon_{ij}\,\epsilon_{hk} = -\,g_{ik}$. Replacing $(ij,\, hk)$ in the formula of type (1′) by $K\epsilon_{ij}\,\epsilon_{hk}$, and using this identity, we get (3).

while for a V_3 the quantities G_{ik} reduce to Ricci's symbols (cf. Chapter VII, p. 199)

$$a^{ik} = \frac{(i+1\ i+2,\quad k+1\ k+2)}{a},$$

the relation being

$$G_{ik} = a_{ik} - \mathcal{M}\, a_{ik}, \qquad . \quad . \quad . \quad . \quad . \quad (4)$$

where $\mathcal{M}$ denotes the mean curvature of the V_3, or in symbols

$$\mathcal{M} = \sum_{1}^{3}{}_{ik}\, a^{ik}\, a_{ik}. \qquad . \quad . \quad . \quad . \quad . \quad (5)$$

For $n = 4$, from the general formula

$$N = \frac{n^2(n^2 - 1)}{12}$$

of Chapter VII, p. 182, it follows that in general the Riemann-Christoffel tensor has 20 algebraically independent components, while the elements G_{ik} of the Einstein tensor provide only 10 linear combinations. This simple arithmetical remark shows that the Einstein tensor cannot exhaust all the curvature properties of the V_4, but, as we shall see, it does suffice to give those of essential physical importance.

Before beginning the examination of this question, we shall find the expression for the divergence of the tensor G_{ik}. From (1′), we have by covariant differentiation

$$G_{ik|l} = \sum_{0}^{3}{}_{jh}\, g^{jh}\, (ij,\, hk)_l,$$

so that the components of the divergence (cf. Chapter VI, p. 153)

$$Y_i = \sum_{0}^{3}{}_{k}\, G_{ik}^{|k} = \sum_{0}^{3}{}_{kl}\, g^{kl}\, G_{ik|l} \qquad . \quad . \quad . \quad (6)$$

become

$$Y_i = \sum_{0}^{3}{}_{jhkl}\, g^{jh}\, g^{kl}\, (ij,\, hk)_l.$$

In virtue of the relations

$$(ij,\, hk) = (hk,\, ij),$$

Bianchi's identities (formula (17′), Chap. VII, p. 183) enable us to substitute

$$-(jl, hk)_i - (li, hk)_j$$

for $(ij, hk)_l$, so that we have

$$Y_i = -\sum_0^3{}_{jhkl}\, g^{jh} g^{kl}\, (jl, hk)_i - \sum_0^3{}_{jhkl}\, g^{jh} g^{kl}\, (li, hk)_j.$$

The first term is merely G_i, as follows from covariant differentiation of (2′), which by interchanging the indices can be written in the form

$$G = -\sum_0^3{}_{jlhk}\, g^{jh} g^{kl}\, (jl, hk).$$

Interchanging j and l, and also h and k, in the second term, it becomes

$$-\sum_0^3{}_{jhkl}\, g^{kl} g^{jh}\, (ji, kh)_l,$$

and in view of the identity

$$(ji, kh) = (ij, hk)$$

it obviously reduces to $-Y_i$. We therefore have

$$Y_i = \tfrac{1}{2} G_i \quad . \quad . \quad . \quad . \quad . \quad . \quad (7)$$

which in virtue of (6) can also be written

$$\sum_0^3{}_k\, G_{ik}^{|k} - \tfrac{1}{2} G_i = 0. \quad . \quad . \quad . \quad . \quad . \quad (7')$$

Since the divergence of the tensor Gg_{ik} (proportional to the fundamental tensor g_{ik}) is

$$\sum_0^3{}_{kl}\, (G\, g_{ik})_l\, g^{kl} = \sum_0^3{}_l\, G_l \sum_0^3{}_k\, g_{ik}\, g^{kl} = G_i,$$

it will be seen that (7), or the equivalent equation (7′), expresses the property that *the divergence of the tensor*

$$G_{ik} - \tfrac{1}{2}\, G\, g_{ik}$$

is zero. This tensor is called the *gravitational tensor*; the name will be justified farther on.

3. Solidarity of physical phenomena. Criteria for the construction of the gravitational equations, and reduction of the inductive proof of their validity to the statical case.

In the immediate vicinity of a point and instant fixed in advance, a mechanical phenomenon is completely determined (at least conceptually) if we know, at the specified point and instant, the density and velocity of the matter (or, which comes to the same thing, of the energy), and the distribution of the specific stress, which includes as a differential consequence the determination of the external force; the latter, however, as already noted (p. 349) in the preceding chapter, can be supposed absorbed into the stresses, the concept of action at a distance being as before excluded. In substance, therefore, the local behaviour of a mechanical phenomenon is completely determined by the knowledge (which is both necessary and sufficient) of the energy tensor T_{ik}.

This remark has a more general scope, since it holds also for phenomena other than mechanical (e.g. electromagnetic phenomena).

Einstein's fundamental view is that the aggregate of physical phenomena influences the metric of V_4; more precisely, that at every point P of the V_4 there must be a local relation between the value of the energy tensor, which may be taken as characteristic of the physical conditions, and the behaviour of the curvatures of the V_4 at the point. As an abstract hypothesis, the possibility of some such influence, limited however to the spatial metric, had already been suggested independently by Riemann and by Clifford. Einstein completed it, applying it not only to the spatial metric, but to the metric of the space-time which includes both space and time and also, as we saw in § 4, p. 291, and § 10, p. 320, when studying the motion of a material particle, the force in the field, which is represented through the coefficient g_{00}.

We have pointed out just above that from the mathematical point of view the external force can be considered as produced by a suitable distribution of stresses. From the point of view of the classical mechanics this principle could also be applied to the particular case of forces of gravitational origin; Einstein, however, assigns a privileged position to these forces, and supposes that all actions of gravitational origin (and only these) are so intimately fused with the geometrical and temporal properties that they are directly determined by the four-dimensional ds^2. Such a possibility is amply justified by the considerations set forth in the preceding chapter (pp. 291–328). All the other non-gravitational forces (in particular, actions of electromagnetic origin), on the contrary, can be absorbed into the energy tensor. In order to put this view in a mathematical form, Einstein had to establish a relation between the ds^2 (i.e. its ten coefficients) and the energy tensor (i.e. the ten functions T_{ik}); he had therefore to determine ten equations. One of these was a necessary consequence, at least approximately, of the Newtonian theory. In the classical mechanics space is considered rigorously Euclidean, and by Newton's law the density ρ of the attracting matter determines the field of force by means of the Newtonian potential

$$U = f\int \frac{\rho\, dS}{r},$$

where f is the gravitation constant, and the meaning of the other symbols is as usual. From this expression for U Poisson's equation

$$\Delta_2 U = -4\pi f\rho$$

follows in the ordinary way for every point of the field. Since the density ρ differs from the element T_{00} of the energy tensor only by a constant multiplier (pp. 349, 354, §§ 22, 24, and 25), while to a first approximation (cf. § 4, p. 291) we have

$$g_{00} = 1 - \frac{2U}{c^2},$$

it follows that Poisson's equation establishes a relation between the energy tensor and a sum of second derivatives of g_{00}.

The differential equations expressing the relation between the coefficients g_{ik} of ds^2 and the quantities T_{ik} must therefore include this relation, at least to a first approximation. A reasonable induction suggests that in order to construct the ten required equations we must equate the ten components T_{ik} of the energy tensor to ten differential expressions of the second order in the coefficients g_{ik}, which, the system being invariant, must themselves constitute a tensor. Now a double tensor of the second order is given by those combinations of the Riemann-Christoffel tensor which we considered in the preceding section. Accordingly, the procedure which would first occur to one would be to assume that the G_{ik}'s were equal or proportional to the T_{ik}'s; and this was in fact what at his first attempt Einstein did. But immediately afterwards he reflected that the fundamental equations must not impose on the metric properties of space-time any *a priori* limitation, in this sense that any value whatever of ds^2 must be capable of being regarded as theoretically possible provided there is a suitable energy tensor. This property would be inconsistent with the condition that the G_{ik}'s and T_{ik}'s are to be proportional, since the latter tensor, from its physical origin, satisfies four differential conditions expressing the vanishing of its divergence (cf. pp. 351, 359, §§ 23 and 25), so that the G_{ik}'s, would have to be connected by corresponding equations. The idea of a linear relation between the two tensors can however be retained without imposing any differential relation on the g_{ik}'s, since the divergence of the tensor

$$G_{ik} - \tfrac{1}{2} G g_{ik}$$

is identically zero, as we saw in the preceding section. If in fact we put

$$G_{ik} - \tfrac{1}{2} G g_{ik} = -\kappa T_{ik} \quad . \quad . \quad . \quad . \quad (8)$$

where κ denotes a constant (to be subsequently connected with the constant f in Poisson's equation), there will be no resulting differential relations between the g_{ik}'s. These are the celebrated gravitational equations. The foregoing considerations serve merely to give them plausibility from the purely formal point of view; their physical justification follows *a posteriori* from arguments of two kinds, which we shall now explain.

For the moment we consider only a first approximation; i.e. we suppose that ds^2 differs from the pseudo-Euclidean value by a small amount. As we saw in Chapter XI, p. 320, we may on this hypothesis assume

$$\left.\begin{aligned} g_{00} &= 1 - 2\gamma \\ g_{0i} &= -\gamma_i \\ g_{ik} &= -\delta_i^k - \gamma_{ik} \end{aligned}\right\} (i, k = 1, 2, 3),$$

where the γ's are small quantities of the second order. We also saw that (still with the same hypothesis) the equations of motion of a material particle, to a first approximation, depend neither on the γ_i's nor on the γ_{ik}'s, but only on the coefficient g_{00}, or, which is the same thing, on the function γ, and that they in fact reduce to the classical Newtonian equations

$$\ddot{x}_i = \frac{\partial U}{\partial x_i} \quad (i = 1, 2, 3),$$

since
$$\gamma = \frac{U}{c^2}. \quad . \quad . \quad . \quad . \quad . \quad . \quad . \qquad (9)$$

In view of this, the problem of justifying the gravitational equations to a first approximation reduces to that of proving:

(*a*) that one of these equations (the one corresponding to $i = k = 0$) involves only γ (i.e. U) and is identical with Poisson's equation;

(*b*) that the other nine are consistent with values of the functions γ of the assumed order of magnitude: their precise values in this first approximation are a matter of complete indifference, since whatever they may be we in any case arrive back at the Newtonian formulæ.

We can therefore limit the scope of (*a*) and (*b*) to the statical case, for the reasons indicated at the end of § 1.

The passage to a further approximation in the equations of motion of a material particle involves (cf. Chapter XI, p. 320) either the values to a first approximation of γ_i and γ_{ik} or the third-order correction ψ in the expression for g_{00}. It is this difference from the results of the Newtonian laws which, being within the range of astronomical observation, provides a means of testing whether Einstein's hypothesis is or is not superior to its classical predecessor.

At this point we are, so to speak, in conditions analogous to those in which Newton found himself when he substituted for Kepler's kinematical laws the dynamical principle of universal attraction, which was capable not only of including Kepler's laws as a first approximation, but also of predicting, and that on a magnificent scale, new facts which have since found marvellous confirmation. When the relativity theory is substituted for the Newtonian, the phenomena predicted by it are much more minute, but even with present experimental resources, some at least of them are within the reach of experiment. This experimental control provides the second line of argument alluded to above in support of the gravitational equations.

4. General equations of Einsteinian statics. Empty space.

When we are dealing with statical phenomena (cf. Chapter XI, p. 326), the ds^2 of space-time has the form

$$ds^2 = V^2\,dx_0^2 - dl^2 \quad \quad . \quad . \quad . \quad . \quad (10)$$

with

$$dl^2 = \sum_{ik}^{3}{}_{1}\, a_{ik}\,dx_i\,dx_k. \quad . \quad . \quad . \quad . \quad (10')$$

The coefficients a_{ik}, like V, are to be functions of x_1, x_2, x_3, only; V is interpreted (cf. Chapter XI, p. 339) as the velocity of light, and is therefore considered essentially positive.

With obvious meanings for the symbols, we have

$$\left.\begin{aligned} &g_{ik} = -a_{ik},\quad g_{0i} = 0,\quad g_{00} = V^2,\quad g = -a\,V^2, \\ &g^{ik} = -a^{ik},\quad g^{0i} = 0,\quad g^{00} = \frac{1}{V^2},\qquad (i, k = 1, 2, 3). \end{aligned}\right\} . \quad (11)$$

We shall use a dash (′) to denote Christoffel's symbols and the components of the Riemann-Christoffel and Einstein tensors relative to the quaternary form (10), and shall keep the ordinary notation without a dash for the analogous symbols and components relative to (10′).

From the definitions and (11) we get

$$\left.\begin{aligned} &\{ik, l\}' = \{ik, l\}, \\ &\{ik, 0\}' = \{0i, k\}' = \{00, 0\}' = 0, \\ &\{i0, 0\}' = \frac{V_i}{V}, \\ &\{00, i\}' = V\,V^i, \end{aligned}\right\} . \quad . \quad (12)$$

where i, k, l, can take any of the values 1, 2, 3, $V_i = \dfrac{\partial V}{\partial x_i}$, and $V^i = \sum_{j}^{3}{}_{1}\, a^{ij}\,V_j$ is the reciprocal system with respect to the purely spatial dl^2.

We shall next express Riemann's symbols of the second kind for the quaternary ds^2 in terms of the analogous symbols for dl^2 and of V. We have by definition, from formula (3) of Chapter VII, p. 175,

$$\{ir, hk\}' = \frac{\partial}{\partial x_k}\{ih, r\}' - \frac{\partial}{\partial x_h}\{ik, r\}'$$
$$- \sum_{0}^{3}{}_l [\{lh, r\}'\{ik, l\}' - \{lk, r\}'\{ih, l\}'].$$

We shall examine separately the various cases which may occur, according to the number of the indices i, r, h, k which are zero.

(1) No index zero. The first group of (12) gives immediately

$$\{ir, hk\}' = \{ir, hk\}. \quad . \quad . \quad . \quad . \quad (13)$$

(2) A single index zero. Riemann's symbols being anti-symmetrical with respect to the last two indices, we need only examine the three cases in which the zero index is i, r, or h. In each case, from the second group of the formulæ (12) it follows immediately that Riemann's symbols of this type are all zero, or

$$\{0r, hk\}' = \{i0, hk\}' = \{ir, 0k\}' = 0. \quad . \quad (14)$$

(3) Two indices zero. From the general properties of the Riemann-Christoffel tensor the symbols of the type $\{ir, 00\}'$ vanish identically (for any ds^2), and those of the type

$$\{00, hk\}' = \sum_{0}^{3}{}_j g^{j0}(0j, hk)$$

vanish whenever $g^{j0} = 0$ (for $j > 0$), as in our case. There remain therefore to be considered the two types $\{0r, 0k\}'$ and $\{i0, 0k\}'$.

From (12) and the fundamental formula of covariant differentiation with respect to the purely spatial dl^2 we find

$$\left.\begin{aligned} \{0r, 0k\}' &= V(V^r)_k, \\ \{i0, 0k\}' &= \frac{V_{ik}}{V} \end{aligned}\right\} \quad . \quad . \quad . \quad . \quad (14')$$

(4) Three or four indices zero. It will be seen immediately from (12) that these symbols are all zero.

We are now in a position to evaluate explicitly the symmetrical double tensor G'_{ik}, the elements of which, as we know (§ 2), are

$$G'_{ik} = \sum_{0}^{3}{}_h \{ih, hk\}' = \sum_{1}^{3}{}_h \{ih, hk\}' + \{i0, 0k\}'.$$

Introducing the analogous system

$$G_{ik} = \sum_{1}^{3}{}_h \{ih, hk\}$$

relative to the ternary form dl^2, we find at once, using the expressions obtained for the symbols $\{ir, hk\}'$,

$$\left.\begin{aligned} G'_{ik} &= G_{ik} + \frac{V_{ik}}{V}, \\ G'_{0k} &= 0, \\ G'_{00} &= -V\sum_{1}^{3}{}_h V_h^h = -V\Delta_2 V. \end{aligned}\right\} \quad . \quad . \quad . \quad (15)$$

From these formulæ and (11) we get for the linear invariant of the system G'_{ik},

$$\begin{aligned} G' &= \sum_{0}^{3}{}_{ik} g^{ik} G'_{ik} = g^{00} G'_{00} - \sum_{1}^{3}{}_{ik} a^{ik} G'_{ik} \\ &= -2\frac{\Delta_2 V}{V} - \sum_{1}^{3}{}_{ik} a^{ik} G_{ik}. \quad . \quad . \quad . \quad . \quad . \quad (16) \end{aligned}$$

We have already seen (Chapter VII, p. 200) that for a three-dimensional manifold we can with advantage replace the tensor G_{ik} by Ricci's tensor α_{ik}, the linear invariant

$$\mathcal{M} = \sum_{1}^{3}{}_{ik} a^{ik} \alpha_{ik}$$

of which (cf. Chapter VII, p. 203) represents the mean curvature (the sum of the three principal curvatures).

The G_{ik}'s and α_{ik}'s are connected by the linear relations

$$G_{ik} = \alpha_{ik} - \mathcal{M} a_{ik};$$

from this, multiplying by a^{ik} and summing with respect to i, k, there follows in particular

$$G = \mathcal{M} - 3\mathcal{M} = -2\mathcal{M}.$$

Applying these results, (15) and (16) become

$$\left.\begin{aligned} G'_{ik} &= a_{ik} + \frac{V_{ik}}{V} - \mathcal{M} a_{ik}, \\ G'_{i0} &= 0, \qquad (i, k = 1, 2, 3) \\ G'_{00} &= -V\Delta_2 V; \end{aligned}\right\} \quad . \quad . \quad (15')$$

$$\tfrac{1}{2} G' = \mathcal{M} - \frac{\Delta_2 V}{V}, \quad . \quad . \quad . \quad . \quad (16')$$

which provide convenient expressions for the components of the Einstein tensor and its linear invariant in statical conditions.

We can now return to the gravitational equations (8) of the preceding section. We note in the first place that since in statical conditions there is no energy flux, the components T_{0i} vanish. Hence from (11) and (15′) three of these equations reduce to pure identities, and there remain seven; six of these, corresponding to non-zero values of the indices, have the form

$$a_{ik} + \frac{V_{ik}}{V} - \frac{\Delta_2 V}{V} a_{ik} = -\kappa T_{ik} \qquad (i, k = 1, 2, 3) \quad (17)$$

in virtue of (15′), (16′), and (11), while the seventh, for $i = k = 0$, is

$$-V\Delta_2 V - \tfrac{1}{2} G' g_{00} = -\kappa T_{00},$$

or, from (15′) and (16′),

$$\mathcal{M} = \kappa \frac{T_{00}}{V^2}. \quad . \quad . \quad . \quad . \quad . \quad . \quad (18)$$

These seven equations[1] (17) and (18), as is naturally to be expected, reduce the Einsteinian statics to the three dimensions of the associated space. Their form is invariant with respect to the metric of this space, which has the dl^2 in question as its fundamental quadratic form. They also involve, in association with the fundamental form, the two invariant functions V and T_{00}, and the covariant double system T_{ik} $(i, k = 1, 2, 3)$. The latter characterizes the distribution of the stresses, while $\dfrac{T_{00}}{V^2}$

[1] Cf. LEVI-CIVITA: *Rend. della R. Acc. dei Lincei,* Vol. XXVI (first half-year, 1917), p. 458.

is to be interpreted as the energy density (cf. Chapter XI, p. 357), V representing the velocity of light, as was said at the outset.

With regard to the energy density it is to be observed that no example of a negative density exists,[1] at least within the range of the better-known phenomena to-day, whether material, or electromagnetic in the broad sense. Hence we may assume that the right-hand side of (18) $\geqslant 0$, and we get the following geometrical corollary: *The mean curvature $\mathcal{M}$, determined in physical space as the effect of purely statical phenomena, is in every case either positive or zero.*

An important consequence of the equations (17) is obtained on multiplying them by a^{ik} and summing with respect to the two indices. Using the definition of $\mathcal{M}$ and (18) we get

$$\frac{\Delta_2 V}{V} = \tfrac{1}{2}\kappa\left(\boldsymbol{T} + \frac{T_{00}}{V^2}\right), \quad . \quad . \quad . \quad . \quad (19)$$

where

$$\boldsymbol{T} = \sum_{1}^{3}{}_{ik}\, a^{ik}\, T_{ik} \quad . \quad . \quad . \quad . \quad . \quad (20)$$

and obviously represents the linear invariant of the system of stresses with respect to our dl^2 (of the associated space). It may be remarked incidentally that this invariant must not be confused with the scalar invariant of the four-dimensional tensor, namely,

$$T = \sum_{0}^{3}{}_{ik}\, g^{ik}\, T_{ik},$$

the value of which, from (11), is on the contrary

$$T = \frac{T_{00}}{V^2} - \boldsymbol{T}.$$

Consider in particular a region of space in which all the components of the energy tensor vanish (empty space). From the physical point of view, this condition can be considered satisfied when the region in question contains neither ordinary

[1] In fact, if at a given point there is matter at rest distributed with density ρ, this implies an energy $c^2\rho$ of material origin, which in normal conditions enormously outweighs all other possible contributions to the total. Moreover, the electromagnetic contribution to the energy density also ≥ 0. Hence even when there is no matter it does not seem possible for the energy density to have a negative value.

matter nor electromagnetic energy, since in this case it follows from the mechanics of material media that the stresses of material origin vanish, and, from Maxwell's theory, that the Maxwellian field of force vanishes, and therefore also the Maxwellian stress.[1]

With this hypothesis the equations (17), in view of (19), plainly reduce to the form

$$\Delta_2 V = 0, \quad . \quad . \quad . \quad . \quad . \quad . \quad . \quad . \quad (21)$$

$$a_{ik} + \frac{V_{ik}}{V} = 0 \qquad (i, k = 1, 2, 3), \quad . \quad . \quad (22)$$

the first of which shows that not the timelike coefficient $g_{00} = V^2$ itself, but its square root, is a harmonic function. Also, (18) gives at once

$$\mathcal{M} = 0. \quad . \quad . \quad . \quad . \quad . \quad . \quad (21')$$

If the energy tensor were zero throughout *all* space, it is intuitive from the physical standpoint that the Einsteinian ds^2 would be rigorously pseudo-Euclidean, and therefore the associated space rigorously Euclidean. This in fact represents the starting-point of Einstein's speculative construction, which assigns any deviation from a pseudo-Euclidean metric to those physical actions which are included in the energy tensor. Serini [2] too has given a rigorous proof of the hypothesis, based on equations (21) and (22).

5. First approximation. Connexion with Poisson's equation.[3]

If we suppose that the expression (10) for ds^2 differs by very little from the Euclidean type referred to Cartesian space coordinates and Römerian time

$$ds^2 = dy_0^2 - \sum_{1}^{3}{}_i \, dy_i^2$$

we can put (cf. § 3, and Chapter XI, § 10, p. 320)

$$V = 1 - \gamma \quad . \quad . \quad . \quad . \quad . \quad (23)$$

$$a_{ik} = \delta_i^k + \gamma_{ik} \qquad (i, k = 1, 2, 3). \quad . \quad . \quad (24)$$

[1] See e.g. JEANS: *The Mathematical Theory of Electricity and Magnetism*, fifth edition, 1925, Chap. VI, Cambridge University Press.

[2] *Rend. della R. Acc. dei Lincei*, Vol. XXVII (first half-year, 1918), p. 285.

[3] LEVI-CIVITA, loco cit., and *ibidem* (second half-year, 1917), pp. 307–317.

We thus have

$$dl^2 = \sum_{1}^{3}{}_{ik}\, a_{ik}\, dy_i\, dy_k = dl_0{}^2 + \sum_{1}^{3}{}_{ik}\, \gamma_{ik}\, dy_i\, dy_k, \qquad (24')$$

where $dl_0{}^2$ is the line element of ordinary Euclidean space referred to Cartesian co-ordinates.

The quantities γ_{ik} are pure numbers, like γ, and the qualitative property we have assigned to ds^2 is equivalent, to a first approximation, to treating all these seven quantities as infinitesimals.

It follows that Christoffel's symbols

$$[ih, j] = \tfrac{1}{2}\left(\frac{\partial a_{ij}}{\partial y_h} + \frac{\partial a_{jh}}{\partial y_i} - \frac{\partial a_{ih}}{\partial y_j}\right)$$

are also infinitesimal. Since to the same order of approximation the quantities a_{jr} keep their Euclidean values δ_j^r, it will be seen that the symbols of the second kind

$$\{ih, r\} = \sum_{1}^{3}{}_j\, a^{jr}\, [ih, j]$$

do not differ appreciably from the homologous symbols $[ih, r]$ of the first kind. It follows that from the definition of Riemann's symbols (p. 175, formula (3)) we have, neglecting terms of higher order,

$$\begin{aligned} \{ir, hk\} &= \frac{\partial}{\partial y_k}\{ih, r\} - \frac{\partial}{\partial y_h}\{ik, r\} \\ &= \frac{\partial}{\partial y_k}[ih, r] - \frac{\partial}{\partial y_h}[ik, r] \\ &= \tfrac{1}{2}\left\{\frac{\partial^2 a_{ik}}{\partial y_r \partial y_h} + \frac{\partial^2 a_{rh}}{\partial y_i \partial y_k} - \frac{\partial^2 a_{ih}}{\partial y_r \partial y_k} - \frac{\partial^2 a_{rk}}{\partial y_i \partial y_h}\right\}. \end{aligned}$$

Hence it follows, from the definition of the G_{ik}'s (cf. § 4) and from (24), that

$$\begin{aligned} G_{ik} &= \sum_{1}^{3}{}_h \{ih, hk\} \\ &= \tfrac{1}{2}\sum_{1}^{3}{}_h \left\{\frac{\partial^2 \gamma_{ik}}{\partial y_h{}^2} + \frac{\partial^2 \gamma_{hh}}{\partial y_i \partial y_k} - \frac{\partial^2 \gamma_{ih}}{\partial y_h \partial y_k} - \frac{\partial^2 \gamma_{hk}}{\partial y_i \partial y_h}\right\}. \end{aligned}$$

We now return to the statical equations (17) and (18). We

have already made them contain Ricci's symbols a_{ik} instead of the G_{ik}'s, the relation between the two being

$$a_{ik} = G_{ik} + \mathcal{M}\, a_{ik}.$$

It is to be noted that $\mathcal{M}$ is now to be considered infinitesimal, like the G_{ik}'s and their linear invariant, so that, from (18), T_{00} is also infinitesimal. Replacing $\mathcal{M}$ by its value (18), the explicit expressions which represent the a_{ik}'s to a first approximation take the form

$$\left.\begin{array}{c} a_{ik} = \frac{1}{2}\sum_1^3{}_h \left\{\frac{\partial^2 \gamma_{ik}}{\partial y_h^2} + \frac{\partial^2 \gamma_{hh}}{\partial y_i \partial y_k} - \frac{\partial^2 \gamma_{ih}}{\partial y_h \partial y_k} - \frac{\partial^2 \gamma_{hk}}{\partial y_i \partial y_h}\right\} + \kappa\, T_{00}\,\delta_i^k \\ (i, k = 1, 2, 3) \end{array}\right\}. \quad (25)$$

Using this result, and noting further that, neglecting infinitesimals of higher order, the covariant derivatives of $V = 1 - \gamma$ do not differ from the ordinary derivatives, so that in particular

$$\Delta_2 V = -\Delta_2^0 \gamma = -\sum_1^3{}_i \frac{\partial^2 \gamma}{\partial y_i^2},$$

we find that (17) and (19) can be written as

$$a_{ik} - (\gamma)_{ik} + \Delta_2^0 \gamma\, \delta_i^k = -\kappa\, T_{ik}, \quad . \quad . \quad . \quad (26)$$

$$\Delta_2^0 \gamma = -\tfrac{1}{2}\kappa\,(\boldsymbol{T} + T_{00}), \quad . \quad . \quad . \quad . \quad (27)$$

where the symbols $(\gamma)_{ik}$ denote covariant derivatives of γ; and in particular, in empty space, since the terms on the right vanish, they become

$$a_{ik} = (\gamma)_{ik}, \quad . \quad . \quad . \quad . \quad . \quad . \quad (26')$$

$$\Delta_2^0 \gamma = 0, \quad . \quad . \quad . \quad . \quad . \quad . \quad . \quad (27')$$

which to a first approximation, as is naturally to be expected, are identical with (22) and (21).

At this point we must consider the mechanical significance of the function γ, or, better, of the product $c^2\gamma$. On p. 293, Chap. XI, when dealing with Einstein's modification of Hamilton's principle, we saw that, when ds^2 is very close to the pseudo-Euclidean form, the difference between g_{00} and unity is $-2\frac{U}{c^2}$

to a first approximation, U being the potential of the field of force in which the motion takes place. In the present case this difference is -2γ, so that we have

$$\gamma = \frac{U}{c^2}. \quad . \quad . \quad . \quad . \quad . \quad . \quad (9)$$

This conclusion could of course also have been deduced from the general proposition in § 12 of Chapter XI, p. 328, that $-\frac{1}{2}c^2 V^2$ (together with a non-essential additive constant) constitutes the potential function of the force exerted in the field in statical conditions. In our case $V^2 = 1 - 2\gamma$, and therefore

$$-\tfrac{1}{2}c^2 V^2 = -\tfrac{1}{2}c^2(1-2\gamma) = -\frac{c^2}{2} + c^2\gamma,$$

which proves the required result.

Now let us for the moment again take the standpoint of the classical mechanics, and consider the field of force due to a generic distribution of matter of density $\rho = \frac{\epsilon}{c^2}$, where ϵ is the corresponding energy density. If U is the Newtonian potential of this field, we know that Poisson's equation

$$\Delta_2 U = -4\pi f\rho = -\frac{4\pi f}{c^2}\epsilon$$

holds, f being the coefficient of universal attraction. If on the other hand we take the standpoint of general relativity, the same distribution of matter gives a ds^2 for which $\gamma = \frac{U}{c^2}$, and an energy tensor whose component T_{00} coincides with ϵ, while the components T_{0i} vanish in statical conditions, so that the remaining components T_{ik} represent stresses (cf. Chapter XI, p. 358). If we are dealing with discrete matter, the components T_{ik}, and therefore also their invariant T, are zero, and (27) becomes

$$\Delta_2 U = -\tfrac{1}{2}c^2\kappa\epsilon.$$

In order that this may be identical with Poisson's equation, it is necessary and sufficient that the constant κ of the gravitational equations and the universal constants $f = 6{\cdot}7 \times 10^{-8}$

and $c = 3 \times 10^{10}$ (in C.G.S. units) of the classical mechanics should be connected by the relation

$$\kappa = \frac{8\pi f}{c^4} \quad . \quad . \quad . \quad . \quad . \quad . \quad (28)$$

which gives in round numbers (C.G.S. units)

$$\kappa = 2 \times 10^{-48}.$$

For the remainder of the argument we shall adopt this value of κ, and shall definitely take up the standpoint of relativity. In relation to the remarks in § 3 we can at this point consider that the preliminary justification of the gravitational equations is terminated. In fact, their first approximation is represented in statical conditions by (26) and (27). The equation (27), as we have now proved, is identical with Poisson's equation; the equations (26), as we shall see in the following section, serve to determine the quantities γ_{ik}, which to a first approximation, as we have already said, do not influence the motion, but will become essential when we come to discriminate on a more refined scale between the Newtonian mechanics and the relativity theory. Here we have referred specifically to the statical case, but the justification of the gravitational equations obtained in this case also holds good, as already pointed out in § 3, in the general case, provided the coefficients γ_i of the product terms in $dx_0\, dx_i$ $(i = 1, 2, 3)$ are of order higher than the first. We have arrived at this condition by a process of induction from experimental facts, and have used it to reduce the ten gravitational equations to the seven of (17) and (18). We are now so to speak at the deductive stage, and must first show that the gravitational equations contain in synthesis all the facts to a first approximation; and at this stage we must point out that in ordinary conditions of material motion (i.e. with velocities which are small compared with that of light) the three gravitational equations

$$G_{0i} - \tfrac{1}{2}\, G\, g_{0i} = -\kappa\, T_{0i} \qquad (i = 1, 2, 3), \quad . \quad (29)$$

which are rigorously true in statical conditions, continue to hold to a first approximation if we suppose the quantities γ_i of order higher than the second (that of γ and of the γ_{ik}'s). In fact, the left-hand side of these equations, as we have already seen (cf. § 4), becomes identically zero when we put $g_{0i} = 0$: this means that if the three g_{0i}'s are treated as quantities γ_i of a certain order of smallness, the left-hand side of (29) will be of at least the same order.[1] If therefore we suppose that the γ_i's are of order higher than the second, the left-hand side of (29) will also be of the same order, and therefore zero to a first approximation. As regards the right-hand side, we know (cf. Chapter XI, p. 356) that in a pseudo-Euclidean metric, and therefore (neglecting terms of higher order) also in the case we are considering,

$$T_{0i} = -\epsilon\beta_i = -T_{00}\beta_i;$$

and hence, from the presence of the factor β_i, it follows that T_{0i} is of higher order of smallness than T_{00}, and therefore that the right-hand side of (29) is of higher order than $-\kappa T_{00}$, and is therefore zero to a first approximation. Hence, in these conditions, the equation (29) is satisfied.

6. The Einsteinian ds^2 which corresponds to a first approximation to an assigned Newtonian field.

Suppose a Newtonian field and its potential U given. From the remark made in § 1, we can ignore the possibility (consequent on the motion of the material masses) that U may depend explicitly on the time, and treat U only as a function of the space co-ordinates, as if the masses were at rest in the positions they occupy at the instant considered. Consider a region not occupied by attracting masses, in which region $\Delta_2 U = 0$. In order to characterize the corresponding Einsteinian ds^2 to a first approximation we have to determine (cf. § 5) the functions γ and γ_{ik}, where γ is given by $\gamma = \frac{U}{c^2}$, and is therefore harmonic (i.e. a solution of (27′)), and the γ_{ik}'s have to satisfy (26′), which can also be written in the simpler form

$$a_{ik} = \frac{\partial^2 \gamma}{\partial y_i \partial y_k}, \quad . \quad . \quad . \quad . \quad . \qquad (26'')$$

[1] The quickest way of showing this is to suppose that the g_{0i}'s are of the form $h\gamma_i^*$, where h is a numerical coefficient determining the order of magnitude, and the γ_i^*'s are functions of position and of the time, to be treated as finite quantities together with their first derivatives. It is clear in this case that the left-hand side of (29) contains h as a factor.

since the covariant derivatives which would occur on the right would differ from the ordinary derivatives by terms of higher order, and can therefore be replaced by these ordinary derivatives.

For the integration of these equations we note in the first place that we get a particular solution by taking

$$\gamma_{ik} = 2\,\delta_i^k\,\gamma. \quad . \quad . \quad . \quad . \quad . \quad (30)$$

The proof follows immediately from the expression (25) for the a_{ik}'s, in which T_{00} is of course put equal to zero. Substituting for a_{ik} in (26″) the values

$$\tfrac{1}{2}\sum_{1}^{3}{}_h \left\{ \frac{\partial^2 \gamma_{ik}}{\partial {y_h}^2} + \frac{\partial^2 \gamma_{hh}}{\partial y_i \, \partial y_k} - \frac{\partial^2 \gamma_{ih}}{\partial y_h \, \partial y_k} - \frac{\partial^2 \gamma_{hk}}{\partial y_i \, \partial y_h} \right\}, \quad . \quad (29')$$

and remembering that γ is harmonic, the required result follows.

Since then the equations (26″) constitute a linear non-homogeneous system in the γ_{ik}'s, the general integral is obtained by adding the solution (30) to the most general solution of the equations with the right-hand side zero, i.e.

$$a_{ik} = 0.$$

The general integral of this system could easily be constructed by using the result (cf. Chapter VII, p. 200) that for a three-dimensional manifold the vanishing of Ricci's symbols a_{ik} implies that all Riemann's symbols are likewise zero, or in other words that the quantities

$$a_{ik} = \delta_i^k + \gamma_{ik}$$

are the coefficients of a Euclidean dl^2 (referred to any curvilinear co-ordinates whatever). But, as it happens, the addition to the particular solution (30) of the general integral of the homogeneous system has no interest, since, as we shall see shortly, this corresponds merely to a change of the co-ordinates of reference.

In fact, the vanishing of the symbols a_{ik}, as we have just pointed out, expresses the necessary and sufficient condition that dl^2 should be Euclidean, i.e. reducible, with a suitable choice of parameters, to the form $\sum_{1}^{3}{}_i \, dy_i^2$. Hence, if x_1, x_2, x_3 denote the co-ordinates of reference in their most general form, the most general method of defining a Euclidean dl^2, with respect to these co-ordinates x, will evidently be to introduce a transformation of any kind

$$y_i = y_i(x_1, x_2, x_3) \qquad (i = 1, 2, 3)$$

between the y's and the x's, and to take for the coefficients a_{ik} those which result from expressing $\sum_1^3{}_i\, dy_i^2$ in terms of the differentials of the x's.

Assuming the functions $y_i(x_1, x_2, x_3)$ in the form

$$x_i + \xi_i(x_1, x_2, x_3),$$

as is always legitimate, and inserting the corresponding differentials in the trinomial $\sum_1^3{}_i\, dy_i^2$, we get

$$dl^2 = \sum_1^3{}_{ik}\, a_{ik}\, dx_i\, dx_k$$

where $$a_{ik} = \delta_i^k + \left(\frac{\partial \xi_i}{\partial x_k} + \frac{\partial \xi_k}{\partial x_i}\right) + \sum_1^3{}_j \frac{\partial \xi_j}{\partial x_i}\frac{\partial \xi_j}{\partial x_k}.$$

In order to take account of the condition that the difference $a_{ik} - \delta_i^k = \gamma_{ik}$ is limited to the first order, together with the further condition that the difference between the Cartesian co-ordinate system of the y's and the curvilinear system of the x's is to be of the same order,[1] it is sufficient (and necessary) that we should be able to treat the functions ξ and their derivatives as infinitesimals. It follows that

$$\gamma_{ik} = \left(\frac{\partial \xi_i}{\partial x_k} + \frac{\partial \xi_k}{\partial x_i}\right), \quad . \quad . \quad . \quad . \quad (31)$$

which constitutes the formal expression for the general integral of the homogeneous system $a_{ik} = 0$ (the a_{ik}'s, in the form (29′), being linearly dependent on the γ_{ik}'s).

[1] If this condition is not imposed, the only necessary condition is that the six numerical quantities

$$\gamma_{ik} = \left(\frac{\partial \xi_i}{\partial x_k} + \frac{\partial \xi_k}{\partial x_i}\right) + \sum_1^3{}_j \frac{\partial \xi_j}{\partial x_i}\frac{\partial \xi_j}{\partial x_k}$$

should be infinitesimal, and this can be secured, as Prof. ALMANSI has shown (cf. "L'ordinaria teoria dell' elasticità e la teoria delle deformazioni finite", in *Rend. della R. Acc. dei Lincei*, Vol. XXVI (second half-year, 1917), pp. 3–8), even when the quantities ξ are not themselves infinitesimal.

But it is not this formal expression with which we are concerned, but rather the circumstance that the term (31), which is to be added to (30) in order to get the general integral of the system (26″) with the right-hand side not zero, can always be made equal to zero by a suitable change of co-ordinates; this change being the substitution for the x's of the combinations

$$y_i = x_i + \xi_i(x_1, x_2, x_3), \quad . \quad . \quad . \quad (32)$$

the result of which is that the expression for dl^2 reduces, by construction, to $\sum_1^3 {}_i\, dy_i^2$, all the differences $a_{ik} - \delta_i^k$ vanishing.

When the y's are chosen as variables, the transformation (32) must naturally be applied also to the particular solution (30). But since the ξ's are to be considered infinitesimal equally with γ, (32) reduces, so far as (30) is concerned, to the mere substitution of the y's for the x's. The expression (30) for the particular solution, which alone is of any interest for our purpose, thus remains unaltered when the system of reference is changed to the y's.

It is further to be noted that the elementary form $\Delta_2^0\gamma$ (the sum of the second derivatives) of the parameter also remains unaltered.

From the foregoing arguments we see that *in an empty field the statical potential* U (Newtonian to a first approximation) *is associated with a metric modification of the associated three-dimensional space.* With a suitable choice of the co-ordinates of reference (the y's just defined) we have

$$\gamma = \frac{U}{c^2},$$

with γ harmonic (in the y's as well as in the x's); the values of the coefficients a_{ik} of the square of the line element are given (to the same degree of approximation) by the expression $\delta_i^k(1 + 2\gamma)$, so that

$$dl^2 = (1 + 2\gamma)(dx_1^2 + dx_2^2 + dx_3^2).$$

It will be seen that in general the space does not remain Euclidean even to a first approximation, but, to this degree of approximation, can only be conformally represented in a Euclidean space.

To sum up, remembering that $g_{00} = 1 - 2\gamma$ and that $g_{0i} = 0$, the ds^2 of the Einsteinian space-time belonging to an assigned Newtonian field of force with potential $U = c^2\gamma$ is given by

$$ds^2 = (1 - 2\gamma)\, dx_0{}^2 - (1 + 2\gamma)\, dl_0{}^2, \quad . \quad . \quad (33)$$

where dl_0 is the line element of a Euclidean space.

In the case of a single point-mass m_0 we need of course only take

$$\gamma = \frac{fm_0}{c^2} \frac{1}{r},$$

where r represents the distance between the mass and the point at which the attraction acts.

7. Further approximation for the coefficient $g_{00} = V^2$ in statical conditions.

In the preceding chapter (p. 320) we saw that if the ds^2 of space-time is not far removed from being pseudo-Euclidean, then the motion of a material particle is affected only to a first approximation by the second-order difference $2\gamma = \frac{2U}{c^2}$ between g_{00} and unity, so that the results are the same as for the Newtonian theory. If, however we wish to proceed to a further approximation, i.e. to calculate the principal part of the Einsteinian correction to be applied to the laws of the classical mechanics, we must not only find the second-order quantities γ_{ik}, which are the differences between the a_{ik}'s and the Euclidean values (the γ_i's being of higher order), but we shall also need an evaluation of $g_{00} = V^2$ carried to the fourth order.

This is easily found if we limit the investigation to the statical case and to a portion of empty space (with the energy tensor zero). The differential equation (21) of § 4 is then rigorously true, i.e.

$$\Delta_2 V = 0, \quad . \quad . \quad . \quad . \quad . \quad . \quad (21'')$$

it being of course understood that Δ_2 refers to the spacelike dl^2.

To a first approximation, as has already been seen, we have

$$V^2 = 1 - 2\gamma,$$

where γ is proportional to the potential of the field, by (9). We shall accordingly have to put

$$V = 1 - \gamma - \psi, \quad . \quad . \quad . \quad . \quad (34)$$

where ψ is to be of higher order than the second. The explicit expression of Δ_2 in generic co-ordinates (cf. Chapter VI, p. 154)

$$\Delta_2 V = \frac{1}{\sqrt{a}} \sum_1^3{}_l \frac{\partial}{\partial x_l} (\sqrt{a}\, V^l)$$

gives in the first place, by (21″) and (34),

$$\sum_1^3{}_l \frac{\partial}{\partial x_l} (\sqrt{a}\, \psi^l) = - \sum_1^3{}_l \frac{\partial}{\partial x_l} (\sqrt{a}\, \gamma^l). \quad . \quad (21''')$$

From this we have to find ψ to a fourth-order approximation. As γ is already of the second order, in calculating γ^l $\left(\text{from } \gamma_l = \frac{\partial \gamma}{\partial x_l}\right)$ and $\sqrt{a}$ we need only consider terms as far as the second order, i.e. we can use the form (cf. formula (33))

$$dl^2 = (1 + 2\gamma) dl_0{}^2.$$

This gives

$$\sqrt{a} = (1 + 2\gamma)^{\frac{3}{2}},$$

$$\gamma^l = \frac{\gamma_l}{1 + 2\gamma},$$

whence, neglecting terms of higher order than the fourth,

$$\sqrt{a}\, \gamma^l = \gamma_l (1 + \gamma) = \gamma_l + \tfrac{1}{2}(\gamma^2)_l.$$

A priori we do not yet know the order (by hypothesis certainly higher than the second) of the additional term ψ, which we have to calculate not only as far as its principal part of order ν, but also so as to include additional terms, if any, up to the fourth order inclusive. For the moment we shall consider the part of order ν. On the left-hand side of (21‴) we can substitute 1 for $\sqrt{a}$, the difference between these two quantities being of the second order, which is equivalent to neglecting terms of order $\nu + 2$. As γ is harmonic, it follows that

$$\Delta_2^0 (\psi + \tfrac{1}{2}\gamma^2) = 0$$

where Δ_2^0 as before represents Laplace's operator $\sum_{1}^{3} l \dfrac{\partial^2}{\partial x_l^2}$, whence

$$\psi = -\tfrac{1}{2}\gamma^2 + \text{a harmonic function.}$$

With suitable hypotheses as to qualitative behaviour, it will be seen that the additional harmonic function must vanish, and there remains

$$\psi = -\tfrac{1}{2}\gamma^2$$

as the principal term of the function ψ. As this is already of the fourth order, we can take $-\tfrac{1}{2}\gamma^2$ as the expression for ψ correct to the fourth order inclusive.

To the same order of approximation we get

$$\begin{aligned} g_{00} &= V^2 = (1 - \gamma + \tfrac{1}{2}\gamma^2)^2 \\ &= 1 - 2\gamma + 2\gamma^2. \qquad (35) \end{aligned}$$

8. **A theorem of mechanical equivalence.**[1]

From the two preceding sections it follows that to a sufficient degree of approximation the Einsteinian ds^2 which corresponds to a statical Newtonian field of potential U, fixed in advance, is given by

$$ds^2 = (1 - 2\phi)dy_0^2 - (1 + 2\gamma)dl_0^2, \qquad (36)$$

where

$$\gamma = \frac{U}{c^2}, \qquad (37)$$

$$\phi = \gamma - \gamma^2 \qquad (37')$$

(cf. formula (35) in the preceding section).

In (36) we are satisfied with the first approximation for the coefficients of the spacelike dl_0^2, while for V^2 the part which is of the fourth order is also given. This formula is a particular case of the ds^2 considered on p. 320 of Chapter XI (formula (27)). In order to define the motion of a material particle, i.e. the geodesics of a space-time of this kind, in accordance with the criteria of § 10, p. 320, we note first of all that (36) gives us

$$\frac{ds^2}{dy_0^2} = 1 - 2\phi - (1 + 2\gamma)\left(\frac{dl_0}{dy_0}\right)^2;$$

[1] Cf. Levi-Civita, *Rend. Acc. Lincei*, Series VI, Vol. IV, 1926, pp. 3–5.

comparing this with the equations (28) on p. 321, and noting that $\left(\frac{dl_0}{dy_0}\right)^2$ is identical with $\beta^2 = \frac{v^2}{c^2}$, we see that it corresponds to the particular case in which the linear form T_1 vanishes and the quadratic form T_2 reduces to $\gamma\beta^2$. This brings us back to the case considered in § 11, p. 323, the necessary values for the symbols then used being

$$\chi = 2\gamma = \frac{2U}{c^2}, \quad \psi = -\gamma^2 = -\frac{U^2}{c^4}.$$

Equation (31) on p. 325 gives

$$U_1 = \left(1 + 4\frac{E}{c^2}\right)U + 3\frac{U^2}{c^2}, \quad . \quad . \quad . \quad (38)$$

which leads to the following theorem: *The trajectories of the Einsteinian motion coincide to a second approximation with those of a Newtonian motion in ordinary Euclidean space for which the total energy is still* E *and the force is derived from the potential* U_1.

If t_1 is the ordinary time in this auxiliary Newtonian problem, the corresponding integral of *vis viva* is

$$\tfrac{1}{2}\left(\frac{dl_0}{dt_1}\right)^2 - U_1 = E.$$

This integral can be put in a more convenient form for the purpose we have in view. From equation (31) on p. 325, neglecting terms of higher order, $U_1 + E$, which we shall call U^*, can be written in the form

$$U^* = (U + c^2\psi + E)\left(1 + \frac{2U}{c^2} + \chi\right),$$

whatever may be the values of ψ and χ. In our case, since $\chi = \frac{2U}{c^2}$, $\psi = -\frac{U^2}{c^4}$, we shall have

$$\tfrac{1}{2}\left(\frac{dl_0}{dt_1}\right)^2 = \left(U - \frac{U^2}{c^2} + E\right)\left(1 + \frac{4U}{c^2}\right). \quad . \quad (39)$$

Further, we saw in the section referred to that for the

Einsteinian motion, to the assigned degree of approximation, there exists the integral

$$\tfrac{1}{2}v^2\left(1 + \frac{2U}{c^2} + \chi\right) - (U + c^2\psi) = E.$$

If t is the variable which acts as the time in this problem, $v^2 = \left(\frac{dl_0}{dt}\right)^2$. Substituting for χ and ψ their values, we can write this as

$$\tfrac{1}{2}\left(\frac{dl_0}{dt}\right)^2\left(1 + \frac{4U}{c^2}\right) = U - \frac{U^2}{c^2} + E.$$

From this and from (39) we get the differential relation

$$dt = dt_1\left(1 + \frac{4U}{c^2}\right);$$

when the Newtonian problem is completely solved, this relation enables us to find also the law of the time in the Einsteinian motion.

9. Motion of the planets according to Einstein, to a second approximation. Displacement of perihelion.

The most striking application of the foregoing result is to the problem of the motion of the planets round the sun. If we treat the planets (as is in fact usually done as a first approximation) as material particles with mass so small compared with the sun that they do not perceptibly affect the field (or more generally the four-dimensional metric associated with the field), then our problem is essentially that solved in the preceding section, for the particular case in which the function U is the potential of a single mass m_0 (the sun) which can be taken as the origin O of the co-ordinates. We have therefore, as in § 6,

$$U = \frac{fm_0}{r},$$

where r is the distance between the sun and the planet, measured as if the space between the two were rigorously Euclidean. We know from the preceding section that as regards the trajectory everything happens as if the ordinary mechanics held and the planet were acted on by a unitary central force derived from the potential (38). This consists of two terms, the first of which, $\left(1+\frac{4E}{c^2}\right)U$, corresponds to an attraction inversely proportional to r^2, of radial component

$$\left(1+\frac{4E}{c^2}\right)\frac{dU}{dr}=-\frac{k}{r^2},$$

where for brevity we have put

$$k=\left(1+\frac{4E}{c^2}\right)fm_0; \quad . \quad . \quad . \quad . \quad . \quad (40)$$

and the second to a disturbing force, also central, but inversely proportional to r^3, of radial component

$$\frac{3}{c^2}\frac{dU^2}{dr}=-\frac{k_1}{r^3},$$

where

$$k_1=6\left(\frac{fm_0}{c}\right)^2. \quad . \quad . \quad . \quad . \quad . \quad (41)$$

There are thus two modifications of the Newtonian law: (1) a change in the coefficient of proportionality, which becomes $fm_0\left(1+\frac{4E}{c^2}\right)$ instead of fm_0; (2) a disturbing force (of the second order relative to the Newtonian force) inversely proportional to the cube of the distance, and therefore of the type already considered by Newton. Now it is known from the theory of central forces[1] that for motion in a plane under a force whose radial component is

$$-\frac{k}{r^2}-\frac{k_1}{r^3},$$

the equation of the orbit in polar co-ordinates r, θ can be put in the form

$$r=\frac{p}{1+e\cos\alpha\theta} \quad . \quad . \quad . \quad . \quad (42)$$

[1] See e.g. LEVI-CIVITA and AMALDI: *Lezioni di Meccanica Razionale*, Vol. II, p. 200 (Bologna, Zanichelli, 1926); or LAMB: *Dynamics*, second edition, Chap. XI, § 91 (Cambridge University Press, 1923).

by a suitable choice of the direction of the polar axis, where, G being the area-constant,

$$\alpha^2 = 1 - \frac{k_1}{G^2}, \quad \frac{1}{p} = \frac{k}{G^2 - k_1},$$

and e is a constant of integration, which can always be supposed positive, θ being if necessary replaced by $\theta + \pi$.

All this holds generally. Now suppose in particular that $e < 1$, denoting elliptic motion as a first approximation (i.e. for $k_1 = 0$, so that α reduces to unity). We can also suppose $e > 0$, which means that we exclude the case of the circular orbit. With this limitation, θ in equation (42) can be made to vary without restriction, and the equation shows that when θ increases by $\frac{2\pi}{\alpha}$, r again takes the same value. This holds in particular for the minimum value of r (i.e. perihelion); and therefore, for two successive passages through perihelion, the anomalies differ by $\frac{2\pi}{\alpha}$. In the particular case $\alpha = 1$ (elliptic orbit with fixed perihelion), the value of this difference is precisely 2π, so that the difference

$$\sigma = \frac{2\pi}{\alpha} - 2\pi = 2\pi\left(\frac{1}{\alpha} - 1\right)$$

represents, in magnitude and sign, the angular displacement of perihelion in one revolution. With the value of α given above, taking into account the smallness of k_1, we have

$$\sigma = \pi \frac{k_1}{G^2}.$$

Since for σ, which is already a correction, we need only a first approximation, we can take for G^2 its Newtonian value [1]

$$G^2 = fm_0 p = fm_0 a(1 - e^2),$$

where a and e denote respectively the semi-major axis and the eccentricity of the orbit. Using the value (41) of k_1, we get for the displacement of perihelion the expression

$$\sigma = \frac{6\pi}{1 - e^2} \frac{fm_0}{ac^2}, \quad \quad (43)$$

which was first calculated by Einstein.

[1] Cf. LEVI-CIVITA and AMALDI: *op. cit.*, p. 212.

In order to adapt the formula to numerical calculation for any planet, we introduce the mean radius a_0 of the earth's orbit, and write (43) in the form

$$\sigma = \frac{6\pi}{1 - e^2} \frac{fm_0}{a_0 c^2} \frac{a_0}{a}.$$

The eccentricity of any planetary orbit being small, we can at once put $e^2 = 0$. The radius of the orbit being a, we know that the velocity v in the orbit is given by

$$\frac{v^2}{a} = \frac{fm_0}{a^2},$$

which expresses the equality of the attraction and the centripetal acceleration. For the earth we have in particular

$$v_0^2 = \frac{fm_0}{a_0},$$

and accordingly (43) becomes

$$\sigma = 6\pi \frac{v_0^2}{c^2} \frac{a_0}{a}. \quad . \quad . \quad . \quad . \quad . \quad (43')$$

The velocity v_0 of the earth in its orbit being practically 30 km. per second, and c being 300,000 km. per second, we have approximately $\frac{v_0}{c} = 10^{-4}$, and therefore

$$\sigma = 6\pi \,.\, 10^{-8} \frac{a_0}{a}.$$

For Mercury, the planet nearest the sun, and therefore evidently showing the most perceptible effect, $\frac{a}{a_0} = 0{\cdot}39$, which gives for σ a little more than one-tenth of a second. Since Mercury completes about 420 revolutions in a century, we thus find for the perihelion of its orbit the centennial displacement of 42″, which corresponds exactly to the difference between the total observed displacement and the amount predicted by ordinary celestial mechanics from the Newtonian theory of the perturbations due to the other planets. It was precisely this residual shift of about 42″ per century which before the birth of the relativity theory could only be explained by introducing hypothetical disturbing forces with constants determined *ad hoc.*

For the other planets, the corresponding calculation naturally gives a much smaller centennial shift, hardly 8·6″ for Venus, 3·8″ for the earth, 1·35″ for Mars, and still less for the others, and the results of observation which are at present available are not accurate enough to provide any basis of comparison with these figures.

10. Displacement of the spectral lines. Deflection of light.

In this section we propose to examine the effect of a field of force on the frequency and the path of light rays. We suppose, as in the preceding section, that the field is statical, with a Newtonian potential U, and we consider regions of the field external to the attracting masses. The effect to a first approximation will be sufficient for our purpose, and we can consequently assume that the expression (33) of § 6

$$ds^2 = (1 - 2\gamma)dx_0{}^2 - (1 + 2\gamma)dl_0{}^2, \quad . \quad . \quad (33)$$

where γ stands for $\dfrac{U}{c^2}$, holds for the four-dimensional ds^2.

Now suppose that a phenomenon which is predominantly timelike (e.g. the vibration of an atom) takes place at a specified point T. If dt is an elementary interval of time in which this phenomenon is considered, and if within this interval the variations dy_i of the space co-ordinates are assumed to be negligible, we shall have from (33) (since $x_0 = ct$)

$$ds_T^2 = (1 - 2\gamma_T)c^2\,dt_T^2,$$

where the suffix T denotes that the values in question are those belonging to the phenomenon at the point T. If the phenomenon takes place instead at another point S we have analogously

$$ds_S^2 = (1 - 2\gamma_S)c^2\,dt_S^2.$$

Now suppose that we have two identically similar phenomena at different points, e.g. the emission of light from two atoms chemically alike and in identical physical conditions. If we admit that in such a case the space-time interval ds^2 will be the same for both, the foregoing formulæ will give

$$\frac{dt_S}{dt_T} = \frac{1-\gamma_T}{1-\gamma_S} = 1-(\gamma_T-\gamma_S).$$

This differential relation between corresponding times of the two phenomena under discussion, expressing the constancy of the ratio $\frac{dt_S}{dt_T}$, naturally implies that the same ratio exists between any finite pair whatever of corresponding intervals, Δt_S and Δt_T; in particular, if the phenomenon considered is periodic, between the respective periods or between the reciprocals of the frequencies ν_S and ν_T. We thus have, neglecting terms of the second order,

$$\frac{\nu_S-\nu_T}{\nu_T} = \gamma_T-\gamma_S = \frac{1}{c^2}(U_T-U_S),$$

which shows that in a gravitational field the variation of the frequency is of sign opposite to that of the potential; hence, in particular, there will be a reduction of the frequency for a given spectral line (and therefore a shift of the line towards the red end of the spectrum) on passing to a region of higher potential.

By way of example, let us compare two monochromatic light rays emitted in the same conditions on the earth T and on the sun S. We can neglect U_T in comparison with U_S and take for U_S (cf. the preceding section) the value

$$U_S = \frac{fm_0}{r_0} = \frac{fm_0}{a_0}\frac{a_0}{r_0},$$

where r_0 denotes the sun's radius. As we saw in the preceding section, we now have

$$\frac{fm_0}{a_0} = v_0^2;$$

the (relative) variation of the frequency, if $\Delta\nu = \nu_T - \nu_S$, is therefore given by

$$\frac{U_S}{c^2} = \frac{\Delta\nu}{\nu} = \frac{v_0^2}{c^2}\frac{a_0}{r_0}, \quad . \quad . \quad . \quad . \quad . \quad (44)$$

and since in round numbers

$$\frac{v_0}{c} = 10^{-4}, \quad \frac{a_0}{r_0} = 200,$$

we get

$$\frac{\Delta\nu}{\nu} = 2 \times 10^{-6}.$$

It was uncertain for some years whether there did in fact exist a shift of this kind towards the red for the solar rays, as compared with corresponding rays emitted from a source on the earth. The most recent measurements by Perot, Fabry, and St. John tend to confirm its existence.

A more remarkable verification has recently been provided by St. John, who, following up a suggestion of Eddington's, has observed analogous displacements in the spectrum of the Companion of Sirius.

We now pass to the consideration of the path of a light ray in a field of force. Along any ray we shall have in the first place (cf. Chapter XI, p. 336) $ds^2 = 0$, and further, the field being statical (Chapter XI, p. 340), Fermat's principle

$$\delta\int dx_0 = 0$$

will also hold.

Since $ds^2 = 0$, the expression (33) for ds^2 gives

$$dx_0{}^2 = \frac{1+2\gamma}{1-2\gamma}\, dl_0{}^2,$$

and therefore, neglecting squares of γ,

$$dx_0 = (1 + 2\gamma)\, dl_0.$$

The rays are therefore defined by the variational equation

$$\delta\int(1 + 2\gamma)\, dl_0 = 0. \qquad (45)$$

At this point we note that in an ordinary Euclidean medium, isotropic but not homogeneous, of refractive index $\mu(y_1, y_2, y_3)$, the geometric path of a ray, by Fermat's principle, is characterized by the variational formula

$$\delta\int\mu\, dl_0 = 0;$$

comparing this with (45) we see that in our field of force, with its ds^2 given by (33), light is propagated as if the space were Euclidean and filled with a medium of refractive index

$$\mu = 1 + 2\gamma.$$

This remark becomes even more expressive if we refer once more to the trajectories of a dynamical problem. In fact, as we have already had occasion to show in § 11, pp. 323–325, the principle of least action leads to the result that the curves (45), or, what comes to the same thing (multiplying by c^2 and remembering the meaning of γ), those for which

$$\delta\int c^2(1 + 2\gamma)\, dl_0 = \delta\int (c^2 + 2U)\, dl_0 = 0, \quad . \quad (45')$$

can be considered as the trajectories of a material particle in ordinary space in a field of potential $\frac{1}{2}c^2(1 + 4\gamma) = \frac{c^2}{2} + 2U$ and with total energy zero, or, if we prefer, in a field of potential $2U$ and with total energy $\frac{c^2}{2}$.

It is interesting to observe that even in the classical mechanics the mere hypothesis of the materialization of energy leads us to predict a curved path for rays in a gravitational field. If in fact we admit that light rays, regarded as lines of flux of energy, are effectively trajectories of material particles, then each of these rays—their mutual reactions being supposed negligible—ought to behave like a free material particle moving under the action of the force in the field (of potential U) with a velocity which tends to c at an infinite distance from the attracting masses (i.e. for $U = 0$), or, which comes to the same thing, with total energy $\frac{1}{2}c^2$ per unit mass. It will be seen that general relativity implies, to a first approximation, solely the substitution of $2U$ for U. Now apply these considerations to the path of the rays in the sun's gravitational field. In accordance with the above remarks, these rays are to be considered the trajectories in the problem of the motion of a point attracted by a fixed centre of force, the potential, with the same notation as before, being

$$2U = \frac{2fm_0}{r}$$

and the total energy

$$E = \tfrac{1}{2}c^2.$$

These trajectories are obviously conics with a focus at the centre of force. The species will depend on the sign of the constant E; in our case $E > 0$, so that the curves are hyperbolas. Since the divergence from a rectilinear path must be very small, it is self-evident that these hyperbolas will be only very slightly curved; this can also be proved analytically from the differential equations. To show this, let $\mathbf{n}$ and $\frac{1}{\rho}$ denote the direction of the principal normal and the curvature at any point of a ray. Equating the centripetal acceleration to the centripetal force per unit mass, we have

$$\frac{v^2}{\rho} = 2\frac{dU}{dn}.$$

The derivative $\frac{dU}{dn}$ represents the force in the field in the direction $\mathbf{n}$, and cannot therefore be greater than the intensity $\frac{fm_0}{r^2} = U\frac{1}{r}$ of this force. Further, the integral of *vis viva*

$$\tfrac{1}{2}v^2 = \tfrac{1}{2}c^2 + 2U = \tfrac{1}{2}c^2\left(1 + \frac{2U}{c^2}\right)$$

shows that if we neglect terms of the second order, v may be taken as equal to c. Consequently we have

$$\frac{1}{\rho} \leqslant \frac{1}{c^2}\,\frac{2fm_0}{r^2}.$$

If r_0 is the sun's radius, the maximum possible value for the force $\frac{2fm_0}{r^2}$ in the space traversed by the light rays is evidently that given by $r = r_0$; the above inequality can therefore be written

$$\frac{1}{\rho} \leqslant \frac{2fm_0}{c^2\,r_0^{\,2}}.$$

As $\frac{fm_0}{r_0}$ is the value U_s of the potential at the surface of the sun, and the value of the ratio $\frac{U_S}{c^2}$ is 2×10^{-6} (cf. formula (44)), we get finally

$$\frac{1}{\rho} \leqslant 4 \times 10^{-6} \frac{1}{r_0}.$$

In other words, if the radius of curvature ρ is not infinite as for a straight line, it is at any rate of the order of a million times the sun's radius.

It is therefore perfectly legitimate to assume that the rays are in any case only very slightly bent, even if they pass very close to the sun; in every case, therefore, the hyperbola in question will have its asymptotes OA', OT' (cf. fig. 4) almost in one straight line.

Consider in further detail the hyperbolic ray which grazes the solar sphere at V. Let O be the centre of the hyperbola, S the centre of the sun and therefore the focus of the given branch of the hyperbola. V will be its vertex, and, if a denotes the transverse semi-axis and e the eccentricity, we shall have by definition

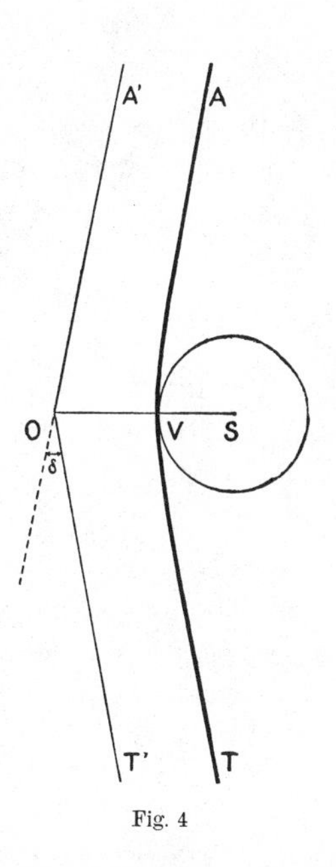

Fig. 4

$$OV = a, \quad OS = ae, \quad SV = r_0 = a(e-1).$$

We know also from analytical geometry that if δ represents the exterior angle between the two asymptotes

$$\sin \frac{\delta}{2} = \frac{1}{e}.$$

In the case we are considering, δ must be very small; hence, from this formula, e is very large. Our results will be quite sufficiently accurate if we take the sine of the angle δ as equal to the arc, and consider $\frac{1}{e}$ negligible in comparison with unity. Thus we can write

$$\delta = \frac{2}{e} = \frac{2}{e} \frac{1}{1 - \frac{1}{e}} = \frac{2}{e-1}.$$

Using the relation $r_0 = a(e-1)$, we get finally as the measure of δ in terms of the two lengths r_0 and a

$$\delta = \frac{2a}{r_0}. \qquad (46)$$

In the classical theory, the transverse semi-axis a in the hyperbolic motion due to the Newtonian attraction of a mass M is connected with the constant E of the *vis viva* by the relation

$$E = \frac{fM}{2a}.$$

Putting for E its value $\frac{1}{2}c^2$, and noting that in our case $M = 2m_0$, this gives a, and (46) becomes

$$\delta = \frac{4}{c^2}\frac{fm_0}{r_0}, \qquad (46')$$

and therefore, using the numerical value already found for this expression,

$$\delta = 8 \times 10^{-6}.$$

The right-hand side is a pure number, which gives the angle δ in radians. In seconds

$$\delta = 1{\cdot}7''. \qquad (47)$$

It will at once be seen that this angle δ gives the measure of the *deflection*, i.e. the maximum angular deviation to which a stellar ray can be subjected by the sun's gravitational action. Suppose in fact that we are considering a ray of light which starts from a star A and arrives at a terrestrial observer after describing an arc of a hyperbola which grazes the solar sphere at V, as in fig. 4. The direction of the hyperbola at T, along which the observer receives the light ray, is indistinguishable from that of the asymptote OT'; the direction in which the light left the star is that of the tangent at A, which in its turn is indistinguishable from the other asymptote $A'O$, so that the deflection is the exterior angle between $A'O$ and OT', i.e. δ.

The direction $A'O$ will naturally be identified with the direction in which T sees the star in normal conditions, i.e. when the sun leaves the earth-star direction and the corresponding gravitational perturbation becomes imperceptible, so that the visual ray again becomes rectilinear (or so nearly rectilinear that the difference is absolutely imperceptible).

It may be well to point out that if the visual ray from a star does not graze the solar sphere but passes at a distance $r > r_0$ from the centre of the sun, the deflection diminishes, being in inverse ratio to the perihelion distance r. This can be seen as follows. The expression (46) for δ naturally holds for any star whatever which is visible from the earth, provided r_0 is replaced by the perihelion distance r. We shall thus have

$$\delta = \frac{2a}{r} = \frac{2a}{r_0}\frac{r_0}{r}.$$

The factor $\frac{2a}{r_0}$ has been calculated above, so that we have finally

$$\delta = 1{\cdot}7'' \times \frac{r_0}{r}.$$

Since r_0 corresponds to an angle of 16′, it will be obvious that if the angular distance from the centre of the sun is even a few degrees δ will not be more than some hundredths of a second, and will therefore be totaliy imperceptible, just as if the ray were rigorously rectilinear.

The angular displacements, if any, due to the sun become capable of observation during a total eclipse. A first attempt in this direction was made by the Lick Observatory in 1918, but the precision of the observations was insufficient for the purpose.

For the total eclipse of 29th May, 1919, two simultaneous expeditions were organized by the Royal Society of London: one for Sobral in the north of Brazil, the other for the island of Principe in the Gulf of Guinea, both localities being within the zone of totality of the eclipse. The results of the observations made by these two expeditions can be summarized as follows. For the deflection of light the mean value of the displacements observed at Sobral gave 1·98″, with probable error $\pm$ 0·12″; at Principe the mean value was 1·61″, with probable error 0·30″. The deflection 1·76″ predicted by Einstein's general relativity lies between these two. This provided a new and striking confirmation of Einstein's theory, as the observed results were definitely incompatible both with the zero deviation of geometrical optics, and with the deviation of half this value (0·88″)

which would be given by the ordinary theory combined with the simple postulate that mass and energy are proportional.

On the occasion of the next total eclipse (21st September, 1922), visible in Western Australia, three further expeditions started for the zone of totality; the American one, organized by the Lick Observatory and conducted by Campbell, was the only one to secure any useful observations. But the available stars were rather far from the limb of the sun, and the deflection was therefore small; the results [1] show a wide dispersion, so that many astronomers do not regard their mean value as a further confirmation of the theory, although it is in almost perfect agreement with the Einsteinian prediction.

11. Three-dimensional metrics with spherical symmetry.

We shall begin by defining what is meant by saying that a metric manifold V_3 has spherical symmetry round one of its points O. We shall follow the geometrical method suggested by Palatini[2], considering along with the V_3 an ordinary Euclidean space V_3' in one-to-one correspondence with it. This correspondence being established, any point-transformation (T) of V_3' into itself (in particular, a rigid motion of V_3') gives rise to an analogous point-transformation of V_3 into itself. There is, however, no *a priori* reason that a rigid motion of V_3 should correspond to a rigid motion of V_3', a rigid motion of a manifold being taken to mean any transformation which leaves dl^2 unchanged, and therefore, in particular, changes geodesics into geodesics.

We shall now say that a metric manifold V_3 has spherical symmetry round one of its points O when each of the ∞^3 rigid rotations of V_3' round the corresponding point O' determines a rigid motion in V_3.

Some important properties of the metric of a V_3 with this property follow easily from the definition, subject naturally to the obvious condition that the metric (i.e. the coefficients of dl^2) is regular in the region round every point, except possibly the point O. It can at once be shown that to any ray j' drawn from

[1] Published in the *Lick Observatory Bulletin*, No. 346, 1923.

[2] Cf. "Lo spostamento del perielio di Mercurio, ecc," in *Nuovo Cimento*, XIV (1917), pp. 12–54.

O' there corresponds in V_3 a geodesic j drawn from O. Thus, let P' be any point on j' which is not O', P the corresponding point (which is therefore not O) in V_3. Let g be the geodesic in V_3 which is tangential to j at P; from the qualitative hypotheses of the case it follows that g exists and is unique. We have to show that g coincides with j.

Consider in V_3' the ∞^1 rotations which have j' for axis: these correspond to ∞^1 rigid motions in the space V_3 which leave fixed all the points of j, and only these. If we suppose that g is distinct from j, the effect of the ∞^1 rotations round j' would be that g would occupy a simple infinity of positions, retaining in each the properties of being geodesic and tangential to j at P; we should therefore have an infinite number of geodesics drawn through P in the same direction, which is impossible; hence g must coincide with j.

An obvious deduction is that to any spherical surface Σ' with centre O' there corresponds in V_3 a geodesic sphere Σ with centre O.

Now consider any pair Σ, Σ' of these surfaces, and the correspondence between the points Q of one and Q' of the other determined by the correspondence between the two spaces. We wish to show that the correspondence between Q and Q' is conformal.

Let $d\sigma'$ be a generic line element in Σ' drawn from Q', $d\sigma$ the homologous element drawn from Q. If we suppose the Euclidean space referred to polar co-ordinates r, θ, ϕ, we shall have

$$d\sigma'^2 = r^2(d\theta^2 + \sin^2\theta\, d\phi^2)$$

where $r = O'Q'$. Further, when r, θ, ϕ are known, they determine Q', and therefore also Q, from the one-to-one correspondence; θ and ϕ can therefore also be regarded as curvilinear co-ordinates of Q on Σ, and the line element $d\sigma$, corresponding to $d\sigma'$ (i.e. to arbitrary differentials $d\theta$ and $d\phi$), will in every case be represented by a quadratic form which we propose to find.

Consider two elementary arcs $d\sigma'$ of equal length, drawn from Q' in two different directions. The two homologous arcs $d\sigma$ will also be equal. For the two arcs $d\sigma'$, being equal in length, can be obtained from one another by a rotation round $O'Q'$; hence we infer that the two arcs $d\sigma$ can also be obtained from one another by a rigid motion in V_3, and are therefore of equal length with respect to the metric of V_3.

It follows that the ratio $\frac{d\sigma}{d\sigma'}$ is the same for the two directions considered, or, in other terms, that this ratio is the same whatever the differentials $d\theta$ and $d\phi$ may be. It is therefore a function H of position alone, i.e. (*a priori*) of r, θ, ϕ; but it will at once be seen that this function must be the same whatever may be the point Q' of Σ' considered, since we can always pass from one Q' to another by a rotation. We can therefore put

$$d\sigma^2 = H^2\, d\sigma'^2,$$

where H denotes a function of r only.

For what follows it is perhaps advantageous to replace the co-ordinate r (the radius vector in V_3') by a function $R(r)$ defined by the equation

$$R(r) = H(r)r. \quad . \quad . \quad . \quad . \quad . \quad (48)$$

The square of the line element of the geodesic sphere Σ thus takes the form

$$d\sigma^2 = R^2(d\theta^2 + \sin^2\theta\, d\phi^2); \quad . \quad . \quad . \quad (49)$$

this gives us the geometrical significance of R, no longer in the auxiliary Euclidean metric, but directly in V_3. In fact, the expression (49) for $d\sigma^2$ is that for a sphere of radius R in ordinary space, and as such (cf. § 7, p. 240) has Gaussian curvature $K = \frac{1}{R^2}$; this curvature, from its intrinsic nature, belongs to any surface whose line element is given by (49), and therefore, in particular, to our surface Σ.

We can therefore attach the following significance to the co-ordinate R: $\frac{1}{R^2}$ represents at any point the Gaussian curvature of the geodesic sphere with its centre at the centre of symmetry O and passing through the point. From the property of symmetry it follows at once that all the geodesics drawn from O cut the sphere Σ orthogonally; hence if we denote by dg the elementary arc of one of these geodesics, the dl^2 of V_3 can be represented in the form

$$dl^2 = dg^2 + d\sigma^2;$$

and since dg depends solely on R (also from symmetry) we can put

$$dg = A(R)dR,$$

where A is a function of R, *a priori* undetermined, so that we get in consequence, with the help of (49),

$$dl^2 = A^2dR^2 + R^2(d\theta^2 + \sin^2\theta\, d\phi^2). \quad . \quad . \quad (49')$$

This is the most general expression for the dl^2 of a V_3 which is symmetrical round a point.[1]

It is not without interest to show that every V_3 of this kind can be conformally represented in Euclidean space. It will be sufficient to show that we can determine two functions $H(r)$ and $r(R)$ such that we have identically

$$A^2dR^2 + R^2(d\theta^2 + \sin^2\theta\, d\phi^2) = H^2\{dr^2 + r^2(d\theta^2 + \sin^2\theta\, d\phi^2)\};$$

the necessary and sufficient conditions for this are

$$Hr = R, \quad Hdr = AdR,$$

and therefore, eliminating H,

$$\frac{dr}{r} = A\frac{dR}{R}. \quad . \quad . \quad . \quad . \quad . \quad (50)$$

When A is a known function of R, this determines r, except for a constant multiplier, which from the strictly geometrical point of view remains arbitrary. The modulus H of the conformal transformation is then defined by

$$H = \frac{R}{r}. \quad . \quad . \quad . \quad . \quad . \quad . \quad . \quad (51)$$

We shall now calculate Ricci's symbols α_{ik} (Chapter VII, p. 199) relative to a metric of this kind. We again make use of the property of symmetry, noting that an obvious consequence of the considerations set out in § 12, pp. 201–208 is that if the quadric which determines the local distribution of curvature has an axis of symmetry, this axis gives one of the three principal directions, while the other two are indeterminate (i.e. may be

[1] This formula had been given as early as 1896, from analytical considerations based on the theory of groups. Cf. *Atti della R. Acc. dei Lincei*, Vol. V (second half-year, 1896), pp. 164–171.

any pair of directions orthogonal to each other and to the axis of symmetry). In our case, a point P distinct from O having been fixed arbitrarily, and the behaviour of every metric property being symmetrical round the geodesic g which joins O and P, it follows that the quadric of curvatures at P is necessarily symmetrical round the direction of g. Hence at every point our co-ordinates r, θ, ϕ give principal directions of curvature, from which it follows at once that in the quadric of curvatures, and therefore in the tensor a_{ik}, the product terms are missing, i.e. that

$$a_{ik} = 0 \text{ for } i \neq k.$$

In addition, if ω_1 is the principal curvature corresponding to g, the other two curvatures ω_2, ω_3 are equal to one another; we shall denote their common value by ω.

We may now recall formula (47) on p. 207, viz.

$$a_{ik} = \sum_1^3{}_h \omega_h \lambda_{h|i} \lambda_{h|k},$$

which gives explicitly all the a's as functions of the curvatures and of the moments of the principal lines. Since these coincide with the co-ordinate lines, along which vary R alone, θ alone, and ϕ alone, respectively, they will have for parameters

$$\lambda_1^1 = \frac{dR}{dl} = \frac{1}{A}, \quad \lambda_1^2 = 0, \quad \lambda_1^3 = 0;$$

$$\lambda_2^1 = 0, \quad \lambda_2^2 = \frac{d\theta}{dl} = \frac{1}{R}, \quad \lambda_2^3 = 0;$$

$$\lambda_3^1 = 0, \quad \lambda_3^2 = 0, \quad \lambda_3^3 = \frac{d\phi}{dl} = \frac{1}{R \sin\theta};$$

and therefore the moments will be

$$\lambda_{1|1} = A, \quad \lambda_{1|2} = 0, \quad \lambda_{1|3} = 0;$$

$$\lambda_{2|1} = 0, \quad \lambda_{2|2} = R, \quad \lambda_{2|3} = 0;$$

$$\lambda_{3|1} = 0, \quad \lambda_{3|2} = 0, \quad \lambda_{3|3} = R \sin\theta.$$

Substituting in the formula quoted above, and putting $\omega_2 = \omega_3 = \omega$, we get

$$a_{11} = A^2\omega_1, \quad a_{22} = R^2\omega, \quad a_{33} = a_{22} \sin^2\theta, \qquad (52)$$

$$a_{ik} = 0 \qquad (i \neq k). \quad . \quad . \quad . \quad . \qquad (53)$$

The equations (53) have already been obtained from the consideration that in our case the principal lines of curvature coincide with the co-ordinate lines.

We shall now calculate explicitly the value of ω at a generic point P from its definition as the Riemannian curvature. From symmetry, it can be considered as belonging to any geodesic surface whatever with pole P and containing the direction R. We shall show that the surface $\phi =$ constant is a particular case of such a surface. Take the differential equations of the geodesics in our V_3 (of line element dl), not, however, in the form given in (47) on p. 134, where they are solved for the variables, which would require the calculation of Christoffel's symbols, but in Lagrange's parametric form, starting from the Lagrangian function (the *vis viva*)

$$T = \tfrac{1}{2}\frac{dl^2}{dt^2}.$$

In the case we are considering

$$T = \tfrac{1}{2}\{A^2\,\dot{R}^2 + R^2(\dot{\theta}^2 + \sin^2\theta\,\dot{\phi}^2)\},$$

(where a dot over a letter denotes differentiation with respect to the parameter t), and therefore

$$\frac{\partial T}{\partial \dot{\phi}} = R^2\sin^2\theta\,\dot{\phi},$$

$$\frac{\partial T}{\partial \phi} = 0.$$

From Lagrange's equation for the angle ϕ, viz.

$$\frac{d}{dt}\frac{\partial T}{\partial \dot{\phi}} - \frac{\partial T}{\partial \phi} = 0,$$

it follows on integrating that one of the equations of the geodesics has the form

$$R^2\sin^2\theta\,\dot{\phi} = \text{constant}.$$

From this it follows that if a geodesic issuing from P touches initially the surface $\phi =$ constant (so that $\dot{\phi} = 0$ at P), $\dot{\phi}$

vanishes along the whole geodesic, which therefore belongs to the surface $\phi =$ constant passing through P, as we wished to prove. To find ω, we have therefore to find the curvature of the binary differential form

$$A^2\,dR^2 + R^2\,d\theta^2, \quad . \quad . \quad . \quad . \quad . \quad (54)$$

which expresses the square of the line element of the surface $\phi =$ constant.

The general expression for this curvature is

$$\omega = \frac{(12,\,12)}{a}$$

(formula (28), p. 194); as our a is A^2R^2, it only remains to calculate Riemann's symbol of the first kind, (12, 12) by means of formulæ (3) and (5) of Chapter VII. The explicit expression for this was formed by Gauss, and is given in all treatises on the subject. We thus get

$$\omega = -\frac{1}{AR}\frac{d}{dR}\left(\frac{1}{A}\right) = -\frac{1}{2R}\frac{d}{dR}\left(\frac{1}{A}\right)^2.$$

For the curvature ω_1 we find

$$\omega_1 = \frac{1}{R^2}\left\{1 - \frac{1}{A^2}\right\}.$$

An independent calculation of these expressions is given in the following section.

12. **Digression on the calculation of curvatures.**

While our specific object is the calculation of ω and ω_1, we may here, for the convenience of the reader, show how the explicit expression for the curvature of a binary form as a function of its coefficients can be obtained without calculating Christoffel's symbols.[1] We shall start from the geometrical property of the curvature expressed by formula (29′) on p. 195, viz.

$$K = \frac{\epsilon}{D\Gamma},$$

[1] Cf. F. Sbrana: *Rend. della R. Acc. dei Lincei*, Vol. XXXIII (second half-year, 1924), pp. 236–238.

where $D\Gamma$ denotes the area of an infinitesimal circuit T containing P, and ϵ represents the angle of parallelism. In order to reduce the calculation to a minimum, we shall calculate ϵ with reference to a dl^2 of orthogonal form, of the type

$$E\,dx_1{}^2 + G\,dx_2{}^2. \quad . \quad . \quad . \quad . \quad . \quad . \quad (55)$$

If on leaving P the direction $\boldsymbol{\lambda}$ which is being displaced makes an angle α with the co-ordinate line x_1, its parameters λ^1, λ^2 are plainly given by

$$\lambda^1 = \frac{\cos\alpha}{\sqrt{E}}, \quad \lambda^2 = \frac{\sin\alpha}{\sqrt{G}} \quad . \quad . \quad . \quad . \quad (56)$$

(cf. §§ 4 and 7, pp. 92, 98). Now consider an infinitesimal displacement δP, of contravariant components δx_1, δx_2; we know that when $\boldsymbol{\lambda}$ is given a parallel displacement along δP the increments $\delta\lambda^i$ of its parameters are given by

$$\delta\lambda^i = -\sum_1^2{}_{jl}\,\{jl, i\}\,\lambda^j\,\delta x_l \qquad (i = 1, 2) \qquad (57)$$

(formula (23), p. 110). In order to avoid the necessity of calculating the coefficients on the right-hand side (Christoffel's symbols), we note that the equations

$$\ddot{x}_i = -\sum_1^2{}_{jl}\{jl, i\}\,\dot{x}_j\,\dot{x}_l$$

of the geodesics have on the right-hand side quadratic forms whose coefficients are precisely the symbols we need. Further, the first of the Lagrangian equations of the geodesics corresponding to the form (57) (the equation relative to x_1) is

$$\frac{d}{dt}\frac{\partial T}{\partial \dot{x}_1} - \frac{\partial T}{\partial x_1} = 0,$$

where
$$T = \tfrac{1}{2}(E\dot{x}_1{}^2 + G\dot{x}_2{}^2),$$

or, performing the differentiations and solving for $\ddot{x}_1$,

$$\ddot{x}_1 = -\left\{\frac{E_1}{2E}\,\dot{x}_1{}^2 - \frac{G_1}{2E}\,\dot{x}_2{}^2 + \frac{E_2}{E}\,\dot{x}_1\,\dot{x}_2\right\},$$

where E_1, E_2, G_1 represent derivatives of E and G with respect to x_1 and x_2. Comparing this with the first of the equations (57), we see that the latter can be written

$$\begin{aligned}\delta\lambda^1 &= -\left\{\frac{E_1}{2E}\lambda^1\,\delta x_1 - \frac{G_1}{2E}\lambda^2\,\delta x_2 + \frac{E_2}{2E}\lambda^1\,\delta x_2 + \frac{E_2}{2E}\lambda^2\,\delta x_1\right\}\\ &= -\lambda^1\,\delta\log\sqrt{E} + \lambda^2\left(-\frac{E_2}{2E}\,\delta x_1 + \frac{G_1}{2E}\,\delta x_2\right).\end{aligned}$$

But from (56) we get

$$\delta\lambda^1 = -\sqrt{\frac{G}{E}}\,\lambda^2\,\delta\alpha - \lambda^1\,\delta\log\sqrt{E},$$

and substituting from this in the preceding equation there results

$$\delta\alpha = \frac{1}{2\sqrt{EG}}\,(E_2\,\delta x_1 - G_1\,\delta x_2).$$

The angle of parallelism is obtained by integrating $\delta\alpha$ round the circuit T. Replacing the line integral by a surface integral in the usual way (for the signs, cf. footnote, p. 190, Chapter VII) we get

$$\epsilon = -\int_\Gamma\left[\frac{\partial}{\partial x_1}\left(\frac{G_1}{2\sqrt{EG}}\right) + \frac{\partial}{\partial x_2}\left(\frac{E_2}{2\sqrt{EG}}\right)\right]dx_1\,dx_2.$$

Noting that the field of integration reduces to the infinitesimal element

$$D\Gamma = \sqrt{EG}\,dx_1\,dx_2,$$

we can write (neglecting infinitesimals of higher order)

$$\epsilon = -\frac{D\Gamma}{2\sqrt{EG}}\left[\frac{\partial}{\partial x_1}\left(\frac{G_1}{\sqrt{EG}}\right) + \frac{\partial}{\partial x_2}\left(\frac{E_2}{\sqrt{EG}}\right)\right].$$

This gives the required expression for the curvature, viz.

$$K = -\frac{1}{2\sqrt{EG}}\left[\frac{\partial}{\partial x_1}\left(\frac{G_1}{\sqrt{EG}}\right) + \frac{\partial}{\partial x_2}\left(\frac{E_2}{\sqrt{EG}}\right)\right].$$

For $E = 1$ (for which the lines x_1 are geodesics) we get in particular the formula

$$K = -\frac{1}{\sqrt{G}} \frac{\partial^2 \sqrt{G}}{\partial x_1^2}$$

which is frequently used in the theory of surfaces.

For the line element given by (54), putting $x_1 = R$, $x_2 = \theta$, so that $E = A^2(R)$, $G = R^2$, the curvature K becomes

$$\omega = -\frac{1}{AR} \frac{d}{dR}\left(\frac{1}{A}\right), \quad . \quad . \quad . \quad (58)$$

as stated in the preceding section.

We now come to the calculation of ω_1, the curvature corresponding to the section normal to the lines R. It is to be noted that the spheres R = constant, unlike the surfaces ϕ = constant, are not geodesic surfaces, so that ω_1 does not coincide with the Gaussian curvature $\frac{1}{R^2}$ (cf. § 11) of these spheres. To calculate it, instead of using the direct definition it will be more convenient to use the property that dl^2 can be conformally represented in a Euclidean space, with

$$dl^2 = H^2\, dl_0^2,$$

as we have already seen.

In § 4, p. 228, we found the explicit form of the relations between homologous Riemann's symbols for two line elements ds and ds' for which

$$ds'^2 = e^{2\tau}\, ds^2.$$

We shall identify ds with our dl, and ds' with the Euclidean dl_0; we can then apply formulæ (18) of p. 231 by making the symbols marked with a dash vanish (since they refer to the Euclidean dl_0^2) and putting

$$\tau = -\log H. \quad . \quad . \quad . \quad . \quad . \quad (59)$$

The formulæ then become

$$\begin{aligned}(ij, hk) = a_{ih}(\tau_{jk} - \tau_j \tau_k) - a_{ik}(\tau_{jh} - \tau_j \tau_h) - a_{jh}(\tau_{ik} - \tau_i \tau_k) \\ + a_{jk}(\tau_{ih} - \tau_i \tau_h) + (a_{ih}\, a_{jk} - a_{ik}\, a_{jh})\Delta\tau,\end{aligned}$$

where the coefficients, the covariant derivatives, and the parameter Δ are all taken to refer to

$$dl^2 = A^2\, dR^2 + R^2(d\theta^2 + \sin^2\theta\, d\phi^2).$$

Multiplying these formulæ by $a^{jh}\, a^{ik}$ and summing with respect to the four indices, the left-hand side, by formulæ (1) and (2) of § 2, gives the linear invariant G relative to our V_3, which after some obvious reductions is thus expressed by

$$G = -4\Delta_2\tau - 2\Delta\tau,$$

where $\Delta\tau = \sum_1^3{}_l\, \tau_l\, \tau^l = \sum_1^3{}_{ik}\, a^{ik}\, \tau_i\, \tau_k$ (Chapter VIII, p. 231); and $\Delta_2\tau = \sum_1^3{}_{ik}\, a^{ik}\, \tau_{ik}$ (Chapter VI, p. 154).

But for a V_3 the linear invariant G is equal to $-2\mathcal{M}$ (cf. § 4), so that the mean curvature $\mathcal{M}$ is in our case given by

$$\mathcal{M} = 2\Delta_2\tau + \Delta\tau. \quad . \quad . \quad . \quad . \quad (60)$$

As $\mathcal{M}$ (the sum of the three curvatures) $= \omega_1 + 2\omega$, and ω has already been calculated, this formula will give the required expression for ω_1; it remains to find the values of the quantities $\Delta_2\tau$ and $\Delta\tau$ on the right, using for this purpose the formulæ (50), (51), and (59).

From the general expression

$$\Delta\tau = \sum_1^3{}_{ik}\, a^{ik}\, \tau_i\, \tau_k,$$

we have for our dl^2, and for a function τ which depends only on R,

$$\Delta\tau = \frac{1}{A^2}\tau'^2,$$

the dash here also denoting the derivative with respect to R. Further, from the general expression (18) on p. 154 we have

$$\Delta_2\tau = \frac{1}{\sqrt{a}} \sum_1^3{}_i \frac{\partial}{\partial x_i}(\sqrt{a}\, \tau^i),$$

and since now $\sqrt{a} = AR^2 \sin\theta$, it follows that

$$\Delta_2\tau = \frac{1}{AR^2} \frac{d}{dR}\left(\frac{R^2}{A}\tau'\right).$$

In our case, from (59) and (51),

$$\tau = -\log H = \log \frac{r}{R},$$

and by (50)
$$\frac{d}{dR}\log r = \frac{A}{R};$$

hence
$$\tau' = \frac{A-1}{R}.$$

It follows that

$$\Delta\tau = \frac{(A-1)^2}{A^2 R^2} = \frac{1}{R^2}\left(1-\frac{1}{A}\right)^2,$$

$$\Delta_2\tau = \frac{1}{AR^2}\frac{d}{dR}\left[R\left(1-\frac{1}{A}\right)\right]$$

$$= \frac{1}{AR^2}\left(1-\frac{1}{A}\right) - \frac{1}{AR}\frac{d}{dR}\left(\frac{1}{A}\right);$$

using the expression (58′) for ω, this can also be written in the form

$$\Delta_2\tau = \frac{1}{AR^2}\left(1-\frac{1}{A}\right) + \omega.$$

Substituting in (60) the expressions just found for $\Delta\tau$ and $\Delta_2\tau$ and for $\mathcal{M}$ its value $\omega_1 + 2\omega$, 2ω cancels out on both sides, and we get for ω_1 the value stated at the end of § 11, viz.

$$\omega_1 = \frac{1}{R^2}\left(1-\frac{1}{A^2}\right). \quad . \quad . \quad . \quad . \quad (61)$$

13. **The gravitational equations in the case of spherical symmetry. Schwarzschild's rigorous solution.**

We shall now apply the equations of the Einsteinian statics to the particular case of a single attracting mass, or more generally of a distribution of masses having spherical symmetry round a point O. Using the terminology of § 11 we shall deal with matter distributed in accordance with any law dependent only on R in layers bounded by geodesic spheres of centre O. The Einsteinian ds^2 will have the statical form

$$ds^2 = V^2\,dx_0{}^2 - dl^2,$$

where dl^2 will necessarily be of the type (49′), and V, from symmetry, will also depend only on R. We shall agree to consider only regions outside the field occupied by the attracting masses. In these regions the statical equations (21′), (22) of § 4 for empty space will hold, i.e.

$$\mathcal{M} = 0, \quad . \quad . \quad . \quad . \quad . \quad . \quad . \quad (21')$$

$$a_{ik} + \frac{V_{ik}}{V} = 0. \quad . \quad . \quad . \quad . \quad . \quad . \quad (22)$$

Since $\mathcal{M}$ denotes the mean curvature of the symmetrical V_3, i.e. the sum $\omega_1 + 2\omega$ of the three principal curvatures, (21′), together with (58) and (61) of the preceding section, gives

$$\omega_1 + 2\omega = \frac{1}{R^2}\left(1 - \frac{1}{A^2}\right) - \frac{1}{R}\frac{d}{dR}\left(\frac{1}{A}\right)^2 = 0\,;$$

whence on separating variables and integrating

$$A^2 = \frac{1}{1 - \dfrac{a}{R}}, \quad . \quad . \quad . \quad . \quad . \quad . \quad (62)$$

where a denotes a constant of integration.

It is to be noted that whatever the constant a may be the expression found satisfies the physically necessary condition that at an infinite distance from the attracting masses the metric tends towards the Euclidean form. In fact, if $R \to \infty$, $A \to 1$, so that the dl^2 (49′) becomes the ordinary Euclidean expression in polar co-ordinates.

The symbols a_{ik} are then completely defined by the formulæ (52) and (53), where ω and ω_1 have the values (58) and (61).

In order to put the gravitational equations (22) in an explicit form, we must again replace the covariant derivatives V_{ik} by the ordinary derivatives. This can also be done without any preliminary calculations, as follows. Let the x_i's denote generic co-ordinates in a space with a generic metric. Take a function V of the x's, and consider its variation along a geodesic line along which the x's are considered as functions of a parameter t. We shall have in the first place

$$\frac{dV}{dt} = \sum_1^3{}_i\, V_i\, \dot{x}_i.$$

Differentiating again, and substituting for the $\ddot{x}_i$'s their values as given by the equations of the geodesics, we get, as a particular case of the notion of covariant derivatives (Chapter VI, §§ 1 and 2, p. 144),

$$\frac{d^2V}{dt^2} = \sum_{1}^{3}{}_{ik} V_{ik}\, \dot{x}_i\, \dot{x}_k. \quad . \quad . \quad . \quad . \qquad (63)$$

Further, assuming in particular $x_1 = R$, $x_2 = \theta$, $x_3 = \phi$, and remembering that our V is a function of R only, we have

$$\frac{dV}{dt} = V'\dot{R}$$

(dashes denoting derivatives with respect to R), and

$$\frac{d^2V}{dt^2} = V''\dot{R}^2 + V'\ddot{R}.$$

But for our metric, i.e. for

$$T = \tfrac{1}{2}[A^2\dot{R}^2 + R^2\,(\dot{\theta}^2 + \sin^2\theta\,\dot{\phi}^2)],$$

the equation of the geodesics for the co-ordinate R gives

$$\frac{d}{dt}\frac{\partial T}{\partial \dot{R}} - \frac{\partial T}{\partial R} = 0$$

or $$A^2\ddot{R} + \frac{d(A^2)}{dR}\dot{R}^2 - \tfrac{1}{2}\frac{d(A^2)}{dR}\dot{R}^2 - R(\dot{\theta}^2 + \sin^2\theta\,\dot{\phi}^2) = 0,$$

whence we get

$$\ddot{R} = -\frac{A'}{A}\dot{R}^2 + \frac{R}{A^2}(\dot{\theta}^2 + \sin^2\theta\,\dot{\phi}^2).$$

Using this result the foregoing expression for $\frac{d^2V}{dt^2}$ becomes

$$\frac{d^2V}{dt^2} = \left(V'' - \frac{V'A'}{A}\right)\dot{R}^2 + \frac{RV'}{A^2}(\dot{\theta}^2 + \sin^2\theta\,\dot{\phi}^2).$$

This expression, like (63), must hold along a generic geodesic, i.e. for arbitrary values of the quantities $\dot{x}_1 = \dot{R}$, $\dot{x}_2 = \dot{\theta}$, $\dot{x}_3 = \dot{\phi}$. Comparing them we get

$$\left.\begin{aligned} V_{11} = V'' - \frac{V'A'}{A},\quad V_{22} = \frac{RV'}{A^2},\quad V_{33} = V_{22}\sin^2\theta, \\ V_{ik} = 0 \;\; (i \neq k). \end{aligned}\right\} \qquad (64)$$

Substituting in the gravitational equations (22) these values for the V_{ik}'s and the values (52) and (53) for the a_{ik}'s, we see at once that the equations with two distinct indices reduce to identities, those for the pairs of indices 11 and 22 take the form

$$A^2 \omega_1 + \frac{V''}{V} - \frac{V'A'}{VA} = 0, \quad . \quad . \quad . \quad . \quad . \quad (65)$$

$$R^2 \omega + \frac{RV'}{VA^2} = 0,$$

and the remaining equation is the same as this last one. Substituting in this equation for ω its value (58), i.e. $\frac{A'}{RA^3}$, it becomes

$$\frac{A'}{A} + \frac{V'}{V} = 0, \quad . \quad . \quad . \quad . \quad . \quad . \quad (66)$$

or

$$AV = \text{constant}.$$

At an infinite distance from the attracting masses the Einsteinian ds^2 must reduce to the pseudo-Euclidean form, and therefore the coefficient V (the Römerian velocity of light) must tend to 1 like A; hence the constant must have the value 1, and we have

$$AV = 1. \quad . \quad . \quad . \quad . \quad . \quad . \quad (66')$$

This equation and (62) give A and V in finite terms, so that the required ds^2 is now completely determined. The equation (65) remains to be considered, but it will at once be seen that with the values (62) and (66′) it reduces to an identity. In fact, substituting for ω_1 its value $\frac{1}{R^2}\left(1 - \frac{1}{A^2}\right) = \frac{1}{R^2}(1 - V^2)$, and for $\frac{A'}{A}$, by (66), the equivalent $-\frac{V'}{V}$, multiplying by V^2, and remembering once more that $AV = 1$, (65) becomes

$$\frac{1}{R^2}(1 - V^2) + VV'' + V'^2 = 0,$$

or

$$\frac{1}{R^2}(1 - V^2) + \tfrac{1}{2}\frac{d^2}{dR^2}V^2 = 0.$$

On substituting for V^2 the value $\frac{1}{A^2} = 1 - \frac{a}{R}$, we find that this equation is satisfied identically, which proves the required result.

The rigorous form of the Einsteinian ds^2 with spherical symmetry is therefore

$$ds^2 = \left(1 - \frac{a}{R}\right) dx_0{}^2 - dl^2 \quad . \quad . \quad . \quad . \quad (67)$$

with
$$dl^2 = \frac{dR^2}{1 - \frac{a}{R}} + R^2 (d\theta^2 + \sin^2 \theta \, d\phi^2).$$

This expression for ds^2 was first given by Schwarzschild.[1] The metric contains a constant a which is *a priori* arbitrary; its value can be deduced from a consideration of the intensity of the field of force at great distances from the attracting masses. In these regions the spacelike dl^2 tends, as we know, to become Euclidean, R becoming identical with the length of the radius vector drawn from the centre of symmetry; further, the expression

$$-\tfrac{1}{2}c^2 V^2 = -\tfrac{1}{2}c^2 + \tfrac{1}{2}c^2 \frac{a}{R}$$

represents the potential of the field (cf. Chapter XI, p. 328). Comparing this with the classical Newtonian expression $\frac{fM}{R}$ for the potential due to a mass M concentrated at the origin (or symmetrically distributed round it in any way), we see that we must put

$$a = \frac{2fM}{c^2}, \quad . \quad . \quad . \quad . \quad . \quad (68)$$

where M is the sum of the attracting masses.

It follows from § 11 that every dl^2 with spherical symmetry, and therefore in particular the Einsteinian dl^2 (67), can be conformally represented in a Euclidean space, the modulus of the

[1] *Sitzungsberichte der Preuss. Akad. der Wiss.*, 1916, pp. 189–196.

conformal representation being $\frac{R}{r}$, where r is defined by (50), i.e. by

$$\frac{dr}{r} = \frac{AdR}{R}.$$

Using the relations

$$A = \frac{1}{V}, \quad V = \sqrt{1 - \frac{\alpha}{R}},$$

we have to express the right-hand side of this equation in terms of V, which gives

$$\frac{dr}{r} = \frac{2dV}{1 - V^2},$$

whence on integrating

$$r = r_0 \frac{1 + V}{1 - V}$$

$$= r_0 \frac{(1 + V)^2}{1 - V^2} = \frac{Rr_0}{\alpha}(1 + V)^2, \quad . \quad . \quad (69)$$

where r_0 denotes a constant.

If we wish to impose the natural condition that r, like R, shall tend to become identical with the ordinary radius vector at an indefinitely great distance from the attracting masses, we shall have to determine r_0 in such a way that

$$\lim_{R \to \infty} \frac{R}{r} = 1.$$

Since when $R \to \infty$, $V \to 1$, this gives

$$r_0 = \frac{\alpha}{4}.$$

Consequently $$H = \frac{R}{r} = \frac{4}{(1 + V)^2},$$

and therefore $$dl^2 = H^2 dl_0^2 = \frac{16}{(1 + V)^4} dl_0^2.$$

As an instructive example, we shall apply these rigorous

formulæ to calculate over again, for a symmetrical field, the expression (33) of § 6, viz.

$$ds^2 = (1 - 2\gamma)dx_0^2 - (1 + 2\gamma)dl_0^2$$

$\left(\text{where } \gamma \text{ stands for } \frac{U}{c^2}\right)$, which to a first approximation gives the Einsteinian ds^2 corresponding to an assigned Newtonian field. In our case comparison of the coefficients V^2 and $1 - 2\gamma$ of dx_0^2 gives rigorously

$$\gamma = \tfrac{1}{2}\frac{a}{R},$$

so that, from the value (68) of a, U is precisely the expression for the Newtonian potential of a mass M symmetrically distributed round the centre. Comparison of the coefficients of dl_0^2 imposes the condition (at least to a first approximation)

$$1 + 2\gamma = H^2 = \frac{16}{(1 + V)^4}.$$

From the expression $1 - 2\gamma$ for V^2 we have to a first approximation $V = 1 - \gamma$, and therefore

$$(1 + V)^{-4} = (2 - \gamma)^{-4} = \frac{1}{16}(1 + 2\gamma),$$

which ensures that the above condition is effectively satisfied so long as we neglect terms of higher order.

14. Spatially uniform metrics; their cosmological interest.

We shall now examine whether there exist solutions of the gravitational equations in statical conditions, and on the hypothesis that the spacelike dl^2 has a constant curvature K and that the energy tensor is also uniform, meaning by this that it is of the type (66) of p. 358 (applied to the statical case). This is equivalent to assuming for the T_{ik}'s the expressions

$$T_{00} = V^2(\epsilon - p) = V^2\eta, \quad . \quad . \quad . \quad . \quad (70)$$

$$T_{ik} = p\,a_{ik} \qquad (i, k = 1, 2, 3), \quad . \quad . \quad (71)$$

where the a_{ik}'s obviously denote the coefficients of dl^2. The two quantities $\eta(\geqslant 0)$ and p represent respectively (cf. Chapter XI, p. 358) the energy density and the pressure (or pull, if $p < 0$) in the medium.

We next take into account the geometrical hypothesis that the spacelike manifold has constant curvature K. When the three principal curvatures $\omega_1, \omega_2, \omega_3$ all reduce to K, the canonical expressions for Ricci's symbols α_{ik} (formula (47) on p. 207), together with (46) on p. 206, give immediately

$$\alpha_{ik} = K\,a_{ik}, \quad \ldots \ldots \quad (72)$$

while by the definition of the mean curvature we have

$$\mathcal{M} = 3K. \quad \ldots \ldots \quad (72')$$

Using these results, the first of the gravitational equations, (18) of § 4, becomes

$$3K = \kappa\eta. \quad \ldots \ldots \quad (73)$$

We deduce from this that $K \geqslant 0$, which comes within the general observation of § 4 that in statical conditions the mean curvature is always either positive or zero. The equation (73) then shows that η is necessarily constant when K is, or in other words that the medium must have a uniform distribution of energy, or, what is the same thing, of matter.

On account of this circumstance, this type of solution has a particular cosmological interest. It is true that the celestial bodies are separated by distances which are large compared with their dimensions, and therefore the distribution of matter in space is essentially discontinuous; but from a statistical point of view it is natural to ask what are, so to speak, the mean mechanical conditions of the universe; i.e. what would be the nature of the space-time metric on the hypothesis that the whole of the cosmic matter, instead of being concentrated in discrete masses, is uniformly distributed throughout all space, with the mean density $\frac{\eta}{c^2}$ of the actual distribution.

It is important to note that, as we are dealing with a space of constant *positive* curvature, its extension S (in the sense of Chapter VI, p. 160) is finite, as we shall show in a moment. Associating it in the meanwhile with the foregoing cosmological consideration, we reach the conclusion that in this type of solution the total quantity M of matter is finite, and is given by

$$M = \frac{\eta}{c^2} S. \quad . \quad . \quad . \quad . \quad . \quad . \quad (74)$$

In order to find the extension S, we take dl^2 in the canonical form (31) on p. 240, viz.

$$dl^2 = \frac{dl_0^{\,2}}{u^2} = \frac{1}{u^2}(dy_1^{\,2} + dy_2^{\,2} + dy_3^{\,2}), \quad . \quad (75)$$

where $$u = 1 + \frac{K}{4} r^2 \quad \text{and} \quad r^2 = \sum_1^3{}_i \, y_i^2. \quad . \quad . \quad (76)$$

We have in the first place, for the element of volume corresponding to the Euclidean $dl_0^{\,2}$ referred to polar co-ordinates r, θ, ϕ,

$$dS_0 = r^2\, dr \sin\theta \, d\theta \, d\phi,$$

and therefore, for the corresponding element of physical space,

$$dS = \frac{dS_0}{u^3}.$$

The total volume is in consequence given by

$$S = \int \frac{dS_0}{u^3},$$

the integral being extended to the whole of space. The integration with respect to θ and ϕ gives 4π, so that we can write

$$S = 4\pi \int_0^\infty \frac{r^2\, dr}{u^3}.$$

Here we can introduce the radius a of the sphere of Gaussian curvature K, putting $K = \frac{1}{a^2}$, and substitute $x = \frac{r}{2a}$ for r as the variable of integration. This gives

$$S = 32\pi a^3 \int_0^\infty \frac{x^2\, dx}{(1 + x^2)^3} = 2\,\pi^2\, a^3,$$

and therefore, from (74),

$$M = 2\pi^2 a^3 \frac{\eta}{c^2}.$$

In the given conditions, physical space has thus the volume $2\pi^2 a^3$, and is therefore finite, though at the same time unlimited. This latter property holds, as for ordinary two-dimensional spherical surfaces, for any manifold of constant positive curvature in any number of dimensions.

Another general property which calls for mention is that in a variety of the kind specified the geodesics are all closed lines, of length $2\pi a$. Consider specifically the case of three dimensions which corresponds to the physical space of the problem under discussion. It will be seen immediately that without loss of generality we can always refer $dl^2 = \frac{dl_0^2}{u^2}$ to polar co-ordinates r, θ, ϕ in such a way that for a geodesic assigned in any manner $\dot{\phi} = 0$ at one point of it; from the Lagrangian equation relative to the parameter ϕ it then follows, as in § 11, that $\dot{\phi} = 0$ all along the curve, which is therefore a geodesic of one of the surfaces $\phi =$ constant, or in particular, by suitable choice of the ϕ-axis, of $\phi = 0$. In view of the transformation formulæ between Cartesian and polar co-ordinates,

$$y_1 = r \sin\theta \cos\phi,$$
$$y_2 = r \sin\theta \sin\phi,$$
$$y_3 = r \cos\theta,$$

this is equivalent to saying that any geodesic can always be considered as belonging to the co-ordinate plane $y_2 = 0$; but, for $y_2 = 0$, dl^2 assumes the canonical form of a two-dimensional manifold of constant curvature K, i.e. of the ordinary sphere of radius a. The geodesic therefore coincides with a great circle on this sphere, and is therefore a closed curve of length $2\pi a$.

We now pass on to the other six gravitational equations.

Taking account of (71) and (72), the equations (17) become

$$\frac{V_{ik}}{V} + \left(K + \kappa p - \frac{\Delta_2 V}{V}\right) a_{ik} = 0 \qquad (i, k = 1, 2, 3), \quad (77)$$

which can be satisfied in two different ways, according as we suppose V constant (Einstein's cylindrical space-time) or V a function of position (De Sitter's hyperspherical space-time).[1]

15. **Einstein's solution.**

First, suppose V constant. In this case it is necessary and sufficient to add to (73) the condition

$$K + \kappa p = 0. \qquad (78)$$

From this it follows first of all that the normal stress p is necessarily the same at every point, and on comparison with (73) there follows

$$p = -\tfrac{1}{3}\eta, \qquad (73')$$

whence we get the following result:

In a homogeneous medium subjected to a uniform pull of $\frac{1}{3}\eta$, η being the energy density, the space assumes the constant positive curvature $K = \dfrac{\kappa}{3}\eta$, the velocity V of light remaining constant (and being naturally supposed not zero).

Remembering that in statical conditions the potential of the force in the field is $-\frac{1}{2}V^2$ (Chapter XI, p. 328), we see at once that in the present case the force is zero.

16. **De Sitter's solution.**

Now suppose that V is a function of position. Multiplying (77) by a^{ik} and summing with respect to i and k we get in the first place

$$\frac{\Delta_2 V}{V} + 3\left(K + \kappa p - \frac{\Delta_2 V}{V}\right) = 0$$

or

$$\frac{\Delta_2 V}{V} = \tfrac{3}{2}(K + \kappa p).$$

The equations (77) are therefore equivalent to

$$\frac{V_{ik}}{V} + K^* a_{ik} = 0 \qquad (77')$$

where for brevity we have put

$$K^* = K - \tfrac{1}{2}(3K + \kappa p). \qquad (79)$$

[1] Cf. T. Levi-Civita: "Realtà fisica di alcuni spazi normali del Bianchi," in *Rend. della R. Acc. dei Lincei*, Vol. XXVI (first half-year, 1917), pp. 519–531.

It is easy to see that the equations (77′) are mutually consistent for V *not* constant (in fact, they constitute a complete system with respect to V considered as the unknown function) if, and only if, $K^* = K$. To prove this, take the commutation formula (20) on p. 186 for the second covariant derivatives of a simple system V_i, which gives

$$V_{ihk} - V_{ikh} = -\sum_{1}^{3}{}_r \{ir, hk\} V_r.$$

Substituting for Riemann's symbols of the second kind the expressions for a manifold of constant curvature K (formula (19′) on p. 234), viz.

$$K\,(a_{ih}\,\delta^r_k - a_{ik}\,\delta^r_h),$$

we get
$$V_{ihk} - V_{ikh} = K\,(a_{ik}\,V_h - a_{ih}\,V_k).$$

Further, multiplying (77′) by V and taking the covariant derivative, we get

$$V_{ikh} = -K^*\,a_{ik}\,V_h; \quad . \quad . \quad . \quad . \quad (77'')$$

substituting in the preceding equation, we get the conditions of integrability

$$(K^* - K)\,(a_{ik}\,V_h - a_{ih}\,V_k) = 0$$

for every set of values of the three indices i, h, k. Since by hypothesis V is an effective function, one at least of its derivatives (say V_k) will not be zero. In the above equations take this value of k and a value of h different from k; multiply by a^{ih} and sum with respect to i. This gives

$$(K^* - K)V_k = 0$$

whence
$$K^* - K = 0,$$

Q. E. D.

Using this result, we get from (79)

$$3K + \kappa p = 0,$$

which leads to the same qualitative statements with regard to the stresses as those made above for the cylindrical space-time.

For the integration of the equations (77′), in which from now onwards we put $K^* = K$, we must again take dl^2 in the canonical form (75).

The covariant derivatives V_{ik} of V with respect to our dl^2 can be found explicitly as functions of the ordinary derivatives, without direct calculation, from the considerations in Chapter VIII, pp. 222–232. In fact, considering our dl^2 and the corresponding Euclidean dl_0^2 referred to the same co-ordinates, we have, from formula (9) on p. 224,

$$V_{ik} - V_{ik}^0 = -\sum_1^3{}_l \rho_{ik}^l V_l,$$

where, by (16) on p. 230,

$$\rho_{ik}^l = \delta_i^l \tau_k + \delta_k^l \tau_i - a_{ik} \tau^l,$$

with in our case $e^{-2\tau} = u^2$, u having the value given in (76).

Noting that for dl_0^2 referred to Cartesian co-ordinates the derivatives V_{ik}^0 are identical with the ordinary second derivatives, and that $\tau^l = \tau_l$, $a_{ik} = \delta_i^k$, we have the required expressions in the form

$$V_{ik} = \frac{\partial^2 V}{\partial y_i \partial y_k} + \frac{1}{u}(u_k V_i + u_i V_k) - \frac{\delta_i^k}{u} \sum_1^3{}_l u_l V_l.$$

Substitute these expressions in the equations (77′), which on multiplying by uV take the form

$$uV_{ik} + \frac{KV}{u}\delta_i^k = 0$$

for $K^* = K$ and $a_{ik} = \frac{\delta_i^k}{u^2}$. Putting for brevity

$$W = uV, \quad . \quad . \quad . \quad . \quad . \quad (80)$$

and using the expression (76) for u, we get

$$u\frac{\partial^2 V}{\partial y_i \partial y_k} + u_k V_i + u_i V_k = \frac{\partial^2 W}{\partial y_i \partial y_k} - \frac{KV}{2}\delta_i^k;$$

whence it follows that

$$\frac{\partial^2 W}{\partial y_i \partial y_k} = \delta_i^k\left(\frac{K}{2}\frac{W}{u} - \frac{KW}{u^2} + \sum_1^3{}_l u_l V_l\right).$$

But by (80)

$$V_l = \frac{W_l}{u} - \frac{W}{u^2} u_l;$$

substituting, and taking into account the definition (76) of u and the consequent identity

$$K + \sum_1^3 {}_l u_l^2 = Ku,$$

the foregoing equations become

$$\frac{\partial^2 W}{\partial y_i \partial y_k} = \delta_i^k \frac{K}{2u} \left(\sum_1^3 {}_l y_l W_l - W \right).$$

From this it follows at once that for $i \neq k$ the second derivatives of W vanish, so that W must be a function with the variables separated (the sum of three functions, one of y_1 alone, one of y_2 alone, and one of y_3 alone). Further, for $i = k$, the equations above show that as the terms on the right are the same for all three cases, we must have also

$$\frac{\partial^2 W}{\partial y_1^2}, \quad \frac{\partial^2 W}{\partial y_2^2}, \quad \frac{\partial^2 W}{\partial y_3^2}$$

all equal; their common value must therefore be a constant, which we can denote by $b_0 \frac{K}{2}$. Hence the most general expression for W is of the type

$$W = \frac{b_0 K}{4} r^2 + w + C,$$

where w is a linear homogeneous function, *a priori* undetermined, and C is a constant. The coefficients of this expression are to be so determined that

$$\frac{b_0 K}{2} = \frac{K}{2u} \left(\sum_1^3 {}_l y_l W_l - W \right),$$

i.e. that

$$\sum_1^3 {}_l y_l W_l - W = b_0 u.$$

By Euler's theorem on homogeneous functions the linear

term w contributes nothing to the left-hand side, so that its three coefficients are arbitrary; there thus remains

$$\frac{b_0 K}{4} r^2 - C = b_0 u,$$

which by (76) reduces to

$$C = -b_0.$$

Hence the final expression for W can be written in the form

$$W = b_0 (u - 2) + w, \quad . \quad . \quad . \quad . \quad (81)$$

the constant b_0 and the coefficients of w being still completely arbitrary. This number of constants could of course be predicted from the fact that the system (77′) is completely integrable; as all the second derivatives of the function V are defined by it, it is obviously equivalent (cf. Chapter II, p. 13) to a total differential system in four unknown functions, viz. V itself and its three first derivatives.

It is also to be noted that the three constants of integration which appear in the linear expression

$$w = 2(b_1 y_1 + b_2 y_2 + b_3 y_3)$$

can obviously be reduced to one, since by a suitable orthogonal transformation applied to the y's (for which r^2, u, and $dl_0{}^2$ are all invariant), we can always reduce the trinomial to the form $2by_1$, with $b = \sqrt{b_1{}^2 + b_2{}^2 + b_3{}^2}$.

But we may also suppose $b = 0$; this can be formally proved (though in a less elementary way) by taking account of the homogeneity of a space with constant curvature, which enables us to take a point fixed in advance as the point $y_i = 0$, while retaining the canonical form $\frac{dl_0{}^2}{u^2}$ for dl^2.[1]

[1] This becomes intuitive for the case of two dimensions, in which a manifold of constant positive curvature is an ordinary sphere, and the canonical expression for dl^2 is obtained by stereographic projection of the sphere on a diametral plane (cf. Chapter VIII, p. 241). The assertion in the text reduces in this case to the obvious geometrical fact that any point whatever of the sphere may be chosen as the centre of projection.

Using these results, it follows from (80) that the expression for the spatially uniform ds^2, on the hypothesis of V variable, is

$$ds^2 = \frac{W^2 \, dy_0{}^2 - dl_0{}^2}{u^2},$$

where

$$\left. \begin{aligned} W &= b_0 (u - 2), \\ u &= 1 + \frac{K}{4} r^2 \end{aligned} \right\} \quad . \quad . \quad . \quad . \quad . \quad (82)$$

It is assumed that the constant b_0 is not zero, as otherwise we should have identically $V = 0$, which is not permissible, since we are considering the case of V variable.

In view of the physical significance of V, those points, if any, at which $V = 0$ obviously denote singularities in the field; they remain, so to speak, optically isolated, in a sense which will be explained further on. On the other hand, as r, and therefore u, increases indefinitely, V tends to b_0; further, for finite values of r, u remains essentially finite and > 1, so that the singular points are determined by the equation $W = 0$. This equation, combined with (82) and the relation $K = \frac{1}{a^2}$, becomes

$$r = 2a,$$

which in the representative Euclidean space defines a sphere D_0. The surface D which corresponds to it in the physical space, and which, by § 11, is also a (geodesic) sphere, is called the *horizon*, because it constitutes in a certain sense the limit of the perceptible universe. This follows from the fact that light, and *a fortiori* a material particle, would take an infinite time to reach it. To prove this, let A and B be two generic points; then by the definition of V the time taken by light to pass from A to B is

$$\int \frac{dl}{V} = \int \frac{dl_0}{uV} = \int \frac{dl_0}{W}$$

where the integral is taken along the ray joining A to B. When B tends to the horizon the integrand tends to an infinity of the first order at B, and therefore the integral cannot remain finite.

As we have already several times recalled (in particular in the preceding section), the force in the field is the gradient of $-\frac{1}{2}V^2$;

in consequence it tends to displace the material masses towards the regions of minimum V^2, i.e. towards the horizon. This circumstance was regarded as an incongruence of De Sitter's space-time; but it is to be observed that it must be taken to refer solely to accidental masses (sufficiently small not to modify the field perceptibly), and not to those uniformly diffused masses which constitute it, the equilibrium of which is automatically assured by the gravitational equations.

It is interesting to remark that the problem of spatially uniform metrics (§ 14) admits of a solution which includes both Einstein's and De Sitter's solutions as particular cases.[1]

In fact, in the argument beginning at equation (77′), it was tacitly assumed, at (77″), that K^* is constant. If we drop this supposition we find

$$V_{ihk} - V_{ikh} = K^* (a_{ik} V_h - a_{ih} V_k) + V (a_{ik} K^*_h - a_{ih} K^*_k),$$

which, on combination as before with

$$V_{ihk} - V_{ikh} = K(a_{ik} V_h - a_{ih} V_k),$$

gives

$$(K - K^*) (a_{ik} V_h - a_{ih} V_k) = V(a_{ik} K^*_h - a_{ih} K^*_k).$$

If we put E for $K - K^*$, this becomes

$$E(a_{ik} V_h - a_{ih} V_k) + V(a_{ik} E_h - a_{ih} E_k) = 0,$$

leading, by the same treatment as in the former case, to

$$EV_h + VE_h = 0,$$

or

$$EV = \text{constant} = A, \text{ say.}$$

Equation (77′) may now be written

$$\left(V - \frac{A}{K}\right)_{ik} + Ka_{ik}\left(V - \frac{A}{K}\right) = 0.$$

Thus, if instead of (80) we write

$$W = u\left(V - \frac{A}{K}\right),$$

[1] This extension of the analysis was suggested to me by Dr. John Dougall.

the investigation proceeds exactly as before, and leads to the same value of W, viz. that given in equation (81). The new value of V is therefore

$$V = \frac{A}{K} + \frac{W}{u}$$

$$= \frac{A}{K} + \frac{b_0(u-2)}{u}.$$

If $A = 0$ we have De Sitter's solution; if $b_0 = 0$ we have Einstein's. If both A and b_0 are different from zero, the curvature K is still constant, but the (normal) stress p is variable, being given by

$$K - K^* = E$$

$$= \frac{A}{V},$$

or

$$\tfrac{1}{2}(3K + \kappa p) = \frac{A}{V}.$$

We shall conclude this section by showing that De Sitter's space-time not only, like Einstein's, implies that physical space (i.e. any manifold $x_0 =$ constant) has constant positive curvature K, but has itself, as a four-dimensional manifold, constant *negative* curvature.

To prove this, we start from a known property of every space-like dl^2 which has constant curvature K, namely (Chapter VIII, p. 234), that Riemann's symbols for dl^2 have the form (19′) of p. 234, or

$$\{ir, hk\} = K(a_{ih}\delta_k^r - a_{ik}\delta_h^r) \qquad (i, r, h, k = 1, 2, 3).$$

By (11) and (13) of § 4, these relations can be written in the form

$$\{ir, hk\}' = -K(g_{ih}\delta_k^r - g_{ik}\delta_h^r), \quad . \quad . \quad (83)$$

still for the same values 1, 2, 3 of the indices. Now it is easy to see that these last formulæ, in virtue of the expressions (14′) for Riemann's symbols for our ds^2 and of the equations

$$\frac{V_{ik}}{V} = -Ka_{ik} = Kg_{ik} \qquad (i, k = 1, 2, 3), \quad (77''')$$

will still hold when 0 is included among the values to be assigned to the indices. This is obvious when one, three, or four indices are equal to 0, since then (§ 4) both sides of the equation vanish. In the case of two indices zero we have, as in § 4, to examine the two types $\{0r, 0k\}'$, $\{i0, 0k\}'$. The corresponding values of the left-hand side are respectively $V(V^r)_k = V \sum_1^3{}_i a^{ri} V_{ik}$ and $\frac{V_{ik}}{V}$, i.e. in view of (77‴),

$$- K V^2 \sum_1^3{}_i a^{ri} a_{ik} = - K V^2 \delta_k^r = - K g_{00} \delta_k^r, \quad K g_{ik}.$$

The values of the expression on the right are clearly the same. Thus the equations (83) hold for all values of the indices from 0 to 3, which is precisely equivalent (still by formula (19′) of p. 234) to saying that the ds^2 of space-time has constant negative curvature $- K$.

It may be well to observe that while the notion of a manifold of constant curvature and the measure K of this curvature are by their nature invariant, i.e. independent of the choice of the co-ordinates of reference, this invariance does not persist for multiplication of ds^2 by a constant factor m. In fact, when all the coefficients g_{ik} are multiplied by m, Riemann's symbols of the *second* kind are unchanged, so that, again by formula (19′) of p. 234, the curvature K is divided by m. In particular, for $m = - 1$, it changes sign. This explains the apparent contradiction between our enunciation and that of some writers who take $- ds^2$ as the fundamental form and assign constant positive curvature to De Sitter's space-time.

17. **Einstein's additional term. Indication of other rigorous solutions.**

For Einstein's solution we found in §§ 14 and 15 (formulæ (73) and (78))

$$3K = \kappa\eta, \quad K + \kappa p = 0.$$

We cannot therefore suppose the matter devoid of stresses ($p = 0$) without concluding that $\eta = 0$, which brings us back to the uninteresting case of a totally empty space. Now if we take the cosmologico-statistical point of view (in the sense indicated in § 14), it seems reasonable to suppose that there must be a solution of the gravitational equations corresponding to the hypothesis of a uniform distribution of matter which shall be so tenuous that the molecular actions between contiguous particles, and therefore the stresses, are imperceptible; such, that is, that $p = 0$, while η is a constant other than zero.

Since the gravitational equations in the original form (8), viz.

$$G_{ik} - \tfrac{1}{2} G g_{ik} = -\kappa T_{ik},$$

have no solution of this type, Einstein was led to modify them (very slightly) by adding a term which maintains the tensorial character of the equations (8), and which in ordinary cases is completely imperceptible while serving to render possible a solution of the type indicated. This term was assumed by Einstein in the particularly simple form λg_{ik}, λ denoting a constant which in most cases is negligible compared with G. The gravitational equations so modified are

$$\left. \begin{array}{c} G_{ik} - \tfrac{1}{2} G g_{ik} + \lambda g_{ik} = -\kappa T_{ik} \\ (i, k = 0, 1, 2, 3). \end{array} \right\} \quad . \ . \quad (84)$$

The statical equations accordingly become

$$\mathbf{M} - \lambda = \kappa\eta,$$

$$a_{ik} + \frac{V_{ik}}{V} - \left(\frac{\Delta_2 V}{V} + \lambda\right) a_{ik} = -\kappa T_{ik} \qquad (i, k = 1, 2, 3).$$

Proceeding as in §§ 14, 15, on the hypothesis that the space-like dl^2 has constant curvature, that the density is constant, and that the stresses are isotropic (i.e. are given by (70) and (71)), we ultimately reach the two equations

$$3K = \kappa\eta + \lambda,$$

$$K + \kappa p = \lambda,$$

between K, η, p, and λ, which take the place of (73) and (78).

Here it plainly becomes possible to put $p = 0$ without η necessarily having to vanish at the same time; we need only take

$$\eta = \frac{2K}{\kappa},$$

$$K = \lambda.$$

To get an idea of the order of smallness of the constant λ, we may note that the mean cosmic density $\frac{\eta}{c^2}$ of matter can certainly be regarded as considerably less than that of the nebulæ, which is of the order of 10^{-17} gm./cm.^{3}. It is therefore legitimate to assume that in any case

$$\lambda = K < \frac{\kappa c^2}{2} 10^{-17}.$$

From the numerical values (in C.G.S. units) $\kappa = 2 \times 10^{-48}$, $c = 3 \times 10^{10}$, we have

$$\lambda = K < 9 \times 10^{-45}.$$

For the radius a of the universe $\left(K = \frac{1}{a^2}\right)$ we thus get a lower limit given by

$$a > 10^{22} \text{ cm.}$$

This radius is therefore certainly considerably greater than 10^{17} km. or 10,000 light-years.

We shall conclude with some bibliographical references concerning the rigorous solutions of the gravitational equations (with or without the cosmological term) in some special cases.

Schwarzschild's solution is supplemented or generalized in various important respects by the original contributions of Birkhoff, De Donder, Eddington, v. Laue,[1] and Weyl, which are given in their respective treatises, and of Signorina Longo[2], Trefftz,[3] Nuyens,[4] and Vanderlinden.[5]

A different type of solution is considered in the researches of Weyl,[6] Levi-Civita,[7] Bach,[8] Chazy,[9] Palatini,[10] and Kasner.[11]

[1] Cf. also *Sitzungsberichte der Preuss. Ak. der Wiss.*, 1923, pp. 27–31.

[2] *Nuovo Cimento*, Vol. XV, 1918, pp. 191–211.

[3] *Math. Annalen*, Vol. 86, 1922, pp. 317–326.

[4] *Comptes Rendus*, Vol. 176, 1923, pp. 1376–1379.

[5] *Bull. de l'Ac. royale de Belgique*, 1921, pp. 260–276.

[6] *Annalen der Physik*, 54 (1918), pp. 117–145; 59 (1919), pp. 185–188.

[7] " ds^2 einsteiniani in campi newtoniani ", Notes I–IX, in *Rend. della R. Acc. dei Lincei*, Vols. XXVI, XXVII, XXVIII, 1917–1919.

[8] *Mathematische Zeitschrift*, Vol. 13, 1922, pp. 134–145.

[9] *Bulletin de la Société Math. de France*, Vol. LII, 1924, pp. 17–37.

[10] *Nuovo Cimento*, Vol. XXVI, 1923, pp. 5–24.

[11] *Trans. of the American Math. Society*, Vol. 27, 1925, pp. 101–105, 155–162.

ADDITIONAL NOTES

P. 122, *line* 9 *from foot.* See a short but substantial article by E. Cartan, who discusses the question exhaustively from the geometrical point of view: *Annales de la Societe Polonaise de Mathematiques*, Vol. VI (1927), pp. 1–7. An earlier paper by M. Janet, *ibidem*, Vol. V (1926), pp. 38–73, may also be consulted.

P. 168, *at the end.* A luminous demonstration of M. Fermi's theorem, as simple as it is intimately related to fundamental principles, has been given recently by Mlle. P. Nalli: *Rend. Acc. Lincei*, Vol. VII (1928), pp. 195–198.

P. 171, *at the end.* The use of locally geodesic co-ordinates enables us to recognize at once *an important property of the ϵ-systems*, which they possess in common with the fundamental tensors a_{ik}, a^{ik} —*their covariant derivative vanishes identically.* For each element of an ϵ-system is either zero or of the form $\pm\sqrt{a}$, $\pm\frac{1}{\sqrt{a}}$. The derivatives of the a_{ik}'s being zero (in geodesic co-ordinates, for the point considered), the same is true for every element of an ϵ-tensor. It follows (p. 71, final paragraph) that the covariant derivative vanishes in any system of co-ordinates whatever.

P. 188, *line* 3. The general case in which the cycle T and consequently the area Γ are not restricted to be infinitely small, can also be treated without great difficulty, as has been shown very ingeniously by J. M. McConnell, *Rend. Acc. Lincei*, Vol. VII (1928), pp. 208–213, 306–309.

P. 209, *end of footnote.* See also J. L. Synge, *On the Geometry of Dynamics*, Phil. Trans. Roy. Soc., A, **226** (1926), pp. 31–106; and various notes by MM. Berwald, Boggio, Cartan, Crudeli, De Mira Fernandes, Onicescu, Vranceanu, *Rend. Acc. Lincei*, Vols. V, VI and VII (1927, 1928).

P. 228, *after formula* (14). Formulæ (13) and (14) can be proved more readily, without any formal development, by means of geodesic co-ordinates, as has been remarked by Mlle. NALLI. See her note *Due dimostrazioni nel calcolo assoluto,* Boll. dell' Unione Mat. Italiana, Vol. VII (1928), pp. 124, 127.

P. 234, *line* 10 *from foot.* A simpler proof, due to Mlle. NALLI, is given in the paper cited in the note to p. 228.

P. 439, *at end of references.* On all these questions, DARMOIS, *Les equations de la gravitation einsteinienne,* Fasc. XX, Mémorial des Sciences Mathématiques (Paris, Gauthier-Villars, 1927) may also be consulted.

NAME INDEX

SUBJECT INDEX

A CATALOGUE OF SELECTED DOVER BOOKS IN ALL FIELDS OF INTEREST

A CATALOGUE OF SELECTED DOVER BOOKS IN ALL FIELDS OF INTEREST

THE NOTEBOOKS OF LEONARDO DA VINCI, edited by J.P. Richter. Extracts from manuscripts reveal great genius; on painting, sculpture, anatomy, sciences, geography, etc. Both Italian and English. 186 ms. pages reproduced, plus 500 additional drawings, including studies for Last Supper, Sforza monument, etc. 860pp. 7⅞ x 10¾. USO 22572-0, 22573-9 Pa., Two vol. set $15.90

ART NOUVEAU DESIGNS IN COLOR, Alphonse Mucha, Maurice Verneuil, Georges Auriol. Full-color reproduction of Combinaisons ornamentales (c. 1900) by Art Nouveau masters. Floral, animal, geometric, interlacings, swashes — borders, frames, spots — all incredibly beautiful. 60 plates, hundreds of designs. 9⅜ x 8 1/16. 22885-1 Pa. $4.00

GRAPHIC WORKS OF ODILON REDON. All great fantastic lithographs, etchings, engravings, drawings, 209 in all. Monsters, Huysmans, still life work, etc. Introduction by Alfred Werner. 209pp. 9⅛ x 12¼. 21996-8 Pa. $6.00

EXOTIC FLORAL PATTERNS IN COLOR, E.-A. Seguy. Incredibly beautiful full-color pochoir work by great French designer of 20's. Complete Bouquets et frondaisons, Suggestions pour étoffes. Richness must be seen to be believed. 40 plates containing 120 patterns. 80pp. 9⅜ x 12¼. 23041-4 Pa. $6.00

SELECTED ETCHINGS OF JAMES A. McN. WHISTLER, James A. McN. Whistler. 149 outstanding etchings by the great American artist, including selections from the Thames set and two Venice sets, the complete French set, and many individual prints. Introduction and explanatory note on each print by Maria Naylor. 157pp. 9⅜ x 12¼. 23194-1 Pa. $5.00

VISUAL ILLUSIONS: THEIR CAUSES, CHARACTERISTICS, AND APPLICATIONS, Matthew Luckiesh. Thorough description, discussion; shape and size, color, motion; natural illusion. Uses in art and industry. 100 illustrations. 252pp. 21530-X Pa. $3.00

TEN BOOKS ON ARCHITECTURE, Vitruvius. The most important book ever written on architecture. Early Roman aesthetics, technology, classical orders, site selection, all other aspects. Stands behind everything since. Morgan translation. 331pp. 20645-9 Pa. $3.75

THE CODEX NUTTALL, A PICTURE MANUSCRIPT FROM ANCIENT MEXICO, as first edited by Zelia Nuttall. Only inexpensive edition, in full color, of a pre-Columbian Mexican (Mixtec) book. 88 color plates show kings, gods, heroes, temples, sacrifices. New explanatory, historical introduction by Arthur G. Miller. 96pp. 11⅜ x 8½. 23168-2 Pa. $7.50

THE BEST DR. THORNDYKE DETECTIVE STORIES, R. Austin Freeman. The Case of Oscar Brodski, The Moabite Cipher, and 5 other favorites featuring the great scientific detective, plus his long-believed-lost first adventure — 31 New Inn — reprinted here for the first time. Edited by E.F. Bleiler. USO 20388-3 Pa. $3.00

BEST "THINKING MACHINE" DETECTIVE STORIES, Jacques Futrelle. The Problem of Cell 13 and 11 other stories about Prof. Augustus S.F.X. Van Dusen, including two "lost" stories. First reprinting of several. Edited by E.F. Bleiler. 241pp. 20537-1 Pa. $3.00

UNCLE SILAS, J. Sheridan LeFanu. Victorian Gothic mystery novel, considered by many best of period, even better than Collins or Dickens. Wonderful psychological terror. Introduction by Frederick Shroyer. 436pp. 21715-9 Pa. $4.50

BEST DR. POGGIOLI DETECTIVE STORIES, T.S. Stribling. 15 best stories from EQMM and The Saint offer new adventures in Mexico, Florida, Tennessee hills as Poggioli unravels mysteries and combats Count Jalacki. 217pp. 23227-1 Pa. $3.00

EIGHT DIME NOVELS, selected with an introduction by E.F. Bleiler. Adventures of Old King Brady, Frank James, Nick Carter, Deadwood Dick, Buffalo Bill, The Steam Man, Frank Merriwell, and Horatio Alger — 1877 to 1905. Important, entertaining popular literature in facsimile reprint, with original covers. 190pp. 9 x 12. 22975-0 Pa. $3.50

ALICE'S ADVENTURES UNDER GROUND, Lewis Carroll. Facsimile of ms. Carroll gave Alice Liddell in 1864. Different in many ways from final Alice. Handlettered, illustrated by Carroll. Introduction by Martin Gardner. 128pp. 21482-6 Pa. $2.00

ALICE IN WONDERLAND COLORING BOOK, Lewis Carroll. Pictures by John Tenniel. Large-size versions of the famous illustrations of Alice, Cheshire Cat, Mad Hatter and all the others, waiting for your crayons. Abridged text. 36 illustrations. 64pp. 8¼ x 11. 22853-3 Pa. $1.50

AVENTURES D'ALICE AU PAYS DES MERVEILLES, Lewis Carroll. Bué's translation of "Alice" into French, supervised by Carroll himself. Novel way to learn language. (No English text.) 42 Tenniel illustrations. 196pp. 22836-3 Pa. $3.00

MYTHS AND FOLK TALES OF IRELAND, Jeremiah Curtin. 11 stories that are Irish versions of European fairy tales and 9 stories from the Fenian cycle — 20 tales of legend and magic that comprise an essential work in the history of folklore. 256pp. 22430-9 Pa. $3.00

EAST O' THE SUN AND WEST O' THE MOON, George W. Dasent. Only full edition of favorite, wonderful Norwegian fairytales — Why the Sea is Salt, Boots and the Troll, etc. — with 77 illustrations by Kittelsen & Werenskiöld. 418pp. 22521-6 Pa. $4.50

PERRAULT'S FAIRY TALES, Charles Perrault and Gustave Doré. Original versions of Cinderella, Sleeping Beauty, Little Red Riding Hood, etc. in best translation, with 34 wonderful illustrations by Gustave Doré. 117pp. 8⅛ x 11. 22311-6 Pa. $2.50

HOUDINI ON MAGIC, Harold Houdini. Edited by Walter Gibson, Morris N. Young. How he escaped; exposés of fake spiritualists; instructions for eye-catching tricks; other fascinating material by and about greatest magician. 155 illustrations. 280pp. 20384-0 Pa. $2.75

HANDBOOK OF THE NUTRITIONAL CONTENTS OF FOOD, U.S. Dept. of Agriculture. Largest, most detailed source of food nutrition information ever prepared. Two mammoth tables: one measuring nutrients in 100 grams of edible portion; the other, in edible portion of 1 pound as purchased. Originally titled Composition of Foods. 190pp. 9 x 12. 21342-0 Pa. $4.00

COMPLETE GUIDE TO HOME CANNING, PRESERVING AND FREEZING, U.S. Dept. of Agriculture. Seven basic manuals with full instructions for jams and jellies; pickles and relishes; canning fruits, vegetables, meat; freezing anything. Really good recipes, exact instructions for optimal results. Save a fortune in food. 156 illustrations. 214pp. 6⅛ x 9¼. 22911-4 Pa. $2.50

THE BREAD TRAY, Louis P. De Gouy. Nearly every bread the cook could buy or make: bread sticks of Italy, fruit breads of Greece, glazed rolls of Vienna, everything from corn pone to croissants. Over 500 recipes altogether. including buns, rolls, muffins, scones, and more. 463pp. 23000-7 Pa. $4.00

CREATIVE HAMBURGER COOKERY, Louis P. De Gouy. 182 unusual recipes for casseroles, meat loaves and hamburgers that turn inexpensive ground meat into memorable main dishes: Arizona chili burgers, burger tamale pie, burger stew, burger corn loaf, burger wine loaf, and more. 120pp. 23001-5 Pa. $1.75

LONG ISLAND SEAFOOD COOKBOOK, J. George Frederick and Jean Joyce. Probably the best American seafood cookbook. Hundreds of recipes. 40 gourmet sauces, 123 recipes using oysters alone! All varieties of fish and seafood amply represented. 324pp. 22677-8 Pa. $3.50

THE EPICUREAN: A COMPLETE TREATISE OF ANALYTICAL AND PRACTICAL STUDIES IN THE CULINARY ART, Charles Ranhofer. Great modern classic. 3,500 recipes from master chef of Delmonico's, turn-of-the-century America's best restaurant. Also explained, many techniques known only to professional chefs. 775 illustrations. 1183pp. 6⅝ x 10. 22680-8 Clothbd. $22.50

THE AMERICAN WINE COOK BOOK, Ted Hatch. Over 700 recipes: old favorites livened up with wine plus many more: Czech fish soup, quince soup, sauce Perigueux, shrimp shortcake, filets Stroganoff, cordon bleu goulash, jambonneau, wine fruit cake, more. 314pp. 22796-0 Pa. $2.50

DELICIOUS VEGETARIAN COOKING, Ivan Baker. Close to 500 delicious and varied recipes: soups, main course dishes (pea, bean, lentil, cheese, vegetable, pasta, and egg dishes), savories, stews, whole-wheat breads and cakes, more. 168pp. USO 22834-7 Pa. $2.00

EARLY NEW ENGLAND GRAVESTONE RUBBINGS, Edmund V. Gillon, Jr. 43 photographs, 226 rubbings show heavily symbolic, macabre, sometimes humorous primitive American art. Up to early 19th century. 207pp. 8³/₈ x 11¼.
21380-3 Pa. $4.00

L.J.M. DAGUERRE: THE HISTORY OF THE DIORAMA AND THE DAGUERREOTYPE, Helmut and Alison Gernsheim. Definitive account. Early history, life and work of Daguerre; discovery of daguerreotype process; diffusion abroad; other early photography. 124 illustrations. 226pp. 6¹/₆ x 9¼. 22290-X Pa. $4.00

PHOTOGRAPHY AND THE AMERICAN SCENE, Robert Taft. The basic book on American photography as art, recording form, 1839-1889. Development, influence on society, great photographers, types (portraits, war, frontier, etc.), whatever else needed. Inexhaustible. Illustrated with 322 early photos, daguerreotypes, tintypes, stereo slides, etc. 546pp. 6¹/₈ x 9¼. 21201-7 Pa. $6.00

PHOTOGRAPHIC SKETCHBOOK OF THE CIVIL WAR, Alexander Gardner. Reproduction of 1866 volume with 100 on-the-field photographs: Manassas, Lincoln on battlefield, slave pens, etc. Introduction by E.F. Bleiler. 224pp. 10¾ x 9.
22731-6 Pa. $6.00

THE MOVIES: A PICTURE QUIZ BOOK, Stanley Appelbaum & Hayward Cirker. Match stars with their movies, name actors and actresses, test your movie skill with 241 stills from 236 great movies, 1902-1959. Indexes of performers and films. 128pp. 8³/₈ x 9¼. 20222-4 Pa. $3.00

THE TALKIES, Richard Griffith. Anthology of features, articles from Photoplay, 1928-1940, reproduced complete. Stars, famous movies, technical features, fabulous ads, etc.; Garbo, Chaplin, King Kong, Lubitsch, etc. 4 color plates, scores of illustrations. 327pp. 8³/₈ x 11¼. 22762-6 Pa. $6.95

THE MOVIE MUSICAL FROM VITAPHONE TO "42ND STREET," edited by Miles Kreuger. Relive the rise of the movie musical as reported in the pages of Photoplay magazine (1926-1933): every movie review, cast list, ad, and record review; every significant feature article, production still, biography, forecast, and gossip story. Profusely illustrated. 367pp. 8³/₈ x 11¼. 23154-2 Pa. $7.95

JOHANN SEBASTIAN BACH, Philipp Spitta. Great classic of biography, musical commentary, with hundreds of pieces analyzed. Also good for Bach's contemporaries. 450 musical examples. Total of 1799pp.
EUK 22278-0, 22279-9 Clothbd., Two vol. set $25.00

BEETHOVEN AND HIS NINE SYMPHONIES, Sir George Grove. Thorough history, analysis, commentary on symphonies and some related pieces. For either beginner or advanced student. 436 musical passages. 407pp. 20334-4 Pa. $4.00

MOZART AND HIS PIANO CONCERTOS, Cuthbert Girdlestone. The only full-length study. Detailed analyses of all 21 concertos, sources; 417 musical examples. 509pp. 21271-8 Pa. $6.00

EAST O' THE SUN AND WEST O' THE MOON, George W. Dasent. Considered the best of all translations of these Norwegian folk tales, this collection has been enjoyed by generations of children (and folklorists too). Includes True and Untrue, Why the Sea is Salt, East O' the Sun and West O' the Moon, Why the Bear is Stumpy-Tailed, Boots and the Troll, The Cock and the Hen, Rich Peter the Pedlar, and 52 more. The only edition with all 59 tales. 77 illustrations by Erik Werenskiold and Theodor Kittelsen. xv + 418pp. 22521-6 Paperbound **$4.00**

GOOPS AND HOW TO BE THEM, Gelett Burgess. Classic of tongue-in-cheek humor, masquerading as etiquette book. 87 verses, twice as many cartoons, show mischievous Goops as they demonstrate to children virtues of table manners, neatness, courtesy, etc. Favorite for generations. viii + 88pp. 6½ x 9¼. 22233-0 Paperbound **$2.00**

ALICE'S ADVENTURES UNDER GROUND, Lewis Carroll. The first version, quite different from the final *Alice in Wonderland,* printed out by Carroll himself with his own illustrations. Complete facsimile of the "million dollar" manuscript Carroll gave to Alice Liddell in 1864. Introduction by Martin Gardner. viii + 96pp. Title and dedication pages in color. 21482-6 Paperbound **$1.50**

THE BROWNIES, THEIR BOOK, Palmer Cox. Small as mice, cunning as foxes, exuberant and full of mischief, the Brownies go to the zoo, toy shop, seashore, circus, etc., in 24 verse adventures and 266 illustrations. Long a favorite, since their first appearance in St. Nicholas Magazine. xi + 144pp. 6⅝ x 9¼. 21265-3 Paperbound **$2.50**

SONGS OF CHILDHOOD, Walter De La Mare. Published (under the pseudonym Walter Ramal) when De La Mare was only 29, this charming collection has long been a favorite children's book. A facsimile of the first edition in paper, the 47 poems capture the simplicity of the nursery rhyme and the ballad, including such lyrics as I Met Eve, Tartary, The Silver Penny. vii + 106pp. (USO) 21972-0 Paperbound $2.00

THE COMPLETE NONSENSE OF EDWARD LEAR, Edward Lear. The finest 19th-century humorist-cartoonist in full: all nonsense limericks, zany alphabets, Owl and Pussycat, songs, nonsense botany, and more than 500 illustrations by Lear himself. Edited by Holbrook Jackson. xxix + 287pp. (USO) 20167-8 Paperbound **$3.00**

BILLY WHISKERS: THE AUTOBIOGRAPHY OF A GOAT, Frances Trego Montgomery. A favorite of children since the early 20th century, here are the escapades of that rambunctious, irresistible and mischievous goat—Billy Whiskers. Much in the spirit of *Peck's Bad Boy,* this is a book that children never tire of reading or hearing. All the original familiar illustrations by W. H. Fry are included: 6 color plates, 18 black and white drawings. 159pp. 22345-0 Paperbound **$2.75**

MOTHER GOOSE MELODIES. Faithful republication of the fabulously rare Munroe and Francis "copyright 1833" Boston edition—the most important Mother Goose collection, usually referred to as the "original." Familiar rhymes plus many rare ones, with wonderful old woodcut illustrations. Edited by E. F. Bleiler. 128pp. 4½ x 6⅜. 22577-1 Paperbound **$1.50**

CATALOGUE OF DOVER BOOKS

EGYPTIAN MAGIC, E.A. Wallis Budge Foremost Egyptologist, curator at British Museum, on charms, curses, amulets, doll magic, transformations, control of demons, deific appearances, feats of great magicians. Many texts cited. 19 illustrations. 234pp. USO 22681-6 Pa. $2.50

THE LEYDEN PAPYRUS: AN EGYPTIAN MAGICAL BOOK, edited by F. Ll. Griffith, Herbert Thompson. Egyptian sorcerer's manual contains scores of spells: sex magic of various sorts, occult information, evoking visions, removing evil magic, etc. Transliteration faces translation. 207pp. 22994-7 Pa. $2.50

THE MALLEUS MALEFICARUM OF KRAMER AND SPRENGER, translated, edited by Montague Summers. Full text of most important witchhunter's "Bible," used by both Catholics and Protestants. Theory of witches, manifestations, remedies, etc. Indispensable to serious student. 278pp. 6⁵/₈ x 10. USO 22802-9 Pa. $3.95

LOST CONTINENTS, L. Sprague de Camp. Great science-fiction author, finest, fullest study: Atlantis, Lemuria, Mu, Hyperborea, etc. Lost Tribes, Irish in pre-Columbian America, root races; in history, literature, art, occultism. Necessary to everyone concerned with theme. 17 illustrations. 348pp. 22668-9 Pa. $3.50

THE COMPLETE BOOKS OF CHARLES FORT, Charles Fort. Book of the Damned, Lo!, Wild Talents, New Lands. Greatest compilation of data: celestial appearances, flying saucers, falls of frogs, strange disappearances, inexplicable data not recognized by science. Inexhaustible, painstakingly documented. Do not confuse with modern charlatanry. Introduction by Damon Knight. Total of 1126pp.
23094-5 Clothbd. $15.00

FADS AND FALLACIES IN THE NAME OF SCIENCE, Martin Gardner. Fair, witty appraisal of cranks and quacks of science: Atlantis, Lemuria, flat earth, Velikovsky, orgone energy, Bridey Murphy, medical fads, etc. 373pp. 20394-8 Pa. $3.50

HOAXES, Curtis D. MacDougall. Unbelievably rich account of great hoaxes: Locke's moon hoax, Shakespearean forgeries, Loch Ness monster, Disumbrationist school of art, dozens more; also psychology of hoaxing. 54 illustrations. 338pp. 20465-0 Pa. $3.50

THE GENTLE ART OF MAKING ENEMIES, James A.M. Whistler. Greatest wit of his day deflates Wilde, Ruskin, Swinburne; strikes back at inane critics, exhibitions. Highly readable classic of impressionist revolution by great painter. Introduction by Alfred Werner. 334pp. 21875-9 Pa. $4.00

THE BOOK OF TEA, Kakuzo Okakura. Minor classic of the Orient: entertaining, charming explanation, interpretation of traditional Japanese culture in terms of tea ceremony. Edited by E.F. Bleiler. Total of 94pp. 20070-1 Pa. $1.25

Prices subject to change without notice.
Available at your book dealer or write for free catalogue to Dept. GI, Dover Publications, Inc., 180 Varick St., N.Y., N.Y. 10014. Dover publishes more than 150 books each year on science, elementary and advanced mathematics, biology, music, art, literary history, social sciences and other areas.